Advanced Nanomaterials for Aerospace Applications

Advanced Nanomaterials for Aerospace Applications

edited by

Carlos R. Cabrera
Félix A. Miranda

PAN STANFORD PUBLISHING

Published by

Pan Stanford Publishing Pte. Ltd.
Penthouse Level, Suntec Tower 3
8 Temasek Boulevard
Singapore 038988

Email: editorial@panstanford.com
Web: www.panstanford.com

British Library Cataloguing-in-Publication Data
A catalogue record for this book is available from the British Library.

Advanced Nanomaterials for Aerospace Applications

Copyright © 2014 Pan Stanford Publishing Pte. Ltd.

All rights reserved. This book, or parts thereof, may not be reproduced in any form or by any means, electronic or mechanical, including photocopying, recording or any information storage and retrieval system now known or to be invented, without written permission from the publisher.

For photocopying of material in this volume, please pay a copying fee through the Copyright Clearance Center, Inc., 222 Rosewood Drive, Danvers, MA 01923, USA. In this case permission to photocopy is not required from the publisher.

Cover image courtesy of NASA (modified).

ISBN 978-981-4463-18-8 (Hardcover)
ISBN 978-981-4463-19-5 (eBook)

Printed in the USA

Contents

Preface

The book titled *Advanced Nanomaterials for Aerospace Applications* has been developed for a community interested in aerospace science and nanotechnology. Scientists and engineers of the NASA-University Research Center for Advanced Nanoscale Materials (CANM) at the University of Puerto Rico and from the NASA John H. Glenn and Ames Research Centers, the George C. Marshall Space Flight Center, the Jet Propulsion Laboratory, Pennsylvania State University, and INFN-Laboratori Nazionali di Frascati, Italy, have joined efforts to present the applications of nanomaterials in sensors, life support systems, regenerative fuel cells, Li-ion batteries, robust lightweight materials, nanoelectronics, and electromagnetic shielding.

The CANM at the University of Puerto Rico has been involved in the development of these exciting research areas in collaboration with NASA centers for more than 10 years. Most of the scientists and engineers who have been part of the CANM are authors or co-authors of the chapters in this Book. Their efforts and collaborations on the research, development, and applications of nanomaterials have advanced some of the technology to higher NASA Technology Readiness Levels. The outcome of some of these collaborations and interactions is summarized in the chapters of this book to be shared with the aerospace and nanotechnology communities.

Chapters 1 and 2 are devoted to sensor technology. Chapter 1 offers a thorough overview of a variety of sensors being developed in support of NASA's mission. Among these are sensors for leak detection, high-temperature physical state monitoring, emissions monitoring, fire and environmental detection, and radiation detection, among others. Chapter 2 focuses on the challenges associated with the fabrication and reproducibility of nanostructures into microsensors and microsensor systems. It also discusses the characterization of the basic properties of nanowires, and the sensing mechanisms of nanostructures of different crystal structures. New sensor systems that can be enabled by nanotechnology are also discussed.

Chapters 3 and 4 present the use of nanomaterials for life support system applications. The chapters summarize the challenges faced by the environmental control and life support system engineers and scientists in improving the atmospheric revitalization and water reclamation subsystems currently used in space platforms such as the International Space Station. In this context, the chapters discuss the promising materials developments that may enable solutions to address these challenges for human space exploration missions beyond low earth orbit. These include water-recycling systems in space platforms and the effects of reduced gravity environment on the performance of life support systems (Chapter 4).

Chapter 5 discusses the performance advantages of different nanostructured electroactive materials, as compared to microstructured materials, for the negative electrodes (i.e., anodes) of lithium-ion batteries. Chapter 6 complements Chapter 5 by discussing the advances in the design of high-energy cathode materials for rechargeable lithium-ion batteries. Chapter 7 provides insights on the use of nanomaterials for renewable energy applications, in particular the development of nanomaterials used as catalysts for fuel cells and electrolyzers in regenerative fuel cell systems for NASA applications.

Chapters 8, 9, and 10 present other areas of nanotechnology relevant for aerospace applications. Chapter 8 addresses the use of nanotechnology for the development of nanoelectronic devices and applications. The chapter provides some examples of nanoelectronic devices and characterization techniques, as well as an assessment of the use of nanoelectronics in future NASA communication systems. Chapters 9 and 10 discuss the use of advanced nanomaterials for electromagnetic shielding, a topic of utmost relevance for long-duration human presence in space.

Since one of the roles of the CANM initiative was the development of approaches and mechanisms to foster nanotechnology outreach and education, it seems appropriate to include in this book a chapter on this most relevant subject matter. Accordingly, Chapter 11 discusses nanotechnology educational components based on advanced nanomaterials for aerospace applications and the initiative implemented in this area as part of the CANM educational and outreach efforts.

Finally, Chapter 12 provides an overall NASA perspective on the future trends in nanotechnology and discusses initiatives and mechanisms currently in place to facilitate the advancement of research and technology development delineated by such trends.

There are many people and organizations that have made this book possible. In particular, the editors would like to thank each of the authors of the chapters featured in this book for their hard work and outstanding contributions. We wish to express our gratitude to the students, colleagues, and officers of the University of Puerto Rico, who in many ways contributed to the work featured in this book. We would like to extend our gratitude to Ms. Katrina Emery, former project director of the NASA University Research Centers Program, for her support and encouragement throughout the years. We are grateful to Ms. Kaprice L. Harris and Ms. Deborah A. Szczepinski (Stinger Ghaffarian Technologies, Inc.) of NASA Glenn Research Center's Office of the Chief Counsel, for their help regarding compliance with government regulations and policies.

It is our hope that this book will fill the expectations of the reader and presents an opportunity to learn more about advanced nanomaterials for aerospace applications.

Dr. Carlos R. Cabrera
NASA-URC Center for Advanced Nanoscale Materials
The University of Puerto Rico, Rio Piedras Campus
San Juan, Puerto Rico

Dr. Félix A. Miranda
NASA John H. Glenn Research Center at Lewis Field, Cleveland, OH

May 2014

Chapter 1

Advanced Sensor Nanomaterials for Aerospace Applications

Gary W. Hunter,[a] Jennifer C. Xu,[a] Laura J. Evans,[a] Luis F. Fonseca,[b] Ram Katiyar,[b] María M. Martínez-Iñesta,[c] Wilfredo Otaño,[d] Neeraj Panwar,[b,e] and Randy L. Vander Wal[f]

[a]*NASA John H. Glenn Research Center at Lewis Field,*
21000 Brookpark Rd., Cleveland, OH 44135, USA
[b]*NASA-URC Center for Advanced Nanoscale Materials, University of Puerto Rico,*
Rio Piedras Campus, San Juan, PR 00931, USA
[c]*NASA-URC Center for Advanced Nanoscale Materials, University of Puerto Rico,*
Mayagüez Campus, San Juan, PR 00681, USA
[d]*NASA-URC Center for Advanced Nanoscale Materials, University of Puerto Rico,*
Cayey Campus, San Juan, PR 00736, USA
[e]*Department of Physics, Central University of Rajasthan, Bandarsindari,*
Kishangarh 305801, India
[f]*Pennsylvania State University,*
203 Hosler Building, University Park, PA 16802, USA

ghunter@grc.nasa.gov

Aeronautic and space applications require the development of sensors with capabilities beyond those of commercially available sensors. This chapter reviews a range of aerospace applications and the sensor operational requirements for those applications. These applications include leak detection, high-temperature physical parameter monitoring, emission monitoring, fire/environmental

Advanced Nanomaterials for Aerospace Applications
Edited by Carlos R. Cabrera and Félix A. Miranda
Copyright © 2014 Pan Stanford Publishing Pte. Ltd.
ISBN 978-981-4463-18-8 (Hardcover), 978-981-4463-19-5 (eBook)
www.panstanford.com

monitoring, and ultraviolet radiation monitoring. In response to these needs, a range of nanotechnology-based sensor material work is being developed. This includes transition metal silicide nanowires for physical measurements; palladium nanoshells, ultrathin films, and nanowires for hydrogen detection; semiconducting metal oxide nanostructures relevant to fire/environmental/emissions monitoring; and modified zinc oxide materials for ultraviolet radiation detection. Sensor development for each application involves its own challenges in the fields of materials science and fabrication technology. The advent of nano-based sensor materials can have notable effect on aerospace applications. However, these materials must be integrated into complete sensor systems and these advantages must be demonstrated for the specific application

1.1 Introduction

Aerospace (aeronautic and space) applications require the development of sensors that operate in a number of different environments. Future aeronautic systems have increasing requirements for decreased maintenance, improved capability and performance, and increased safety. Space exploration missions will require significantly improved monitoring in the vehicle and crew habitat environments throughout all phases of the mission. Potential problems with the vehicle, habitat or robotic system, or even the spacesuit, must be identified before they threaten the mission or astronaut health. This implies that the inclusion of automated vehicle intelligence into the system design and operation is necessary [1–4]. In particular, improvement is necessary in sensor systems, i.e., sensors and their associated data acquisition systems, packaging, communications, power, etc. The monitoring of a range of parameters are an important component of an overall health and operational monitoring system for aeronautic and space vehicles. This chapter provides a brief overview of the needs of a range of aerospace applications, including relevant operational parameters.

A major point of this chapter is that nanotechnology development can address the needs of aerospace applications; several examples are given of the development of a range of sensors and sensor materials based on nanotechnology to address aerospace applications. Finally, a summary of the future steps needed to transition this core sensor development into sensor systems applicable to aerospace applications will be discussed. The conclusion is that the material advancements enabled by nanotechnology can have a notable impact on aerospace applications, but demonstration of such capabilities is necessary for such potential to be realized.

1.2 Overview of Aerospace Applications

Aerospace sensor application areas include leak detection, high-temperature physical parameter monitoring, emissions monitoring, fire detection/environmental monitoring, and radiation detection. Each of these areas is the subject of effort throughout NASA to, for example, improve safety and decrease the cost of space travel, significantly decrease the amount of emissions produced by aeronautic engines, and improve the safety of commercial airline travel. Each area of application has vastly different operational conditions and challenges associated with performing the measurement. However, the development of a common base technology can address the measurement needs of a number of applications. The following give a brief description of these various applications.

1.2.1 Leak Detection

In launch vehicle safety applications, detection of low concentrations of hydrogen at potentially low temperatures is important for operations involved with, for example, operation of launch vehicles such as the Space Shuttle [2,3]. In 1990, leaks on the Space Shuttle while on the launch pad temporarily grounded the fleet until the leak source could be identified. Other events since then have also impacted launch availability and schedule. Concern continues that fuel leaks are a potential risk for both present flight systems and also for future flights systems, which include in-space and lunar/planetary/asteroid applications [5]. The drive is for miniature, mobile sensor systems that can be used throughout ground, flight, and in-space operations. Detection of low concentrations of hydrogen and other fuels is important to avoid explosive conditions that could harm personnel and damage the vehicle. Dependable vehicle operation also depends on the timely and accurate measurement of these leaks. Further, since an explosive condition depends not only on the amount of fuel present but on the oxygen concentration as well, the simultaneous measurement of both fuel and oxygen (O_2) is an important component of a leak detection system. Thus, the development of multiple sensors (a sensor array) to determine the concentration of fuels such as hydrogen or hydrocarbons, as well as oxygen, is necessary for space propulsion applications [5].

The operational requirements for leak detection sensors are illustrated by the hydrogen sensor. The hydrogen sensor must be able to detect hydrogen from low concentrations through the lower explosive limit (LEL), which is 4% in air. The sensor must be able to survive exposure to 100% hydrogen without damage or change in calibration. Further, the sensor may be exposed to gases emerging from cryogenic sources. Thus, sensor temperature control is necessary. Operation in inert environments is necessary since the sensor may have to operate in areas purged with helium. Ability to operate in a vacuum is needed for in-space applications. Being able to multiplex the signal from a number of sensors so as to "visualize" the magnitude and location of the hydrogen leak is also desired. Size, weight, and power consumption for each sensor is an issue to minimize impact to the vehicle. Commercially available sensors previously available, which often needed oxygen to operate or depended upon moisture [6], did not meet the needs of this application and thus the development of new types of sensors was necessary [7]. While significant progress has been made in establishing miniaturized leak detection systems for a number of applications [2,3,5,8], low-power systems for in-space, cryogenic applications are still a technical challenge.

1.2.2 High-Temperature Physical Parameters

Sensor and measurement technologies are needed to provide a better understanding of aeronautic engine systems and enable safer, lighter, quieter, and more fuel-efficient aircraft with fewer emissions [9–12]. This includes measurement of the physical parameters related to engine operation in harsh environments. For example, component degradation that develops over time in the engine hot section can lead to catastrophic failure. At present, the degradation processes that occur in the harsh hot section environment are poorly characterized, which hinders development of more durable components. Since it is difficult to model turbine blade temperatures, strains, etc., actual measurements are needed [13]. This implies the need for sensor systems that can be integrated into the engine system and operate in high temperature, harsh environments as well as on rotating components and with thermal cycling from high temperature to ambient. Physical sensors required to enable critical vehicle health monitoring (VHM) of future space and air vehicles include those for strain, temperature, heat flux, and surface flow [14].

Current challenges of physical sensor technology for engine applications include further development of specialized sensor systems for specific measurements. This includes development of instrumentation techniques on complex surfaces, improved sensor durability, and ability to perform higher temperature measurements above 1000°C [11,14]. This includes stable operation in both inert and oxidizing environments, minimal disruption to the airflow or operation of the engine or system, and time response and measurement accuracy tailored for the characterization needs of the application. For example, the capability for thin-film sensors to operate in 1500°C environments for 25 h or more is considered critical for ceramic turbine engine development [14–16]. For future space transportation vehicles, temperatures of propulsion system components over 1650°C are expected [14,17]. The overall need is to be able to have reliable harsh environment measurements, often of multiple parameters simultaneously, not only in tests and text conditions, but throughout the life of the propulsion system.

1.2.3 Emissions Monitoring

The control of emissions from aircraft engines is an important component of the development of the next generation of these engines [2,3,18]. The emissions produced by the engine are understood to be indicative of the state of the combustion process and reflective of other engine parameters. The ability to monitor the type and quantity of emissions being generated by an engine is important for qualifying engine technology as well as determining the health status of the engine, and eventually controlling those emissions. Ideally, an array of sensors placed in the emissions stream close to the engine could provide information on the gases being emitted by the engine. However, there are few sensors available commercially that are able to measure the components of the emissions of an engine in situ. The harsh conditions and high temperatures inherent near the reaction chamber of the engine render most sensors inoperable. While commercially available oxygen sensors have been used in automobile engines [19] for years and have been instrumental in decreasing automotive engine emissions, this sensor tends to be larger and with power consumption on the order of watts, rather than milliwatts or lower. The availability of other high-temperature emission sensors other than oxygen sensors has historically been more problematic, although significant progress has been made in qualifying such sensors for tests and applications [18].

The operational requirements for sensors in emission monitoring applications include operation at high temperatures with exposure to low concentrations of the gases to be measured. Measurements close to the engine exit or combustion process are strongly preferred to obtain improved readings of the species generated before mixing with the atmosphere or chemical reactions change the relative ratio of the emission species. Although the measurement of nitrogen oxides is important in these applications, the measurement of other gases present in the emission stream such as carbon dioxide, hydrogen, hydrocarbons, and oxygen is also of interest. The measurement range depends on the gas and the engine, but generally the detection of nitrogen oxides, hydrogen, and hydrocarbons may be necessary at concentrations of less than 200 ppm with corresponding measurements of oxygen from less than 1% to near 20%. The sensors should be small so as not to interfere with the flow of gases in the engine, or to become significant projectiles if dislodged from their measuring site and emitted into the interior of the engine.

1.2.4 Fire Safety and Environmental Monitoring

Fire detection equipment presently used in the cargo holds of many commercial aircraft relies on the detection of smoke [20]. Although highly developed, these sensors are subject to false alarms. These false alarms may be caused by a number of sources, including changes in humidity, condensation on the fire detector surface, and contamination from animals, plants, or other contents of the cargo bay [21,22]. A second method of fire detection to complement existing techniques, such as the measurement of chemical species indicative of a fire, can help reduce false alarms and improve aircraft safety. Further, the detection of the combustion products associated with fires addresses a critical risk for exploration vehicles and habitats due to the fact that a fire is a significant risk to crew safety and health both during and after the fire event [23,24]. The crew must remain in the proximity of the fire throughout the event, and continue to be within the closed environment after the fire. The hazard from a fire can be significantly reduced if fire detection is rapid and occurs in the early stages of fire development. Further, an understanding of the environment after a fire is important for astronaut health as well as for further monitoring of hazardous conditions.

Although monitoring of a broad range of chemical species is of interest for safety and environmental concerns, the targeted detection of a limited number of species can provide notable benefits [23–25]. Chemical species of interest in detecting fire are signature gases such as carbon monoxide (CO), carbon dioxide (CO_2), hydrocarbons, and acid gases such as hydrogen cyanide (HCN) and hydrogen chloride (HCl). Furthermore, in space applications future exploration vehicles are considering a requirement to monitor HCl and HCN in the crew cabin after a fire event to verify that it is safe for the crew to remove their portable breathing apparatus [26]. Monitoring of oxygen (O_2) concentrations as well as CO is of interest in assuring that the environment is safe to breathe. The emphasis for fire detection is on reliable detection and a minimization of false alarms. Long duration between recalibration is needed, especially for space applications, as well as accurate concentration determination of the species present.

1.2.5 Radiation Detection

Ultraviolet (UV) photodetectors, especially those operating in the solar blind (220–280 nm) region (due to not having interference from solar radiation [27] in this region), have a wide range of applications in military and civilian areas, including missile launching detection, flame sensing, UV radiation calibration and monitoring, chemical and biological analysis, optical communications, and astronomical studies. NASA is interested in low-cost technologies to meet programmatic objectives for global environmental observations. There is a particular need for mapping and sounding instruments that can measure a variety of environmental and climatic parameters (e.g., ozone, water vapor, sulfur dioxide (SO_2), temperature, aerosols, and UV radiation). Such low-cost, high-quality, space-borne sensor instruments are possible with emerging technologies for UV detectors.

The operational conditions of an ideal UV detector are as follows: Its band gap should lie in the solar blind region (greater than 4.5 eV) so that it exhibits high UV to visible rejection ratio. Its response time, i.e., the time required to detect the UV radiation should be in the nanoseconds range or lesser. The fall time should also be in the same range. Its quantum efficiency should be more than 90% and the signal to noise ratio should be as high as possible. It should operate at low voltage (a few volts). It should show high gain and photoresponsivity, so that amplifying equipment is not required. The detector should be physically and chemically stable and moreover, during the operation, its conductivity should not change. For space applications, it must perform in harsh environments, e.g., at high or extremely low temperatures. For this purpose, excitonic binding energy should be higher than for conventional materials (for example, in this regard ZnO-based detection systems with an exciton binding energy on the order of 60 MeV have an edge over other UV detectors).

1.3 Sensor Examples

In order to meet the needs of these applications, a new generation of sensor technology must be developed. Nanotechnology can be an integral part of the solution to these aerospace application challenges, as well as a range of other challenges facing industry and society [28,29]. The following provides examples of nanotechnology development relevant to aerospace applications. The common feature is that new materials can enable new capabilities; the potential of these new materials is just beginning to be explored. The following work is a foundation to significant new methods of addressing the aerospace sensor applications challenges described above, as well for other applications.

1.3.1 Palladium Nanoshells for Hydrogen Sensing

Palladium (Pd) metal is one of the most prominent materials studied for the detection of hydrogen gas. Hydrogen rapidly dissociates on its surface and diffuses into subsurface layers forming palladium hydride with consequent changes in optical, mechanical and electrical properties that are easily detected. At low concentrations, <1% hydrogen in air at STP, palladium hydride α phase is formed in bulk materials with H:Pd ratio of 0.015–0.03, while at higher concentrations the β phase is favored [30,31]. The insertion of the small hydrogen atom in the Pd lattice at interstitials produces changes in the properties of the material, including changes in scattering of light and electrons that are used for sensing [32]. The lattice constant of the Pd fcc (face-centered cubic) structure increases with the incorporation of hydrogen atoms at octahedral sites resulting in a volumetric expansion almost linear with concentration of hydrogen [30,33]. In single crystals, the speed of the process depends on the amount of surface exposed, the sticking probability for adsorption, the reactivity for dissociation of the hydrogen molecule on the particular crystalline faces exposed, the presence of interfering gases on the surface, and the bulk diffusion of hydrogen into the subsurface layers. In polycrystalline materials, the diffusion paths include grain boundaries. In both types of materials, voids and imperfections in the microstructure will produce sites with different binding energy for hydrogen and other gases resulting in variations in the kinetics of the sensing process [33,34].

Several technologies can be used for the Pd-based sensing of hydrogen, including resistive elements, field effect transistors, surface acoustic wave devices, quartz crystal microbalances, cantilevers, and optical fibers [2,3,5,32,35 and references within]. Sensors based on resistive elements are typically simple in design, fabrication and incorporation in a chip [2,3,35,36]. They require materials that change their conductance as a result of their exposure to hydrogen. In palladium, the transformation from the metal to the hydride by the incorporation of hydrogen results in rapid and measurable changes in conductance as a function of voltage, temperature, as well as gas concentration and composition. In bulk, the change in resistance, at a fixed voltage and temperature, is dominated by scattering of the conduction electrons resulting in decreased values of current as the hydrogen concentration increases. The formation of the Pd hydride also produces changes in weight, volume, density and reflectivity that are exploited in the sensing technologies mentioned previously. In metal-insulator-semiconductor (MIS) devices, the Pd-insulator interfaces are electrically disturbed by hydrogen that induces the formation of a dipole layer resulting in a shift in the C–V (capacitance–voltage) or I–V (current voltage) response of the device [2,3,37].

Materials with nanoscale morphologies are promising to improve sensor performance. Nanoscale morphological structures, such as nano-rosettes [38], nanowires [39–41], ultraporous networks [42], nanoshells [43], nanotubes [44], and ultrathin films [32,36,42,45–48], provide large surface areas for adsorption and smaller crystallite size reducing the time needed for "bulk" diffusion. The number of sites available for hydrogen adsorption per Pd atom is also higher in the surface and subsurface layers resulting in higher sensitivity [49].

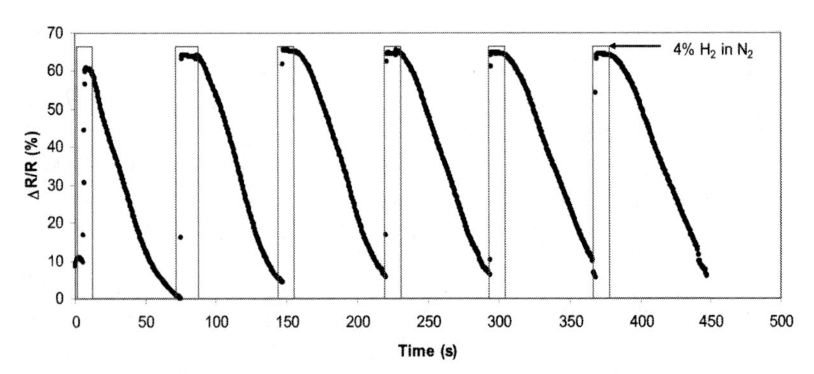

Figure 1.1 The electrical response of the Pd nanoshells is characterized by measuring the current as a function of time while exposed to flows of nitrogen and nitrogen/hydrogen mixtures. The sensitivity, $\Delta R/R \times 100\%$ is extracted from those curves and plotted as a function of time as shown. The Pd nanoshells are about 10–25 nm thick in the middle region of the nanoshells.

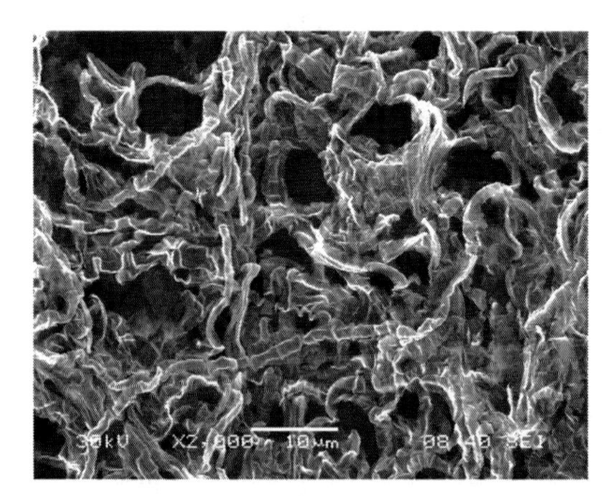

Figure 1.2 Pd shells obtained by sputtering the metal on top of a mat of electrospun polyethylene oxide fibers followed by a heat treatment to gasify the polymer. The thickness of the shell is estimated to be 25 nm in the center and thinner toward the edges [43].

An example of this can be seen in Fig. 1.1, which shows the sensing response of Pd nanoshells deposited on top of a mat of electrospun polymer fibers measured across electrodes. An example of such a Pd nanoshell structure is shown in Fig. 1.2. The synthesis method is described elsewhere [43]. Briefly, the nanoshells are prepared by magnetron sputtering deposition on top of the mat of polymer fibers, which act as a template that shapes the morphology of the palladium being deposited while providing support to the metallic scaffold that is created. Sputtering is a line-of-sight deposition process and the fibers become a variable angle-substrate for the incoming Pd flux. A larger amount of palladium is deposited on top of the fiber where the incoming flux is perpendicular to the surface compared to the sides where the flux is incident at a glancing angle. The top and sides of the fibers shadow their bottom parts closer to the substrate preventing any substantial deposition there. The end result of the deposition is the formation of Pd nanoshells that are thicker in the middle region than at the edges with a larger void network. Process parameters such as deposition time, sputtering pressure, and power can be used to produce nanoshells with different thickness and crystallinity. The high sensitivity and response time shown to 4% hydrogen in nitrogen is understood to result from the reduced dimensions combined with this unique nanostructure. Methods to improve the recovery time of the sensor structure are being investigated. However, this work shows the use of nanotechnology combined with unique processing approaches to produce new sensor geometries with different behavior and morphology than simple thin-film or wire approaches.

1.3.2 Pd Ultrathin Films and Nanowires Using Solid State Reduction and Zeolite Templates

The uses of Pd as a base sensor material can also be advantageously used to enable other sensing structures given that Pd can absorb 900 times its volume in H_2, no surface treatments are needed to achieve this property, and a phase transition from the α to the β phase occurs that causes its resistance to increase, thus resulting in a dramatic change in the measured current [33,50–52].

One example of the use of nanoscale morphologies is in ultrathin films, where the thickness is decreased to increase the surface area/volume ratio. When that thickness is less than about 5 nm, the film is not continuous, consisting of separate islands that could be completely isolated (discontinuous) or partially in contact (semicontinuous) [53]. Both types of non-continuous films have a higher resistance because the current paths are limited, resulting in an increased effective cross-sectional area. The electrons need to move through gaps between clusters by tunneling, thus increasing the resistivity. When hydrogen is adsorbed, the nanoclusters expand and their separation gap is reduced or eliminated. The current in the resistive sensor will then increase with exposure to hydrogen, a response that is contrary to Pd bulk sensors, with excellent sensitivity and cycleability [36,45,46,53–55].

A second approach also involves the phase change associated with the adsorption of hydrogen into Pd. This approach is based on the fact that the Pd phase transition is a function of the morphology of the Pd structures. Narehood et al. [56] showed that the phase transition occurs at a narrower H/Pd range for nanoparticles when compared with the bulk. This property of nanomaterials is potentially of high significance to the design of sensors based on Pd when considering the work of Sakamoto et al. [57], who showed a direct correlation between Pd hydrogen uptake and its changes in resistance.

Several groups have studied the response of Pd nanoparticles to hydrogen [39,40,48]. Favier et al. [58] found an inverse response of their Pd mesowires to hydrogen. They concluded that gaps in the connections of the Pd particles increased the resistance of the Pd while the swelling caused by the phase change when exposed to hydrogen decreased their resistance. This is analogous to the results in thin films mentioned above and suggests that, while it is undisputable that Pd has a large electrical response due to structural changes, the morphology of the Pd affects the response mechanism and should be very well characterized.

Fan Yang et al. [59] investigated the H_2 sensing properties of Pd nanowires of different grain diameters: (1) nano-crystalline Pd nanowires with a grain diameter ≈5 nm (*nc5*-Pd), and (2) nanocrystalline Pd nanowires with a grain diameter of 15 nm (*nc15*-Pd). They found that despite their fundamental similarities, the behavior of these nanowires upon exposure to H_2 was dramatically and reproducibly different: *nc5*-Pd nanowires spontaneously fractured upon exposure to H_2 above 1–2%. Fractured nanowires continued to function as sensors for H_2 concentrations above 2%, actuated by the volume change associated with the α to β-phase transition of PdH_x. *nc15*-Pd nanowires, in contrast, withstood repeated exposures to H_2 up to 10% without fracturing. Therefore, the sensing mechanism is highly dependent on the size of the nanostructure. Their results also showed that smaller nanowires synthesized via lithographically patterned nanowire electrodeposition [60] show accelerated response and recovery times [61].

A new synthesis technique to form Pt and Pd nanowires using solid state reduction and zeolites as templates has been developed that expands on these basic Pd sensing mechanisms [62]. The resulting Pd nanowires are formed on the surface of the zeolite with an approximate average diameter of 6 nm (Fig. 1.3). This method is the only one found in the literature in which metal nanowires are formed inside microporous materials. Molecular mechanics and dynamics simulations are under way to understand the effect of the synthesis conditions such as zeolite structure, zeolite Si/Al ratio, and temperature on the growth of these nanowires. After formation, these nanowires are isolated from the zeolite via hydrogen fluoride (HF) dissolution as shown in Fig. 1.4. It is evident that the morphology of the Pd nanowires changes in the process and that they become more grainy, this despite its generally accepted inertness towards the acid.

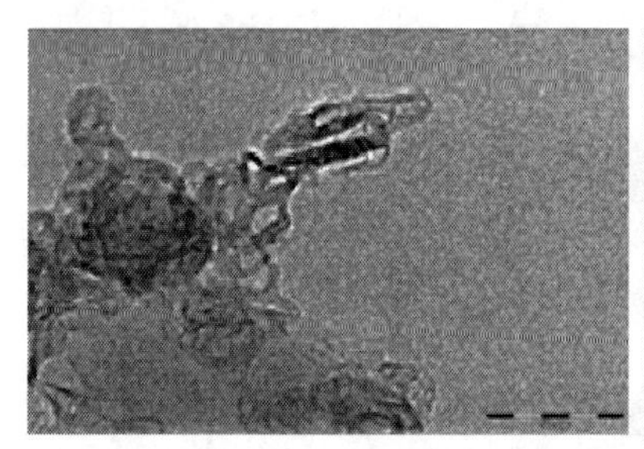

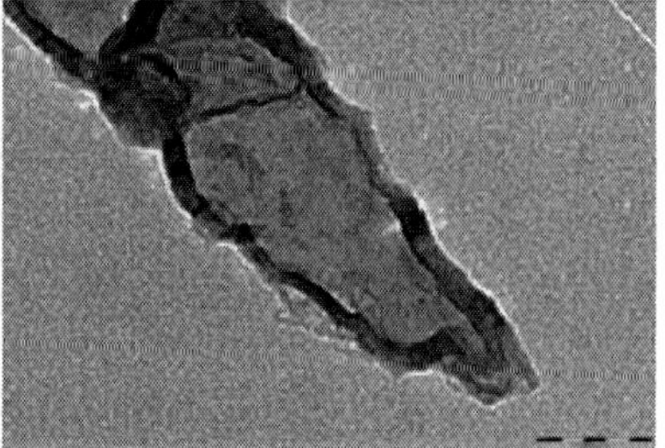

Figure 1.3 Pd nanowires formed on the surface of zeolite mordenite via solid-state reduction.

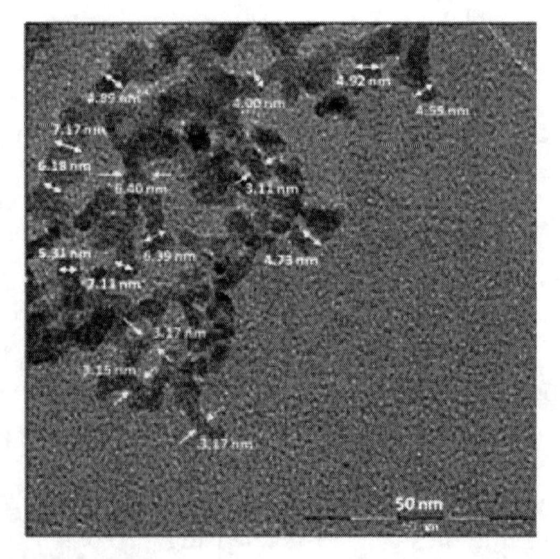

Figure 1.4 Pd nanowires after recovery with Hydrofluoric (HF) acid.

Figurc 1.5 shows the results on the response of the nanowires when deposited on interdigitated electrodes. It shows thcir response to hydrogen is typical with the resistance decreasing in the presence of hydrogen. The sensitivity is approximately 50%

and the response and recovery times are on the order of ~10 s. These are short times compared to minutes observed in some traditional Pd sensors suggesting that the nano-size of the particles does improve the response of the sensor. This could be explained by the phase transition of these materials in the presence of hydrogen shown in Fig. 1.6. Consistent with similar published results, the phase transition occurs in a narrower range of H/Pd, suggesting that it occurs faster than in the bulk when exposed to similar conditions.

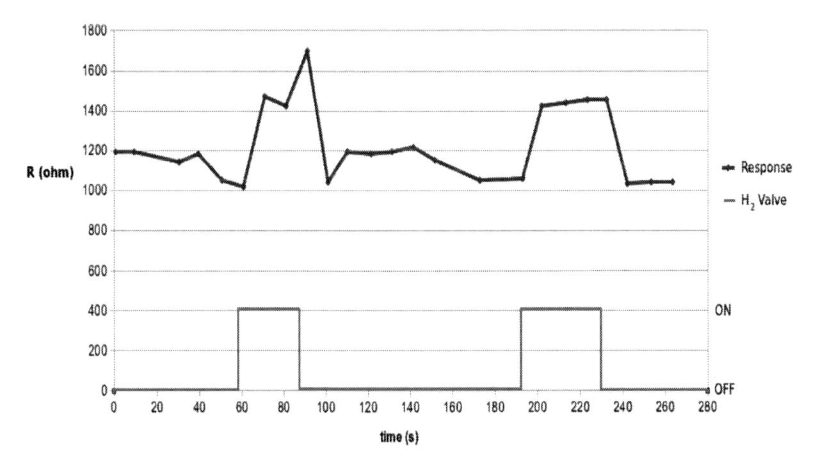

Figure 1.5 Sensor response at 1 V bias and 1% v/v H_2 concentration.

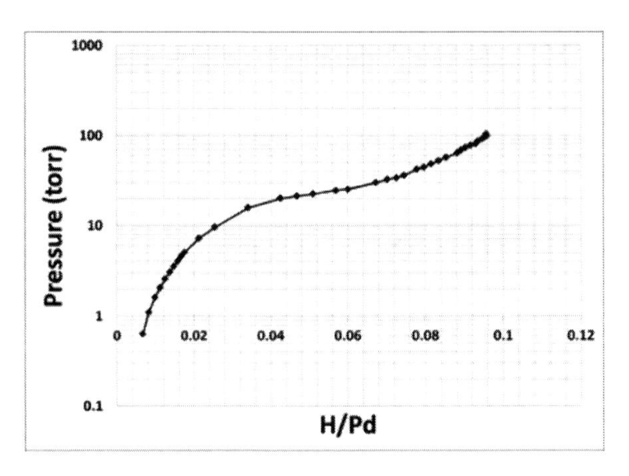

Figure 1.6 Pd nanowire hydrogen loading per Pd atom with presssure at 308 K.

These results agree qualitatively with the results by Yang et al. [59] and Narehood et al. [56], and suggest that this synthesis technique and sensor setup has potential for further development. In the calculation of the H/Pd values reported in Fig. 1.6, it was assumed that this sample was 100% Pd when in fact there was a nonquantified amount of zeolite impurity still present. The crystallinity of the different nanostructures mentioned still needs further study to improve the sensing mechanism. Further, it is well known that using Pd alloys improve sensing performance by stabilizing the crystalline structure [63], which may be used when dealing with these Pd-based structures. Overall, this work shows that controlling the details of the nanomaterial structure has a notable affect on sensor response. New and exciting mechanisms and materials are expected in the future combining in a clever way different nanoscale morphologies and crystallinity, and thus offering new opportunities to develop ultrasensitive sensors.

1.3.3 Transition Metal Silicide Nanowires for Temperature Sensors

The thermoelectrical properties of transition metal silicides materials have been thoroughly studied since 1958 [64–66]. Metal silicides have relative high Seebeck coefficients and good thermal and chemical tolerances that make them convenient materials for developing thermal sensors for aerospace applications. Thin films of transition metal silicides have been investigated for developing temperature sensor devices at NASA Glenn Research Center (GRC). Chromium silicide/tungsten carbide ($CrSi_2$/TC) thermocouples using thin-film technology with a sensitivity of 98 μV/K working with a 600°C temperature gradient was reported by NASA [67,68].

Nanowires (NW) are promising nanostructures for sensing applications where reduced size is required, such as in multi-functional sensors platforms. In many cases, nanowires show improved crystallinity and purity that enhance their sensing properties. The reduced size of the nanostructures makes them ideal to operate within environmental protecting layers that can expand their operational life times when working in harsh environments. The chemical vapor transport (CVT) and the chemical vapor deposition (CVD) methods have resulted in convenient synthesis of free standing nanowires of different transition metal silicides [69–77]. Figure 1.7 shows the SEM (scanning electron microscopy) image of (a) chromium silicide ($CrSi_2$) NWs on a Si substrate synthesized from chromium chloride ($CrCl_2$) precursor powder [78] and (b) cobalt silicide (CoSi) NWs on a silicon (Si) substrate synthesized from cobalt chloride (CoCl) precursor powder. In both cases, high yields are obtained. The formation of the NWs involves a chemical reaction between the evaporated precursor that reaches the silicon wafer and the silicon dioxide (SiO_2) passivating layer. The thickness of the silica layer is a critical parameter to consider during the synthesis [73]. Figure 1.8a shows the superposition of the secondary electron imaging (SEI) SEM image of a $CrSi_2$ NW and the O, Si and Cr (energy-dispersive X-ray spectroscopy (EDS) mappings. The image confirms that the CVD process produces silicide nanowires covered with a SiO_2 layer. Figures 1.8b–d show the Si-KA, O-KA, and Cr-KA EDS mappings, respectively (from Fig. 1.8d the radius of the NW is obtained ~70 nm). Figure 1.9 shows the high-resolution TEM (HRTEM) images of the tip of the NW. These wires grow perpendicular to the *c*-axis of the crystal.

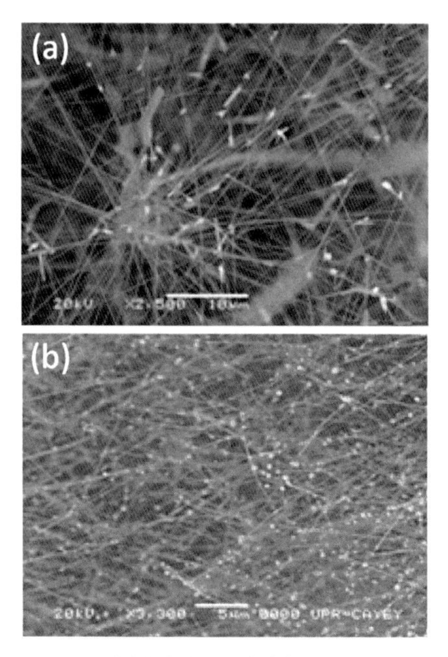

Figure 1.7 SEM images of (a) $CrSi_2$ and (b) CoSi nanowires grown by CVT.

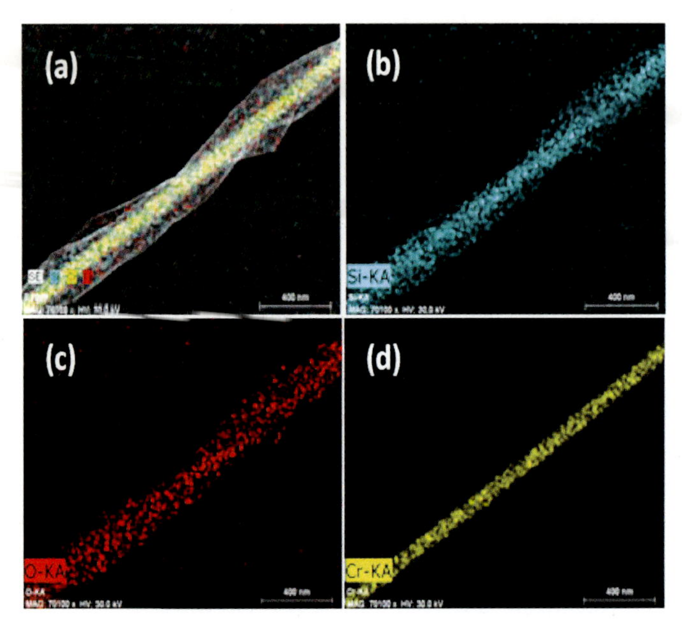

Figure 1.8 (a) Superposition of the SEI image of a $CrSi_2$ nanowire and EDS maps of Si, O and Cr. (b), (c) and (d): EDS distribution maps of Si, O, and Cr, respectively.

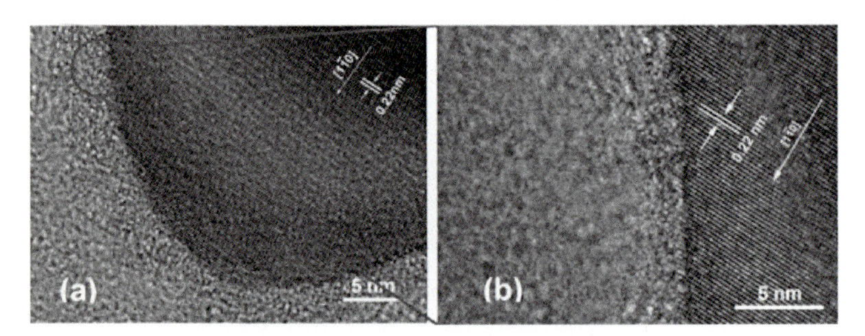

Figure 1.9 (a) HRTEM image of a $CrSi_2$ nanotip. (b) Closeup of a region near the surface showing the lattice fringes.

The thermal tolerance for the $CrSi_2$ NWs was studied by keeping the material at constant temperature in air during 30 min for T = 700, 800, 900, 1000, and 1100°C. Selective Area Electron Diffraction (SAED) and EDS analyses were used to determine changes in crystalline structure and composition, respectively. No significant changes were detected up to 800°C. Figures 1.10 shows the changes for 900 and 1100°C heating temperatures [78]. The

SAED patterns show an increasing lattice disorder as temperature reaches 900°C. Oxidation of the silicide is observed in the EDS mappings where an increasing thickness of the SiO_2 overlayer is confirmed. At 1100°C the SiO_2 amorphous phase is the predominant component of the nanostructure. In effect, while this level of oxidation would have a limited effect on a bulk material, the impact on a NW is more significant.

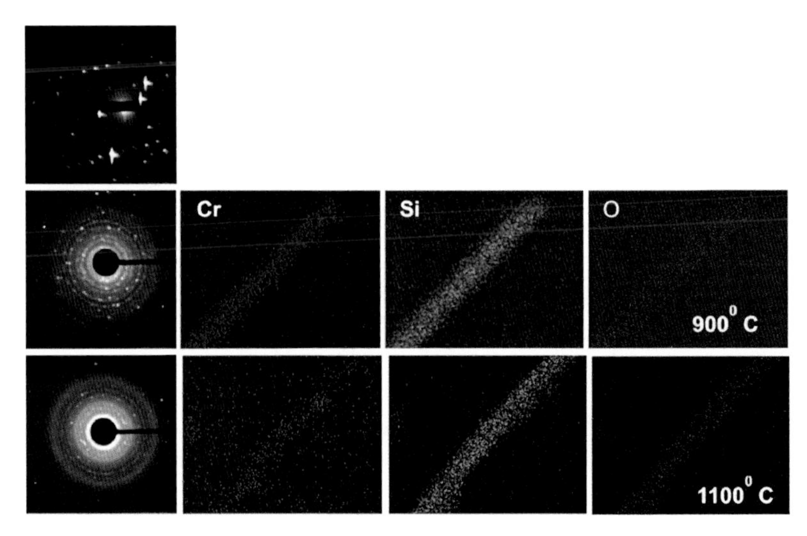

Figure 1.10 The upper image shows the SAED pattern of a $CrSi_2$ nanowire previous to heating. The figures below and in the first column show similar patterns after 30 min annealing in air for 900 and 1100°C, respectively. The images in columns 2 to 4 show the changes in the EDS mappings of Cr, Si, and O elements from 900 to 1100°C [78].

In Zhou et. al. [79], the thermoelectrical properties of $CrSi_2$ NWs grown by CVT along the *c*-axis of the crystalline structure were determined using specialized TE (thermoelectric) microdevices (see Chapter 2) and compared with bulk values. The resulting data show that the TE voltage of the material at the nanoscale compares with single crystal bulk value at room temperature. With Seebeck coefficients larger than 120 μV/K at room temperature and increasing at higher temperatures, this nanomaterial is a good starting choice for developing temperature sensors that can work in high-temperature environments. This work emphasizes the differences between nanomaterials and bulk materials.

This includes differences in physical parameters, (e.g., Seebeck coefficient), and the relative effect of oxidation processes. This core approach is being expanded to include other materials. The aim is to have an assortment of materials whose properties can be targeted for the application environment. As noted above, these silicides show excellent properties at temperatures below 800°C and can be considered for their measurement potential and for energy scavenging purposes provided the thermal conductivity is low. Other materials are being investigated for higher temperatures, using the same core fabrication and characterization approaches.

1.3.4 Combustion Product Sensors Based on Nanostructures

The use of nanocrystalline materials, i.e., materials with grain size on order of nanometers (nm), has reached a high level of maturity, including the capability to reproducibly fabricate the nanocrystalline materials and integrate them into sensor structures [1,2]. The use of these nanocrystalline materials results in improved sensor sensitivity, stability, and response time. Likewise, the ability for fabrication of sensor systems based on microfabrication techniques has allowed the production of a range of sensor structures. However, the ability to fabricate operational sensors from nanostructures is significantly less mature. Nanostructures, e.g., nanowires, nanotubes, and nanobelts, have fundamentally different structural properties than nanocrystalline grains. The fabrication procedures, properties, and possible advantages of nanostructured oxide sensors are just beginning to be explored [80–83] (see Chapter 2).

An example of this work is the use of tin oxide (SnO_2) nanostructures rather than tin oxide film or nanocrystalline materials. This work was aimed toward high-temperature emission monitoring, but SnO_2 and other metal oxide semiconductors have also been used for fire detection and environmental monitoring [8]. In particular, operation of a SnO_2 nanowire on a microsensor platform at high temperatures was demonstrated [8,84]. This example is meant to demonstrate both existing capabilities and the level of maturity of the technology.

These sensors are composed of SnO_2 and are synthesized by a thermal evaporation-condensation (TEC) approach [80]. The tin (IV) oxide (SnO_2) nanostructures are fabricated beginning with evaporating tin (II) oxide from a ceramic growth boat within a small flow of argon plus 5% oxygen in a tube furnace heated to 700–1000°C. Under these conditions, the vapor phase oxide self assembles into coherent rods that are tens of microns in length. Without the presence of a catalyst, growth proceeds by a self-catalyzed vapor–solid (VS) mechanism with the feed boat providing heteroepitaxial nucleation sites. After cooling, the oxide materials are collected from the growth boat for deposition upon the sensor pattern. The SnO_2 nanostructures are aligned on microelectrodes using dielectrophoresis resulting in the sensor structure shown in Fig. 1.11.

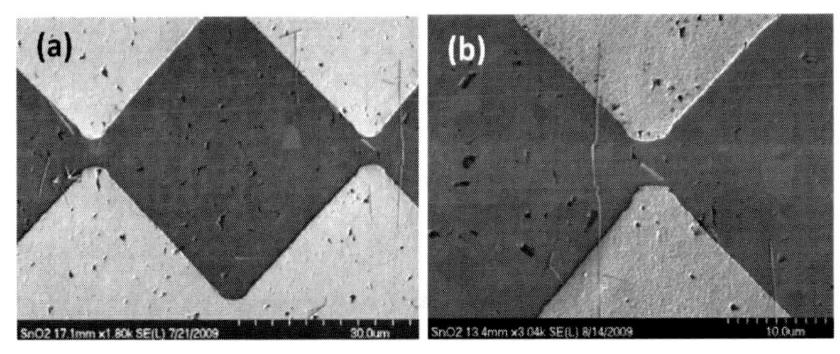

Figure 1.11 Dispersed SnO_2 nanowires deposited on a sensor and aligned with dielectrophoresis. (a) Image of bridging across several pairs of sawteeth. (b) A higher magnification view of a single set of opposing sawteeth with a single oxide nanowire bridging [84].

The gas sensing properties of the sensor structure of Fig. 1.11 were tested at 600°C as part of the development of these structures for high-temperature engine emission applications. The sensors were tested in a chamber with the sensor temperature controlled on a heater stage and electrical contact made with probes. The sensor was exposed to air, nitrogen, and propylene in nitrogen (propylene/nitrogen) at 600°C. The response of the sensor was measured by current output over time (I–t) at a constant voltage of 0.8 V. Current–voltage (I–V) curves from −1 to 1 V were taken periodically to examine sensor response over a range of voltages. The test was conducted over 75 h at 600°C with the overall objective of determining the sensor response at low concentrations of hydrocarbons (propylene).

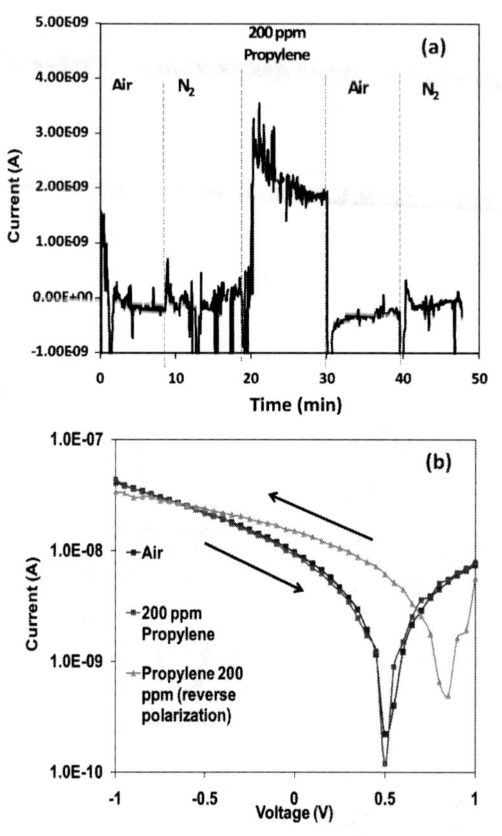

Figure 1.12 The response of the SnO_2 nanowire sensor at 600°C after 75 h to lower concentrations of propylene/ nitrogen. (a) The current (A) vs. time (min) at 0.8 V after exposure to air, N_2, 200 ppm propylene/nitrogen, followed by air and N_2 for 5 min each. (b) The magnitude current (A) vs. voltage (*V*) of the SnO_2 sensor in air as well as 200 ppm propylene/nitrogen taken from −1 V to 1 V and in reverse polarization 1 V to −1 V. While the sensor responds to low concentrations of propylene/nitrogen (as seen in closer examination of the *I*–*V* curves and Fig. 1.12a), the character of the *I*–*V* curves have changed and the mawith extended heating [84].

Figure 1.12 shows sensor response to low concentrations of propylene/nitrogen after 75 h at 600°C. Figure 1.12a shows the *I*–*t* in a linear scale and that the sensor responds to low concentrations of propylene nitrogen (200 ppm). Figure 1.12b shows there is limited shifting of the *I*–*V* curve between air and propylene/nitrogen.

Further, a hysteresis effect is observed [82]. As the *I–V* curve is taken in the reverse polarization, that is, from 1 to −1 V rather than −1 to 1 V, the propylene/nitrogen *I–V* curve actually shifts to the right rather than the left. The reason for this behavior after heating, and the overall sensor response in general, is unclear. Possible contributing factors include changes in the oxide structure, electrical contact between the metal oxide and the microsensor platform, and/or changes in the microsensor platform itself.

Further work is necessary for this sensor structure in particular to improve the (1) Characterization of the properties of the oxide nanostructures to understand their basic sensing properties; (2) Fabrication of the microstructures to allow optimization of the uniformity and density of nanostructure alignment; and (3) Reproducible electronic contact between the nanostructures and the electrodes of the interdigitated fingers. Thus, while the sensor has high sensitivity to low concentrations of hydrocarbons at high temperature and has high potential for emission sensor application, significant work is necessary to understand sensor behavior and optimize sensor response (see Chapter 2). Nonetheless, this work is a core foundation for future sensor systems based on metal oxide nanostructures.

1.3.5 ZnO Bandgap Modulation for Solar Blind UV Detectors

Zinc oxide (ZnO), a wide bandgap semiconductor, is one of the most important materials for the fabrication of UV detectors because of its high chemical stability and radiation hardness in harsh conditions [85]. However, the UV detectors based on ZnO thin films suffer the interference from the visible radiation and therefore, the signal to noise ratio is drastically affected for such type of detectors. In order to increase the bandgap, substitution of manganese (Mg) at Zn sites in ZnO has been done in the last decade [86–94]. However, there exists a miscibility gap between the two and only a certain amount of Mg can substitute at the Zn site, thus limiting the enhancement of the bandgap to deep UV regions [86]. Similar problems were observed with another binary system zinc beryllium oxide (ZnBeO) [95]. Therefore, in order to achieve the modification of the ZnO bandgap value to the regions where the interference from the visible radiation cannot deter the performance of the detector based on ZnO films, recently the co-doping of Mg and Be has been carried out and the bandgap has been successfully modulated, without any change in the wurtzite structure of ZnO, to more than 4.5 eV, i.e., in the solar blind region [96–99]. The large lattice mismatches of ZnO/BeO and ZnO/MgO are expected to be counteracted by each other. Here, we will discuss the progress made in the fabrication of crystalline ZnBeMgO thin films. Two approaches used are: Spin coating and pulse laser deposition (PLD).

1.3.5.1 ZnBeMgO films by spin coating

We fabricated, for the first time, thin films of $Zn_{1-x-y}Be_xMg_yO$ ($0 \leq x \leq 0.10$, $0 \leq y \leq 0.20$) on *c*-axis oriented sapphire substrates by chemical solution spin coating technique (Figure 1.13a) shows the XRD patterns of ZnBeMgO films on sapphire substrates. From the patterns it can be observed that in the co-doped films there are peaks only related to ZnBeMgO and no extra peaks related to either BeO or MgO. The pristine ZnO film has the wurtzite polycrystalline structure, however, it is observed that doped films exhibit the preferential (0002) *c*-axis orientation indicating that *c*-axis of the grains becomes uniformly perpendicular to the substrate surface. The (0002) peak intensity increases with the corresponding increase in dopants concentrations. The *c*-axis lattice parameter value of the pristine ZnO film is 5.189 Å and changes to 5.177 Å for $Zn_{0.8}Be_{0.1}Mg_{0.1}O$ and 5.158 Å for $Zn_{0.7}Be_{0.1}Mg_{0.2}O$ film. The systematic decrease in the lattice parameter value in the co-doped films also suggests the incorporation of smaller ions Be^{2+}/Mg^{2+} at Zn^{2+} sites and an enhanced out-of-plane residual compressive stress. The full width at half maximum (FWHM) values for the co-doped films are 0.31° for $Zn_{0.8}Be_{0.1}Mg_{0.1}O$ and 0.25° $Zn_{0.7}Be_{0.1}Mg_{0.2}O$, which implies that the films deposited by spin coating method have good crystallinity. Figure 1.13(b) shows the transmission spectrum of the films under the same conditions. All the films exhibited almost 80% transmittance in the visible region with sharp and single absorption edges in the UV region.

The bandgap of the films is estimated using Tauc's relationship [100] between the absorption coefficient, α, and the photon energy, $h\nu$, using

$$(\alpha h\nu)^2 = A(h\nu - E_g), \tag{1.1}$$

where A is a constant. This equation gives the bandgap values of ZnBeMgO films in the straight portion of $(\alpha h\nu)^2$ versus $h\nu$ plot, and extrapolated to $\alpha = 0$. The plots are given in Fig. 1.13(c). It can be seen that the bandgap value increases from 3.26 to 3.60 eV.

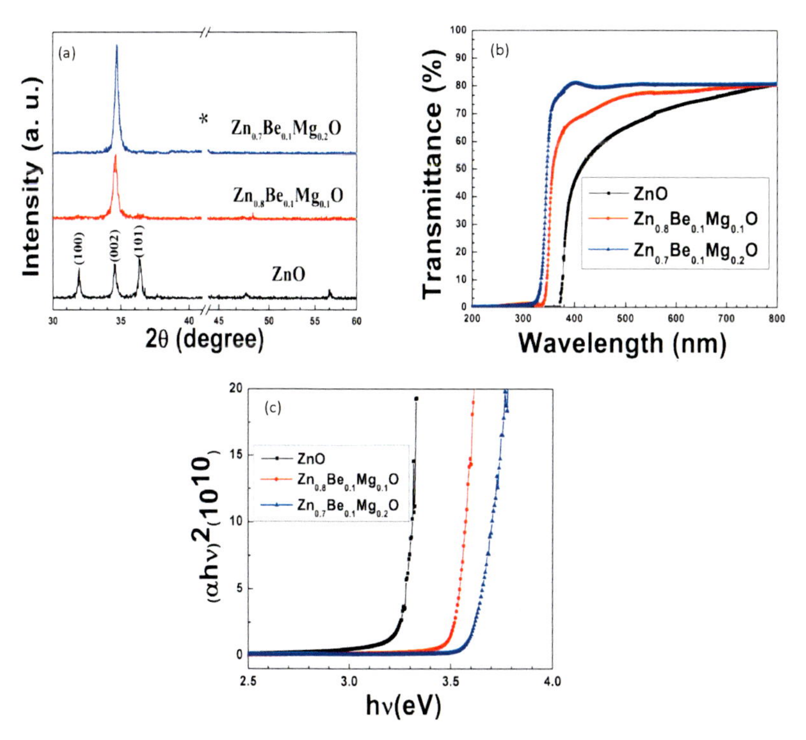

Figure 1.13 (a) X-ray diffraction pattern of $Zn_{1-x-y}Be_xMg_yO$ ($0 \leq x \leq 0.10$, $0 \leq y \leq 0.20$) films, (b) Optical transmittance measurement, (c) Plot for bandgap calculation using Eq. (1.1).

1.3.5.2 ZnBeMgO films by pulsed laser deposition (PLD)

Thin films of $Zn_{1-x-y}Be_xMg_yO$ ($0 \leq x \leq 0.15$; $0 \leq y \leq 0.20$) were deposited from the targets by PLD technique. Figure 1.14(a) shows the XRD patterns (in semi-log scale) of ZnBeMgO films on (0001) sapphire substrates. The pristine and the co-doped films exhibit the preferential *c*-axis orientation indicating that the *c*-axis of the grains is perpendicular to the substrate surface that can be assigned to the self-texturing mechanism in the films. It is worthwhile to note that no other phase related to MgO or BeO was detected in the doped films like those observed in targets. This may be due to the fact that during deposition at lower oxygen pressure some of the Mg^{2+} and Be^{2+} ions (because of their light weights) were taken out from the plume and flushed out of the chamber. In such cases, there may be less ions reaching the substrates. However, in the present case the targets with higher Be and Mg co-doping possess excess MgO and BeO, so during the film growth we get only the single phase film. From the XRD patterns of the films, it is also clear that the 2 theta value of the main peak increases significantly towards higher values. This can be thought due to the incorporation of Be^{2+}/Mg^{2+} smaller ions at Zn^{2+}-site in ZnO. The full width at half maximum (FWHM) of the pristine ZnO film is 0.18° and gets enhanced up to 0.31° in $Zn_{0.7}Be_{0.1}Mg_{0.2}O$ film. The increase in FHWM reveals that the crystalline nature of the films degrades after the incorporation of Be^{2+}/Mg^{2+} ions because these ions produce distortion in the lattice [96].

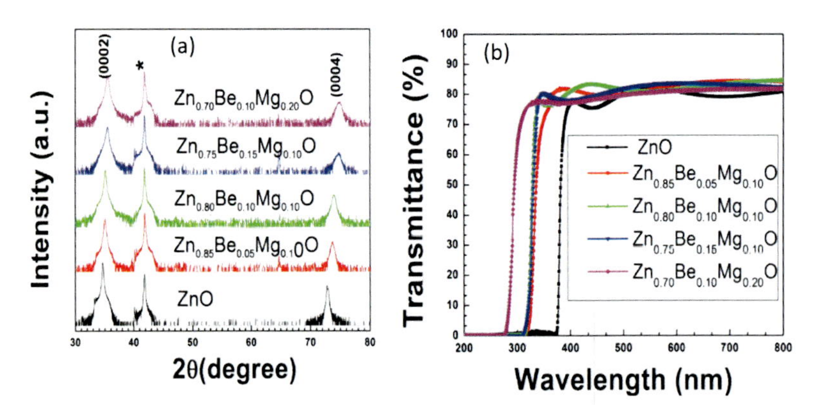

Figure 1.14 (a) X-ray diffraction patterns and (b) optical transmittance measurement of $Zn_{1-x-y}Be_xMg_yO$ ($0 \leq x \leq 0.15$; $0 \leq y \leq 0.20$) thin films by PLD.

Figure 1.14(b) shows the transmission spectrum of the films under the same conditions. The films show 80% transmittance in the visible region. The multi-absorption-edge, a typical characteristic of phase separation frequently observed in doped ZnO films, does not occur here. The band gap for the ZnO film is 3.37 eV, whereas with 10% Be and 20% Mg co-doped film $Zn_{0.7}Be_{0.1}Mg_{0.2}O$, the band gap increases up to 4.51 eV, which lies in the solar blind reason and is helpful in the realization of the deep UV detectors

[98]. Thus, ZnO-based thin films with bandgaps modulated to the solar blind region (UV transmission cutoff wavelength less than 280 nm and bandgap more than 4.5 eV) with Be and Mg co-doping have been realized by the pulsed laser deposition technique.

1.3.6 Long-Term Implementation/Further Necessary Steps

The development of nanotechnology such as that presented in this chapter is a core step to enable a revolution in how measurements are made [101] and the ability to monitor operational aerospace systems. The ability to make sensors with new material properties and enhanced capabilities can enable a next generation of smart sensor systems (see Chapter 2). One version of a smart sensor system is a complete self-contained sensor system that includes the capabilities for logging, processing sensor response, self-contained power, and an ability to transmit or display relevant data. The fundamental idea of a Smart Sensor is that the integration of silicon microprocessors with sensor technology can not only provide interpretive power and customized outputs, but also significantly improve sensor system performance and capabilities. A driving goal in the development of Smart Sensor Systems is the implementation of systems in a nonintrusive manner so that the information is provided to the user wherever and whenever needed, as well as in whatever form is needed for the application. In effect, Smart Sensor technology can enable intelligent systems for aerospace applications.

A long-term vision for an intelligent system is one that is self-monitoring, self-correcting and repairing, and self-modifying or morphing [1,4]. One approach is to build the system bottom-up from smart components. These smart components are independently self-monitoring, self-correcting, and self-modifying. Smart components monitor and adapt their individual status to the mission objectives and local conditions. Information is communicated two-way to local nodes and the collection of these smart nodes encompasses the overall system level operation. Each smart component has the capabilities to be, in a sense, self-aware. In biological terms, the smart component will know its environment (see, feel, hear and smell), think (process information), communicate, and adapt to the environment (move and self-reconfigure). Over all, the approach is self-aware components yielding self-aware intelligent systems.

Nanotechnology has the potential to enable such a vision through a new generation of smart sensor systems, but the realization of such a vision depends not only on developing the technology to enable the vision but also on the successful application of the technology. The application of sensor systems goes beyond development of simply the sensor element itself, but also the integration of these systems into supporting technologies. These supporting technologies include packaging, power, communication, and interfacing with the vehicle system. Nanotechnology enabled solutions are needed in these areas as well.

The implementation of nanotechnology and intelligent system technology in aerospace applications can take place in different stages of vehicle system development and operation. Three envisioned uses of intelligence enabling hardware in aerospace applications are (1) system development and ground testing where the sensor provides information on the state of a system that does not fly; this information could be used for the design and advanced modeling of systems that are in flight or for the design of future systems; (2) vehicle health monitoring (VHM), which involves long-term monitoring of a system in operation to determine the health of the vehicle system (e.g., is the engine increasing fuel burn or increasing emissions); this information could be used to change system parameters to improve vehicle safety or assist in ground-based maintenance; (3) active control of the vehicle in a feedback mode where information from a sensor system is used to change a vehicle parameter in real time (e.g., fuel flow to the engine changed due to system measurements) [2,4]. Each vehicle function or stage of operation (system development, VHM, or active control) has its own technology requirements. Therefore, the requirements for hardware to contribute to intelligent system operation depend on the state of vehicle development and the vehicle function the hardware is performing.

In order to qualify for any of the applications, significant application testing is necessary. An example is an oxygen sensor for the International Space Station (ISS) [23,24]. Attempts to qualify this system for ISS application included 8 years of equivalent pressure cycling testing and 2.5 years of benchtop testing. Similar testing would be needed for future implementation of nanotechnology for future applications. In effect, not only would nanotechnology for sensor applications need to provide new capabilities of value to the aerospace application, but also these capabilities would need to be demonstrated consistently for the specific application being considered. Sensor reliability is core to its acceptance in an application; this reliability must be proven for the sensor system to be implemented in vehicle systems. Overall, while nanotechnology for sensor systems has significant potential, in order for it to be implemented, the technology must be validated and proven for the individual application.

Acknowledgments

The authors would also like to acknowledge the contributions of Prof. C. C. Liu of Case Western Reserve University, Dr. A. Biaggi-Labiosa, Gus Fralick, John Wrbanek, David Litt, Dr. L. Matus, and Dr. M. Zeller, of NASA GRC; Dr. C. Chang and D. Lukco of ASRC Aerospace/NASA GRC; Dr. B. Ward and Dr. D. Makel of Makel Engineering, Inc.; Prof. P. Dutta of Ohio State University; J. González, M. Artale, P. Lampard, C. Hampton of Sierra Lobo/NASA GRC. L. F. Fonseca acknowledges contributions from L. Valentín and J. Carpena from UPR-Rio Piedras, and Prof. M. José-Yacamán (UT-San Antonio) for the use of the facilities at the University of Texas at San Antonio. W. Otaño acknowledges contribution from Diego Rodríguez-Vindas, Dr. Carlos Ortiz, and Dr. Víctor M. Pantojas from UPR-Cayey. Martínez-Iñesta acknowledges the contribution of Héctor Méndez-Colberg from UPR-Mayaguez to this manuscript and the collaboration of Dr. Maxime Guinel from UPR-Rio Piedras to the experimental results. UPR authors acknowledge support from NASA-URC grant #NNX08BA48A and NSF-grant #1002410.

References

1. Hunter, G. W., Oberle, L. G., Baakalini, G., Perotti, J., and Hong, T. (2011). Integrated system health management in exploration applications. In *System Health Management: With Aerospace Applications* (Johnson, S. B., Gormley, T., Kessler, S., Mott, C., Patterson-Hine, A., Reichard, K., Scandura, Jr., P., eds.) Chapter 25, John Wiley & Sons, Ltd, Chichester, West Sussex, UK.

2. Hunter, G. W., Xu, J. C., Liu, C. C., and Makel, D. B. (2006). Microfabricated chemical sensors for aerospace applications. In *The MEMS Handbook, Second Edition: Volume 2 Design and Fabrication* (Gad-El-Hak, M., ed.) Chapter 11, CRC Press, Baton Rouge, LA.

3. Hunter, G. W., Xu, J. C., and Makel, D. B. (2008). Case studies in chemical sensor development. In *BioNanoFluidic MEMS* (Hesketh, P. J., eds.) Chapter 8 Springer Science and Business Media, New York, pp. 197–231.

4. Hunter, G. W. (2003). Morphing, Self-Repairing Engines: A Vision for the Intelligent Engine of the Future, *AIAA/ICAS International Air & Space Symposium, 100th anniversary of Flight* (14–17 July 2003, Dayton, OH, AIAA paper 2003-3045).

5. Hunter, G. W., Xu, J. C., Neudeck, P. G., Makel, D. B., Ward, B., and Liu, C. C. (2006). Intelligent Chemical Sensor Systems For In-Space Safety Applications. *42nd AIAA/ASME/SAE/ASEE Joint Propulsion Conference And Exhibit, Sacramento, CA,* (9–12 July 2006, Paper AIAA-2006-4356).

6. Hunter, G. W. (1992). A Survey and Analysis of Commercially Available Hydrogen Sensors, *NASA Technical Memorandum 105878.*

7. Hunter, G. W. (1992). A Survey and Analysis of Experimental Hydrogen Sensors, *NASA Technical Memorandum 106300.*

8. Hunter, G. W., Xu, J. C., Evans, L., Biaggi-Labiosa, A., Ward, B. J., Rowe, S., Makel, D. B., Liu, C. C., Dutta, P., Berger, G. M., and Vander Wal, R. L. (2010). The Development of Micro/Nano Chemical Sensor Systems for Aerospace Applications, *SPIE Newsroom, June 2010, 10.1117/2.120100 6.002984.*

9. Lei, J. F., Martin, L. C., and Will, H. A. (1997). Advances in Thin Film Sensor Technologies for Engine Applications, *NASA TM–107418, Turbo Expo'97,Orlando, FL, June 2–5.*

10. Martin, L. C., Wrbanek, J. D., and Fralick, G. C. (2001). Thin film sensors for surface measurements, *NASA/TM—2001–211149, Sept. 2001.*

11. Wrbanek, J. D., and Fralick, G. C. (2006). Thin film physical sensor instrumentation research and development at NASA Glenn Research Center, *NASA/TM—2006–214395, Sept. 2006.*

12. Hunter, G. W., Wrbanek, J. D., Okojie, R. S., Neudeck, P. G., Fralick, G. C., Chen, L., Xu, J., and Beheim, G. M. (2006). Development and application of high temperature sensors and electronics for propulsion applications, *Proc. SPIE,* 6222, 622209.

13. Stange, W. (2006). Instrumentation Needs to Reduce Turbine Engine Related Class A Mishaps Within the USAF, *52nd International Instrumentation Symposium*, (Cleveland, OH, May 7–11).

14. Wrbanek, J. D., Fralick, G. C., Gonzalez, J. M., and Laster, K. L., (2008). Thin Film Ceramic Strain Sensor Development for High Temperature Environments, *NASA/TM*—2008-215256, *June 2008.*

15. Anson, D., and Richerson, D. W. (2002). The benefits and challenges of the use of ceramics in gas turbines. In *Progress in Ceramic Gas Turbine Development, Volume 1—Ceramic Gas Turbine Design and Test Experience* (van Roode, M., Ferver, M. K., and Richerson, D. W., eds.) ASME PRESS, New York, pp. 1–10.

16. Schenk B., Easley, M. L., and Rickerson, D. W. (2002). Evolution of ceramic turbine engine technology at Honeywell Engines, Systems & Services. In *Progress in Ceramic Gas Turbine Development, Volume 1—Ceramic Gas Turbine Design and Test Experience* (van Roode, M., Ferver, M. K., and Richerson, D. W., eds.) ASME PRESS, New York, pp. 77–110.

17. Levine S. R., Calomino A. M., Verrilli M. J., Thomas D. J., Halbig M. C., Opila E. J., and Ellis J. R. (2003). Ceramic Matrix Composites (CMC) Life Prediction Development-2003, *NASA/TM*—2003–212493, *Aug. 2003.*

18. Ward, B. J., Wilcher, K., and Hunter, G. (2010). "Gas Microsensor Array Development Targeting Enhanced Engine Emissions Testing," *Infotech@Aerospace, AIAA,* (Reston, VA, 20–22 April, AIAA 2010-3327).

19. Logothetis, E. M. (1991). Automotive oxygen sensors. In *Chemical Sensor Technology* (Yamazoe, N., ed.), Kodansha Ltd. vol. 3, pp. 89–104.

20. Hunter, G. W., Xu, J. C., Greenberg, P., Ward, B., Carranza, S., Makel, D., Liu, C. C., Dutta, P., Lee, C., Akbar, S., and Blake, D. (2004). Miniaturized Sensor Systems for Aerospace Fire Detection Applications, *The Fourth Triennial International Aircraft Fire and Cabin Safety Research Conference, Lisbon, Portugal, Nov. 9.*

21. Proceedings of the 1995 Workshop on Fire Detection Research (1995). (Grosshandler, W. L., June, Gaithersburg, MD ed.), NISTIR 5700.

22. Grosshandler, W. L. (1995). "A Review of Measurements and Candidate Signatures for Early Fire Detection," NISTIR 555, *Nat. Inst. Stand. Tech.*, Gaithersburg, MD.

23. Hunter, G. W., Greenberg, P. S., Xu, J. C., Ward, B., Makel, D., Dutta, P., and Liu, C. C. (2009). Miniaturized sensor systems for early fire detection in spacecraft, *39th International Conference On Environmental Systems, SAE*, Warrendale, PA, 12–16 July, 09ICES-0335.

24. Hunter, G. W., Xu, J. C., Dungan, L., Ward, B., Dutta, P., Adeyemo, A. D., Liu, C. C., and Gianettino, D. P. (2010). Smart chemical sensor systems for fire detection and environmental monitoring in spacecraft, *40th International Conference On Environmental Systems, AIAA,* (Barcellona, Spain, 11–15 July, AIAA766637).
25. Hunter, G. W., Xu, J. C., Biaggi-Labiosa, A. M., Ward, B., Dutta, P., and Liu, C. C. (2011). Smart Sensor Systems for Spacecraft Fire Detection and Air Quality Monitoring, *40th International Conference On Environmental Systems, AIAA,* (Portland, Oregon, 17–21 July, 17–21).
26. Constellation Program Human-Systems Integration Requirements, Rev. C, Document CxP 70024.
27. Walker, D., Kumar, V., Mi, K., Sandvik, P., Kung, P., Zhang, X. H., and Razeghi, M. (2000). Solar-blind AlGaN photodiodes with very low cutoff wavelength, *Appl. Phys. Lett.*, **76**(403), 125768.
28. "Nanotechnology-Enabled Sensing," (2009). Report on the National Nanotechnology Initiative Workshop, Arlington, Virginia, May 5–7, available at http://www.nano.gov/.
29. Stetter, J. R., Hesketh, P. J., and Hunter, G. W. (2006). Sensors: engineering structures and materials from micro to nano, *ECS Interface*, **15**(1), 66–69.
30. Zoltowski, P. (2010). On the importance of equilibrium in studies on the transport of hydrogen in metals, *Electrochem. Act.*, **55**, 6274–6282.
31. Jewell, L. L., and Davis, B. H. (2006). Review of absorption and adsorption in the hydrogen-palladium system, *Appl. Catal. A: Gen.*, **310**, 1–15.
32. Patton, J. F., Hunter, S. R., Sepaniak, M. J., Daskos, P. G., and Barton, D. (2010). Rapid response microsensor for hydrogen detection using nanostructured palladium films, *Sens. Actuators A: Phys.*, **163**(2), 464–446.
33. Lewis, F. A. (1982). The palladium-hydrogen system, *Platinum Metals Rev.*, **26**(2), 70–78.
34. Ravi Prakash, J., McDaniel, A. H., Horn, M., Pilione, L., Sunal, P., Messier, R., McGrath, R. T., and Schweighardt, F. K. (2007). Hydrogen sensors: role of palladium thin film morphology, *Sens. Actuators B*, **120**, 439–446.
35. Buttner, W. J., Post, M. B., Burgess, R. T., and Rivkin, C. (2010). An overview of hydrogen safety sensors and requirements, *Int. J. Hydrogen Energy*, **36**(3), 2462–2470.

36. Xu, T., Zach, M. P., Xiao, Z. L., Rosenmann, D., Welp, U., Kwok, W. K., and Crabtree, G. W. (2005). Self-assembled monolayer-enhanced hydrogen sensing with ultrathin palladium films, *Appl. Phys. Lett.*, **86**(20), 203104.

37. Salomonsson, A., Eriksson, M., and Dannetun, H. (2005). Hydrogen interaction with platinum and palladium metal-insulator-semiconductor devices, *J. Appl. Phys.*, **98**, 014505.

38. Zou, J., Hubble, L. J., Iyer, K. S., and Raston, C. L. (2010). Bare palladium nano-rosettes for real-time high-performance and facile hydrogen sensing, *Sens. Actuators B: Chem.*, **15**(1), 291–295.

39. Walter, E. C., Ng, K., Zach, M. P., Penner, R. M., and Favier, F. (2002). Electronic devices from electrodeposited metal nanowires, *Microelectron. Eng.*, **61–62**, 555–561.

40. Cherevko, S., Kulyk, N., Fu, J., and Cheng, C.-H. (2009). Hydrogen sensing performance of electrodeposited conoidal palladium nanowire and nanotube arrays, *Sens. Actuators B: Chem.*, **136**(2), 388–391.

41. Offermans, P., Tong, H. D., Van Rijn, C. J. M., Merken, P., Brongersma, S. H., and Crego-Calama, M. (2009). Ultralow-power hydrogen sensing with single palladium nanowires, *Appl. Phys. Lett.*, **94**, 223110.

42. Zeng, X. Q., Latimer, M. L., Xiao, Z. L., Panuganti, S., Welp, U., Kwok, W. K., and Xu, T. (2011). Hydrogen gas sensing with networks of ultrasmall palladium nanowires formed on filtration membranes, *Nanoletters*, **11**(1), 262–268.

43. Pantojas, V. M., Rodríguez-Vindas, D., Morell, G., Rivera, A., Ortiz, C., Santiago-Aviles, J. J., and Otaño, W. (2008). Synthesis of palladium with different nanoscale structures by sputtering deposition onto fiber templates, *J. Nanophoton*, **2**, 021920.

44. Yu, S., Welp, U., Hua, L. Z., Rydh, A., Kwok, W. K., and Wang, H. H. (2005). Fabrication of palladium nanotubes and their application in hydrogen sensing, *Chem. Mater*, **17**, 3445–3450.

45. Dankert, O., and Pundt, A. (2002). Hydrogen-induced percolation in discontinuous films, *Appl. Phys. Lett.*, **81**(9), 1618–1620.

46. Kaltenpoth, G., Schnabel, P., Menke, E., Walter, E. C., Grunze, M., and Penner, R. M. (2003). Multimode detection of hydrogen gas using palladium-covered silicon μ-channels, *Anal. Chem.*, **75**(18), 4756–4765.

47. Van Lith, J., Lassesson, A., Brown, S. A., Schulze, M., Partridge, J. G., and Ayesh, A. (2007). A hydrogen sensor based on tunneling between palladium clusters, *App. Phys. Lett.*, **91**(18), 181910.

48. Favier, F. (2009). Nanogaps for sensing, *Procedia Chem.*, **1**(1), 746–749.

49. Jeon, K. J., Jeun, M., Lee, E., Lee, J. M., Lee, K. I., Allmen, P. V., and Lee, W. (2008). Finite size effect on hydrogen gas sensing performance in single Pd nanowires, *Nanotechnology*, **19**, 495501.

50. Lewis, F. A. (1960). The hydrides of palladium and palladium alloys, *Platinum Metals Rev.*, **4**, 132–137.

51. Lewis, F. A. (1961). The hydrides of palladium and palladium alloys, *Platinum Metals Rev.*, **5**, 21–25.

52. Lewis, F. A. (1967). *The Palladium Hydrogen System*, (Academic Press, New York).

53. Kiefer, T., Villanueva, L. G., Fargier, F., Favier, F., and Brugger, J. (2010). The transition in hydrogen sensing behavior in noncontinuous palladium films, *Appl. Phys. Lett.*, **97**, 121911.

54. Dasari, R., and Zamborini, F. P. (2008). Hydrogen switches and sensors fabricated by combining electropolymerization and Pd electrodeposition at microgap electrodes, *J. Am. Chem. Soc. Commun.*, **130**, 16138–16139.

55. Ramanathan, M., Skudlarek, G., Wang, H. H., and Darling, S. B. (2010). Crossover behavior in the hydrogen sensing mechanism for palladium ultrathin films, *Nanotechnology*, **21**, 12550.

56. Narehood, D. G., Kishore, S., Goto, H., Adair, J. H., Nelson, J. A., Gutierrez, H. R., and Eklund, P. C. (2009). X-ray diffraction and H-storage in ultra-small palladium particles, *Int. J. Hydrogen Energy*, **34**, 952–960.

57. Sakamoto, Y., Takai, K., Takashima, I., and Imada, M., (1996). Electrical resistance measurements as a function of composition of palladium–hydrogen(deuterium) systems by a gas phase method, *J. Phys. Condens. Matter*, **8**, 3399–3411.

58. Favier, F., Walter, E., Zach, M. P., Benter, T., and Penner, R. M. (2001). Hydrogen sensors and switches from electrodeposited palladium mesowire arrays. *Science*, **293**, 2227–2231.

59. Yang, F., Taggart, D. K., and Penner, R. M. (2009). Fast, sensitive hydrogen gas detection using single palladium nanowires that resist fracture, *Nano Lett.*, **9**, 2177–2182.

60. Yang, F., Kung, S. H., Taggart, D. K., and Penner, R. M. (2010). Hydrogen sensing with a single palladium nanowire, *Sens. Lett.*, **8**, 534–538.

61. Yang, F., Kung, S. H., Cheng, M., Hemminger, J. C., and Penner, R. M. (2010). Smaller is faster and more sensitive: the effect of wire size on the detection of hydrogen by single palladium nanowires, *ACS Nano*, **4**, 5233–5244.

62. Quiñones, L., Grazul, J., and Martínez-Iñesta, M. M. (2009). Synthesis of platinum nanostructures in zeolite mordenite using a solid-state reduction method. *Mater. Lett.*, **63**, 2684–2686.

63. Wang, M., and Feng, Y. (2007). Palladium–silver thin film for hydrogen sensing, *Sens. Actuators B: Chem.*, **123**, 101–106.

64. Nitkin, E. N. (1958). Study of temperature dependences of electrical conductivity and thermal power of silicides, *Zhurnal Tekhnicheskoj Fiziki*, **28**, 23.

65. Viningn, C. B. (1995). *CRC Handbook of Thermoelectrics.* (Rowe, D. M, ed.), Chapter 23, CRC Press, New York.

66. Federov, M. I., and Zaitsev, V. K. (2006). *Thermoelectrics Handbook, Macro to Nano,* (Rowe, D. M., ed.), Chapter 31, Taylor and Francis, Florida.

67. Wrbanek J. D., and Fralick, G. C. (2006). Thin film physical sensor instrumentation research and development at NASA Glenn Research Center, *NASA/TM-2006-214395, Sept. 2006.*

68. Wrbanek, J. D., Fralick, G. C., Farmer, S. C., Sayir, A., Blaha, C. A., and Gonzalez, J. M. (2004). Development of thin film ceramic thermocouples for high temperature environments, *NASA/TM—2004-213211, AIAA–2004–3549, Aug 2004.*

69. Wu, Y., Xiang, J., Yang, C., Lu, W., and Lieber, C. M. (2004). Single-crystal metallic nanowires and metal/semiconductor nanowire heterostructures, *Nature*, **430**, 61–65.

70. Ouyang, L., Thrall, E. S., Deshmukh, M. M., and Park, H. (2006). Vapor-phase synthesis and characterization of ε-FeSi nanowires, *Adv. Mater.*, **18**, 1437–1440.

71. Schmitt, A. L., Bierman, M. J., Schmeisser, D., Himpsel, F. J., and Jin, S. (2006). Synthesis and properties of single-crystal FeSi nanowires, *Nano Lett.*, **6**, 1617–1621.

72. Song, Y., and Jin, S. (2007). Synthesis and properties of single-crystal β_3-Ni_3Si nanowires, *Appl. Phys. Lett.*, **90**, 173122.

73. Schmitt, A. L., and Jin, S. (2007). Selective patterned growth of silicide nanowires without the use of metal catalysts, *Chem. Mater.*, **19**, 126–128.

74. Szczech, J. R., Schmitt, A. L., Bierman, M. J., and Jin, S. (2007). Single-crystal semiconducting chromium disilicide nanowires synthesized via chemical vapor transport, *Chem. Mater.*, **19**, 3238–3243.

75. Song, Y., Schmitt, A. L., and Jin, S. (2007). Ultralong single-crystal metallic Ni_2Si nanowires with low resistivity, *Nano Lett.*, **7**, 965–969.

76. Schmitt, A. L., Zhu, L., Schmeisser, D., Himpsel, F. J., and Jin, S. (2006). Metallic single-crystal CoSi nanowires via chemical vapor deposition of single-source precursor, *J. Phys. Chem. B*, **110**, 18142–18146.

77. Seo, K., Varadwaj, K. S. K., Cha, D., In, J., Kim, J., Park, J., and Kim, B. (2007). Synthesis and electrical properties of single crystalline $CrSi_2$ nanowires, *J. Phys. Chem. C*, **111**, 9072–9076.

78. Valentin, L. A., Carpena-Nuñez, J., Yang, D., and Fonseca, L. F. (2013). Field emission properties of single crystal chromium disilicide nanowires, *J. Appl. Phys.*, **113**, 014308.

79. Zhou F., Szczech, J., Pettes M. T., Moore A. L., Jin S., and Shi L. (2007). Determination of transport properties in chromium disilicide nanowires via combined thermoelectric and structural characterizations, *Nano Lett.*, **7**, 1649–1654.

80. Hunter, G. W., Xu, J. C., Evans, L. J., Vander Wal, R. L., Berger, G. M., Kulis, M. J., and Liu, C. C. (2006). Chemical sensors based on metal oxide nanostructures, *ECS Trans.*, **3**(9), 199–209.

81. Hunter, G. W., Vander Wal, R. L., Xu, J.C., Evans, L. J., Berger, G. M., and Kulis, M. J. (2008). The development of metal oxide chemical sensing nanostructures, *ECS Trans.*, **16**(14), 73–84.

82. Vander Wal, R. L., Hunter, G. W., Xu, J. C., Kulis, M. J., Berger, G. M., and Ticich, T. M. (2009). Metal-oxide nanostructure and gas-sensing performance, *Sens. Actuators B*, **138**(1), 113–119.

83. Vander Wal, R. L., Berger, G. M., Kulis, M. J., Hunter, G. W., Xu, J. C., and Evans, L. E. (2009). Synthesis methods, microscopy characterization and device integration of nanoscale metal oxide semiconductors for gas sensing, *Sensors*, **9**(10), 7866–7902.

84. Hunter, G. W., Xu, J. C., Evans, L., Biaggi-Labiosa, A., Ward, B. J., Rowe, S., Makel, D. B., Liu, C. C., Dutta P., Berger, G. M., and Vander Wal, R. L. (2010). The development of micro/nano chemical sensor systems for aerospace applications, in *Micro- and Nanotechnology Sensors, Systems, and Applications II*, SPIE Proceedings vol. 7679.

85. Ghosh, T., and Basak, D. (2010). Highly efficient ultraviolet photodetection in nanocolumnar RF sputtered ZnO films: a comparison between sputtered, sol–gel and aqueous chemically grown nanostructures, *Nanaotechnology*, **21**, 375202.

86. Ohtomo, A., Kawasaki, M., Koida, T., Masubuchi, K., Koinuma, H., Sakurai, Y., Yoshida, Y., Yasuda, T., and Segawa, Y. (1998). $Mg_xZn_{1-x}O$ as a II–VI widegap semiconductor alloy, *Appl. Phys. Lett.*, **72**, 2466.

87. Takeuchi, I., Yang, W., Chang, K. S., Aronova, M. A., Venkatesan, T., Vispute, R. D., and Bendersky, L. A. (2003). Monolithic multichannel

ultraviolet detector arrays and continuous phase evolution in $Mg_xZn_{1-x}O$ composition spreads, *J. Appl. Phys.*, **94**, 7336.

88. Hullavarad, S. S., Dhar, S., Varughese, B., Takeuchi, I., Venkatesan, T., and Vispute, R. D. (2005). Realization of $Mg_{(x = 0.15)}Zn_{(1-x = 0.85)}O$-based metal-semiconductor-metal UV detector on quartz and sapphire, *J. Vac. Sci. Technol. A*, **23**, 982–985.

89. Zhang, X., Li, X. M., Chen, T. L., Zhang, C. Y., and Yu, W. D. (2005). p-Type conduction in wide-gap $Zn_{1-x}Mg_xO$ films grown by ultrasonic spray pyrolysis, *Appl. Phys. Lett.*, **87**, 092101.

90. Kong, J. F., Shen, W. Z., Zhang, Y. W., Yang, C., and Li, X. M. (2008). Resonant Raman scattering probe of alloying effect in ZnMgO thin films, *Appl. Phys. Lett.*, **92**, 191910.

91. Ghosh, R., and Basak, D. (2007). Composition dependent ultraviolet photoresponse in $Mg_xZn_{1-x}O$ thin films, *J. Appl. Phys.*, **101**, 113111.

92. Li, C., Meng, F. Y., Zhang, S., and Wang, J. Q. (2010). Effects of Mg content and B doping on structural, electrical and optical properties of $Zn_{1-x}Mg_xO$ thin films prepared by MOCVD, *J. Cryst. Growth*, **312**, 1929–1934.

93. Yoshino, K., Oyama, S., and Yoneta, M. (2008). Structural, optical and electrical characterization of undoped ZnMgO film grown by spray pyrolysis method, *J. Mater. Sci. Mater. Electron.*, **19**(2), 203–209.

94. Sandeep, C. S. S., Philip, R., Satheeshkumar, R., and Kumar, V. (2006). Sol-gel synthesis and nonlinear optical transmission in $Zn_{(1-x)}Mg_{(x)}O$ ($x \leq 0.2$) thin films, *Appl. Phys. Lett.*, **89**, 063102.

95. Ding, S. F., Fan, G. H., Li, S. T., Chen, K., and Xiao, B. (2007). Theoretical study of $Be_xZn_{1-x}O$ alloys, *Phys. B*, **394**, 127–131.

96. Yang, C., Li, X. M., Gu, Y. F., Yu, W. D., Gao X. D., and Zhang, Y. W. (2008). ZnO based oxide system with continuous bandgap modulation from 3.7 to 4.9 eV, *Appl. Phys. Lett.*, **93**, 112114.

97. Panwar, N., Liriano, J., and Katiyar, R. S. (2011). Structural and optical analysis of ZnBeMgO powder and thin films, *J. Alloys Compd.*, **509**, 1222–1225.

98. Yang, C., Li, X. M., Gao, X. D., Cao, X., Yang, R., and Li, Y. Z. (2010). Effects of the oxygen pressure on the structural and optical properties of ZnBeMgO films prepared by pulsed laser deposition, *J. Cryst. Growth*, **312**(7), 978–981.

99. Yang, C., Li, X. M., Yu, W. D., Gao, X. D., Cao, X., and Li, Y. Z. (2009). Zero-biased solar-blind photodetector based on ZnBeMgO/Si heterojunction, *J. Phys. D: Appl. Phys.*, **42**, 152002.

100. Tauc, J. (Ed.) (1974). *Amorphous and Liquid Semiconductors*, Plenum Press, NY.

101. Nanotechnology-Enabled Sensing, Report on the National Nanotechnology Initiative Workshop, Arlington, Virginia, May 5–7, (2009). Available at http://www.nano.gov/.

Chapter 2

Challenges and Possibilities in Nanosensor Technology

Gary W. Hunter,[a] Laura Evans,[a] Jennifer Xu,[a] Azlin Biaggi-Labiosa,[a] Randy L. Vander Wal,[b] Luis F. Fonseca,[c] Gordon M. Berger,[d] and Mike J. Kulis[d]

[a]*NASA John H. Glenn Research Center at Lewis Field,*
21000 Brookpark Rd., Cleveland, OH 44135, USA
[b]*Pennsylvania State University,*
203 Hosler Building, University Park, PA 16802, USA
[c]*NASA-URC Center for Advanced Nanoscale Materials, University of Puerto Rico,*
Rio Piedras Campus, San Juan, PR 00931, USA
[d]*National Center for Space Exploration Research,*
21000 Brookpark Road, Cleveland, OH 44135, USA

ghunter@grc.nasa.gov

Nanotechnology can enable a new generation of sensor systems due to the potentially unique and advantageous properties of these materials. In particular, the properties of nanostructures such as nanowires, nanofibers, nanorods, and nanoribbons are now being investigated to enable new sensing material properties and approaches. In order to achieve the potential of nanotechnology, basic and fundamental capabilities are needed in order to produce and evaluate sensor systems based on these materials. These include the ability to reproducibly fabricate sensors, understand the material properties, and determine their sensing mechanisms. However, the

Advanced Nanomaterials for Aerospace Applications
Edited by Carlos R. Cabrera and Félix A. Miranda
Copyright © 2014 Pan Stanford Publishing Pte. Ltd.
ISBN 978-981-4463-18-8 (Hardcover), 978-981-4463-19-5 (eBook)
www.panstanford.com

very nature of these materials makes the use of traditional sensor fabrication and characterization techniques, such as those used for microsystems, problematic. This chapter describes the challenges associated with the reproducible fabrication of nanostructures into microsensor systems; characterization of the basic properties of a nanowire; and investigations into the sensing mechanism of nanostructures of different crystal structure. These examples suggest that the transition from microsystem technology into those based on nanostructures involve a series of basic challenges beyond that seen in macroscopic materials. However, if these challenges can be met, the advent of nanotechnology into sensor systems enables the possibility of new sensor systems significantly changing how measurements are done. New sensor systems that can be enabled by nanotechnology, such as "Lick and Stick" smart sensor systems, are discussed.

2.1 Overview of Challenges in Nanosensor Technology

Sensor technology allows measurement of a range of parameters within operational systems to provide more complete information on the environment, system operational parameters, or human health. Sensor technology can have impact on a vast range of human activities, including safety, security, medicine, industrial process control, systems operation, and situational awareness in general [1]. This includes the use of nanomaterials for sensors in aerospace applications, which is addressed in another chapter in this book [2]. A significant motivation behind the development and use of nanotechnology in sensor development is that its inherent advantages can potentially enable a paradigm shift in the capabilities afforded by sensor technology and related systems. The vision is that not only can nanotechnology provide comparable capabilities as traditional technologies and even microsystems, but it will enable a revolution in sensor technology and correspondingly have a significant effect on the average person's health and lifestyle [1–3].

In order to achieve such a revolution, nanotechnology must be able to provide capabilities at least comparable, if not superior, to that of conventional sensor systems. Some of the major enabling technical areas for nanotechnology to surpass conventional sensor systems include sensor fabrication reproducibility; significantly reduced power consumption; improved sensor selectivity, sensitivity, and reliability; ease of application and integration; improved redundancy and cross-correlation; and multiparameter, orthogonal detection [4–6]. These technical challenges are consistent with what is necessary regardless of whether the sensor is based on nanotechnology or other approaches; they address basic operational parameters for a sensor system. In the end, a user typically does not care that a sensor system is based on nanotechnology; they first care that the sensor system meets the needs of the application. However, a major technical hurdle associated with achieving the vision of nanotechnology for sensor systems is the ability to produce operational sensors for the targeted application based on these materials.

One notable example of how material structure can affect sensor performance is that of metal oxide semiconductors such as semiconducting tin oxide (SnO_2) [6]. Tin oxide has been used as a chemical sensor for a number of years with wide industrial application. The fundamental sensing mechanism of these metal oxide-based gas sensors relies upon the change in electrical conductivity due to the interaction between the gases in the environment and oxygen in the grain boundaries. Grain boundary growth during long-term heating at high temperatures has previously been noted [7,8]. Given the sensing mechanism of SnO_2, such grain boundary annealing leads to a drift in the sensor response over time. This sensor drift has affected the application of this material in the commercial marketplace.

In order to stabilize the SnO_2 grain structure for long-term operation, the fabrication of nanocrystalline SnO_2, that is material with a grain size on the order of 10 nanometers (nm), has been investigated [5,6,9]. Nanocrystalline materials have several inherent advantages over conventionally fabricated materials, including increased stability and sensitivity at high temperatures [10,11]. These nanocrystalline materials provide a significant increase in the surface/volume ratio for a given material. Given that the sensor response is dependent on grain boundaries, this increased surface/volume ratio results in more sensitive sensors that can operate at lower temperatures. More importantly, the reaction mechanisms that govern the sintering process are different for nanocrystalline materials than for bulk or macrograined material. This results in grains that are more stable and less likely to sinter, yielding a more stable sensor. Therefore, the use of nanocrystalline material results in a decrease of grain growth while improving sensor sensitivity, stability, and response time. However, of high importance is the feature that nanocrystalline materials can be processed by sol-gel processing techniques, and thus manufactured using well-known procedures. If nanotechnology is to become an enabling technology for sensor applications, there should be a capability to fabricate, characterize, and understand the properties of the resulting sensor system.

In contrast to the work done in nanocrystalline materials, the advantages and processing approaches associated with nanostructured oxide sensors, e.g., nanorods, nanofibers, nanoribbons, nanotubes, and nanobelts, are just beginning to be explored. Based on the progress made in nanocrystalline materials, it is envisioned that significant gains in sensor performance and capabilities can be achieved by use of these nanostructures (for examples, see references [12–14]). However, these gains must be demonstrated, and significant technical challenges remain before standard implementation of nanostructured oxides in sensing applications can take place.

Further, notable miniaturization of sensors has been under way using MEMS (micro-electro-mechanical systems) or microsystems technology. Fabrication of sensors using MEMS technology [1,5,6,15,16] has reached a relatively mature stage in that at least the basic methods and principles are well established. That is, MEMS sensor technology is based on the processing approaches of the silicon electronics processing industry. This includes multiple step fabrication approaches based on photolithography using linewidth typically larger than 1 micron (μm), batch fabrication of a number of sensors within a single processing run, and reproducibility of the sensor structure over multiple processing runs [5,6,15,17]. Advantages of these silicon-based processing techniques include the ability to produce micro-sized structures in an identical, highly uniform, and geometrically well-defined manner, as well as the ability to produce three-dimensional structures (microfabrication and micromachining).

An example of a MEMS-based sensor is the hydrogen sensor structure shown in Fig. 2.1 [18]. This hydrogen sensor structure has multiple components: a resistor is used to measure higher concentrations of hydrogen while a Schottky diode is used to measure lower concentrations of hydrogen. Included are a temperature detector and heater all within a compact package. Thus, this sensor structure has the capability to measure a range of hydrogen concentrations as well as internally control the temperature. The fabrication processes used are standard to the silicon industry and include thin film deposition of metals, etching of oxides, mounting on a standard package, and the ability to form interconnects from a package to the sensor die. The sensor die featured at the center of Fig. 2.1b can be mass-produced typically with good uniformity. Further, not only can Schottky diodes and resistors be fabricated using MEMS-based techniques, but so can structures such as calorimeters and electrochemical cells [15]. Changing the sensor materials within a given MEMS structure can allow improved selectivity toward different chemical species. Characterization of individual sensors can be done either using standard commercially available technologies such as a probe station, or as a complete sensor package such as that shown in Fig. 2.1b.

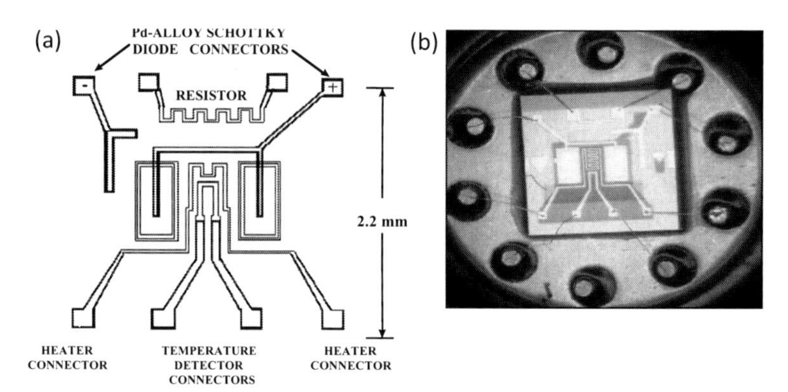

Figure 2.1 (a) Schematic diagram of the silicon-based hydrogen sensor structure produced using MEMS-based technology. Included are two sensors, a heater, and a temperature detector. (b) Picture of the packaged sensor mounted on a standard package and including wire interconnects [18].

Although significant work is still necessary to advance MEMS sensor technology and establish it as the industry standard, the fundamental principles of fabrication and characterization are those that presently exist in the electronics industry. It is suggested that the level of control of processing and the range of tools already existing for MEMS-based structures will be needed in order for nanotechnology to reach its potential. This chapter discusses challenges associated with the fabrication and characterization of sensor technology based on nanotechnology. In particular, this chapter will examine challenges and possible approaches associated with (1) repeatable fabrication of sensors based on nanostructures; (2) characterization of the sensor nanostructures; and (3) determination of the nanostructure's sensing mechanisms. This discussion is not meant to be completely inclusive but presents examples of some of the fundamental issues associated with producing operational sensor systems based on nanotechnology. This chapter will not concentrate on nanomaterial fabrication or details of sensor application; that topic is covered elsewhere within this book [2]. Finally, this chapter will then discuss a specific example of how sensors based on nanotechnology might be integrated into operational sensor systems.

2.2 Fabrication of Sensors Based on Nanostructures

A major issue associated with sensor fabrication using nanostructures is the controlled integration of these nanostructures into a sensor platform in a time-efficient, cost-effective manner, as well as maintaining controlled electrical contact between the nanomaterial and sensor structure. Central to this challenge is that the basic ability to control the orientation and alignment of nanostructures on microstructures is still in its early stages. Nanostructures, e.g., nanowires, nanoribbons, nanotubes, and nanofibers, have fundamentally different structural properties than nanocrystalline grains. While techniques such as sol gel processing can be used for nanostructures, such techniques do not address control and alignment of individual nanostructures, e.g., a single nanorod. The fabrication procedures for sensors using nanostructures are just beginning to be explored. These efforts are fundamental to the ability to produce operational sensor systems since they provide a method to make contact with the sensor structure and fabricate reproducible sensor systems. In effect, no matter how good the sensing material, if one cannot make contact with it or implement it in a sensor structure, its sensing applicability is limited. Thus, in order to control nanotechnology, it is suggested that one first has to have some control of microtechnology in order to interface it with the nanostructures. Significant further work is necessary toward the realization of repeatable, controlled sensor systems composed of oxide-based nanostructures.

Overall, the ability to mass produce sensors in a reproducible way is currently limited with nanostructures such as nanorods or nanowires. Previously, a standard method of deposition of nanostructures onto a sensing structure has been to disperse them in a solvent and deposit the suspension on a substrate (for examples see references [9,19–21].) Such an approach is shown in Fig. 2.2a, where nanorods formed by a thermal evaporation condensation process were put into a suspension and dried on the substrate between two electrodes. This approach has little control over the alignment, orientation, or location of the nanostructures. The density of the nanorod materials, quality of the contact, and baseline sensor properties in general varied with each sensor fabrication, that is, these approaches had limited reproducibility. In effect, this approach relies on the concept that the average distribution will yield repeatable results. However, even if this assumption is valid, it does not take into account the potential advantages that, for example, a specific orientation or properties of the individual nanostructures might provide.

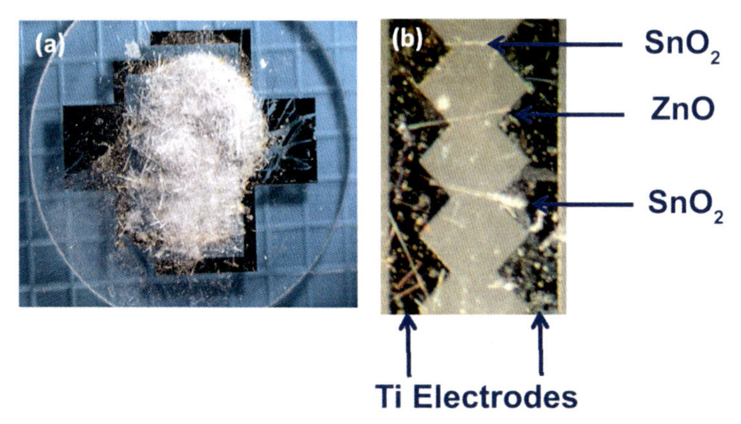

Figure 2.2 (a) Tin oxide nanorods produced using a thermal evaporation condensation process deposited by a solution technique onto a substrate between electrodes [9]. (b) Nanostructures aligned by dielectrophoresis (DEP) on a sawtooth electrode pattern [12].

More complex approaches include atomic force microscopes or laser tweezers [22]. Other work has involved the growth of materials in situ followed by deposition of electrodes [23], high-end processing technologies such as electron beam lithography with nanodimensioned linewidths [24], Langmuir–Blodgett method followed by contacts formed using microfabrication techniques [25–27], and printing approaches [28,29] such as superlattice nanowire pattern transfer (SNAP). These have been performed separate from standard microfabrication processing. In other cases, nanostructures have been buried under metallic contacts on microstructures after random alignment [30]. Depending on the technique, limitations can include that they are labor-intensive approaches, not highly viable for mass fabrication, and the range of materials to which the approach applies is limited. Overall, the use of these techniques are outside of standard microfabrication techniques of larger linewidth resolution.

Other work has involved using techniques such as dielectrophoresis (DEP) [31] to align nanostructures on an existing microplatform. For detailed discussions of DEP, see references [9,12,32–39], which discuss DEP as well as its use in alignment of nanomaterials. To summarize, DEP uses dispersion media such as dimethylformamide (DMF) to align nanomaterials on metal contacts using electric fields. The alignment of nanomaterials between electrodes is dependent on multiple forces resulting from the electric field that are affected by the nanostructure dimension, including (1) an induced dipole interaction from the electric field, which is frequency dependent; (2) hydrodynamic drag forces dictated by rod aspect ratio, fluid viscosity, and temperature; and (3) the induced torque as the rod rotates out of line to the electric field. The emphasis of this previous work is on the nanomaterial alignment between metal contacts.

Figure 2.2b shows an example of nanostructures aligned by DEP on a microsensor platform [9]. To summarize the approach, two different materials, SnO_2 and zinc oxide (ZnO), are assembled at different locations on the same microsensor platform. The microsensor platform is composed of a sawtooth sensor pattern where the "points" and recesses of the sawteeth align with each other. This is done in order to maximize the effect of the electric field on aligning the nanostructures by dielectrophoresis. An alternating current (AC) field is used for alignment; an AC field is used rather than a direct current (DC) field to prevent, in effect, forming an electrochemical cell and the possible resulting electrochemical reactions at electrode surfaces. Alignment of nanostructures between the sawteeth occurs due to the differential hydrodynamic drag force generated by the rod aspect ratio. Further, a torque is induced on the nanostructures within the AC electric field to rotate the nanorod into the line of the electric field. Typically, the solvent used to suspend the nanostructures in standard DEP is either DMF or a light alcohol, and the geometry of the sawtooth pattern helps create the desired electric field gradient to result in movement of the nanorods. The sensing element in Fig. 2.2b is fabricated with titanium (Ti) electrodes having 30 μm spacing between the parallel electrodes. The bridging of the various nanostructures between the sawteeth is visible in the figure. This work is one example of using DEP with microsensor platforms to align nanostructures.

However, while this DEP approach begins to address the issue of control and alignment of nanostructures on a microsensor platform, it does not address some of the notable issues associated with reproducible fabrication of nanostructure-based sensors, and it is still significantly short of the capabilities that enable fabrication of microsensors such as that in Fig. 2.1. For example, the DEP approach above results in the nanostructures laying on the surface of the electrodes. The electrical connection is based on surface contact, and the process does not cover the nanostructures with a top metal layer (burying the electrodes) to insure good electrical contact of the nanostructures to the underlying electrodes. The DEP process was also performed on existing electrode patterns and is not a step in a multistep photoresist-based microfabrication process.

A notable step forward in this processing approach is the use of DEP combined with standard microfabrication or MEMS processing techniques [40,41]. That is, the use of DEP alignment techniques on nanostructures that are mixed into the photoresist that is typically used for microfabrication processing. There are two major features to this processing approach: (1) The use of sawtooth electrodes again using the DEP approach shown in Fig. 2.2b for improved alignment of the nanostructures; and (2) The added feature of including nanostructures within the photoresist itself as part of a series of processing steps. In this advanced approach, rather than DMF, conventional photoresist is used in order to assist with subsequent processing steps. By controlling the amount of nanorods added into the photoresist-based solution, a suspension of nanorods is achieved and shown to be compatible with subsequent DEP and conventional micro-processing steps.

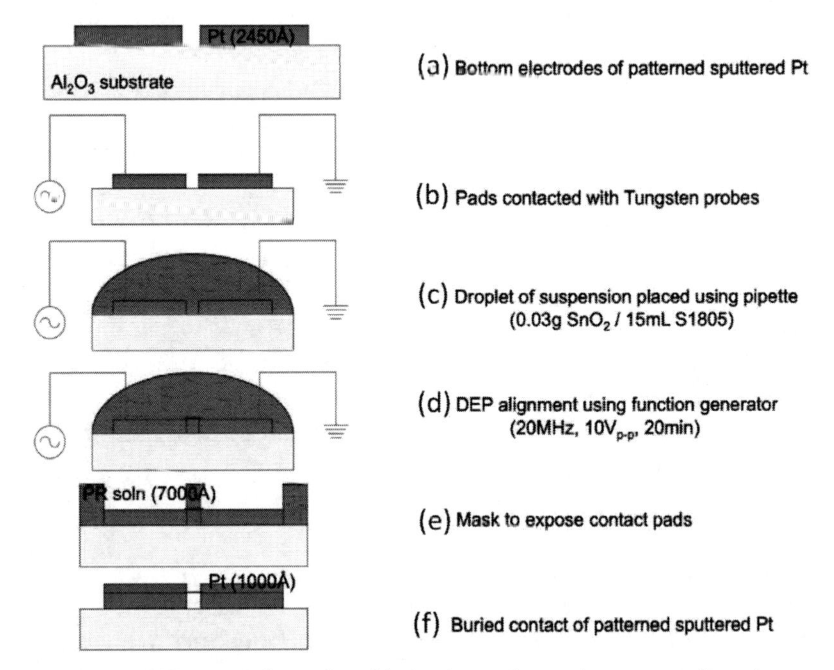

Figure 2.3 Process flow for fabrication of a microsensor based on nanostructures [41].

An outline of the fabrication steps is shown in Fig. 2.3 and is as follows: (a) Define a (bottom) metal electrode pattern using microfabrication techniques on a substrate. (b) Connect function generator to electrode contacts. (c) Add nanostructures to standard photoresist with sufficient concentration to form a dilute suspension within the photoresist. Apply the photoresist suspension to the microstructure. Disperse suspension using spin coating techniques to form a film on the wafer. This film contains aligned nanostructures across the interelectrode gap. (d) Before the photoresist suspension has solidified, apply an alternating electric field across the electrodes. That is, perform DEP on the nanostructures within the photoresist suspension. (e) Expose the photoresist to uncover the bottom electrode layer (previously deposited upon the substrate) to allow the deposition of a second layer of electrodes on top of the ends of the nanostructures and over the bottom electrode. (f) Deposit a second (top) layer of metal through the exposed photoresist directly over the first layer using standard deposition techniques. This step buries the ends of the bridging nanostructures between two layers of metal. Finish processing by completing development of the wafer and removing residual photoresist using solvents and gentle agitation.

The resulting chemical sensor consists of nanorods bridging between two electrodes and secured at both ends between two layers of metal. Figure 2.4 shows the resulting sensor structure as well as several scanning electron microscope images of sensors fabricated using this method. Figure 2.4a shows a wider view schematic of the overall structure. The nanostructures are concentrated between the sawteeth, and contact pads are available to allow mounting of the sensor. Figures 2.4b–d show different materials aligned and characterized using this technique. This includes multiple nanostructures that are aligned between the electrodes, with contact achieved by sandwiching the nanorod ends between two layers of metal. Both metal oxides and carbon nanotubes can be aligned using this approach, and the overall metal sandwich burying the electrodes can simply replicate the sawtooth pattern itself or blanket the region of the sawteeth.

Overall, this approach has allowed the use of microfabrication techniques to produce microsensors using nanostructures. This is a first step toward producing uniform and reproducible sensor structures using batch fabrication and microfabrication techniques. There are several advantages to this approach [40,41]. First, the sawtooth electrode configuration allows dielectrophoresis to align and spatially localize the nanostructures. Through control of photoresist density and nanostructure concentration/dispersion, differing densities of nanostructures can be aligned. Single or multiple nanostructures have been observed to be deposited across the electrode gaps in a given sensor, as demonstrated in Fig. 2.4. This approach allows for the possibility of reproducible manipulation of nanomaterials at a single point or over an array of devices.

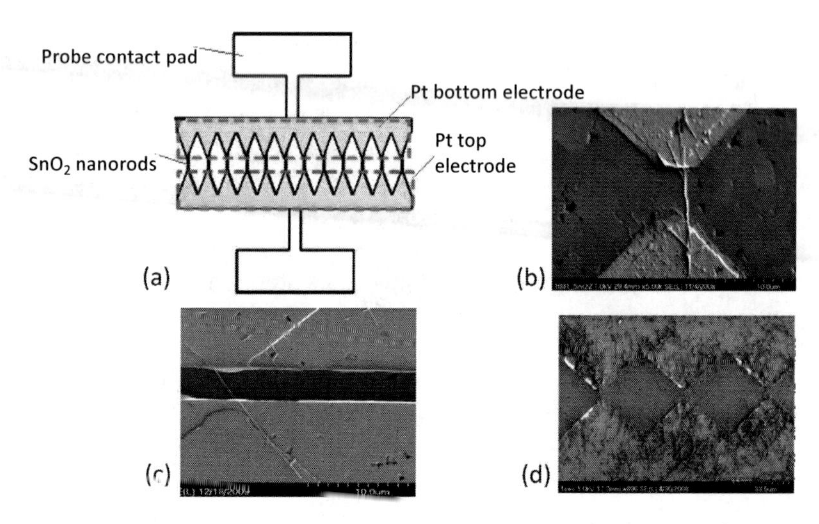

Figure 2.4 (a) Schematic of completed sensor structure. Scanning electron microscope images of three sawtooth pairs with buried electrodes, illustrating the diversity possible ranging from (b) Single SnO_2 nanowire across sawtooth electrodes. (c) Single SnO_2 nanowire across the gap with uniform top metal layer coverage. (d) Carbon nanotubes across sawtooth pattern [40].

Second, this approach allows the possibility of improved electrical contacts to nanostructures. Each nanostructure has two buried electrical contacts on an alumina insulator. This is important for minimizing the variability of necessary electrical interface connections and establishing reproducible and reliable contacts to the nanostructures. Reliable electrical contact increases the potential for reproducible sensor devices using nanostructures.

Third, this approach permits incorporation of nanomaterials into standard photolithographic processing procedures. It does not require highly expensive equipment beyond standard clean room processing facilities. Removal of the devices from a clean room environment is not necessary and alignment of the nanostructures by hand does not take place. This deposition is performed in a continuous set of microfabrication steps without removal from the clean room. This process not only preserves the as-grown nanostructure features but also brings them into contact with a microstructure for integration into existing microplatforms. Further, the use of a nanostructure/photoresist mixture is compatible with other mechanical fabrication methods, including drop coating, spin coating, dip coating, and jet printing. In effect, the combination of including nanostructures in a processing suspension, and then using the suspension in standard processing with the addition of electromechanical nanostructure alignment, has a range of processing impications.

This approach is intended to address the significant barriers of deposition control, contact robustness, and simplified processing to realizing the potential of nanotechnology as applied to sensors. It is suggested that this technique of integrating and aligning nanostructures with microfabrication methods can lead to a standardized approach to chemical sensor processing using nanostructures. This procedure will allow a better understanding of the properties of nanostructures by fabricating reproducible structures and electrical contacts. It can also allow, in principle, nanostructured material-based microsensors to be mass produced and thus applied in broader applications. In a more mature design of this approach, an array of paired patterns on a given wafer will be electrically connected so that a field applied across one set of teeth is simultaneously applied to the full array of multiple sensor paired contact patterns on the wafer. This is considered to be a matter of scale and determination of operating parameters, not a change in the fundamental principle. Further refinement of the dielectrophoresis and photoresist suspension are planned to increase the yield of the bridging for each paired contact pattern. Work is also under way to understand the sensing properties of these sensors based on nanostructured materials produced on these microstructures. Other approaches to understanding the fundamental properties of nanostructures are discussed in the next section.

2.3 Characterization of the Sensor Nanostructures

Nanowires (NW) are promising structures for sensing applications. Besides their large surface-to-volume ratio, the 1D geometry allows easier electrical connectivity than nanoparticles (0D geometry) and can offer novel sensing mechanisms when compared with bulk and thin films. Owing to their reduced size, electron microscopy techniques are standard procedures to characterize NWs: select area electron diffraction (SAED), energy dispersive X-ray spectroscopy (EDS), and electron energy loss spectroscopy (EELS) are commonly used to determine crystalline order and composition. New Auger nanoprobes with lateral resolution below 8 nm are now available.

However, the handling and testing of such structures can be challenging. The task can be done at two levels by testing (a) an ensemble of NWs, or (b) a single NW. The former case is shown in Fig. 2.5a, where a large number of palladium (Pd) NWs are deposited on platinum (Pt interdigitated electrodes) [42]. This method is easier to implement but the measurements will give average values only. The latter involves more difficult procedures and needs to be repeated in a representative number of samples to obtain general conclusions, but can shed more light about the sensing mechanisms in general.

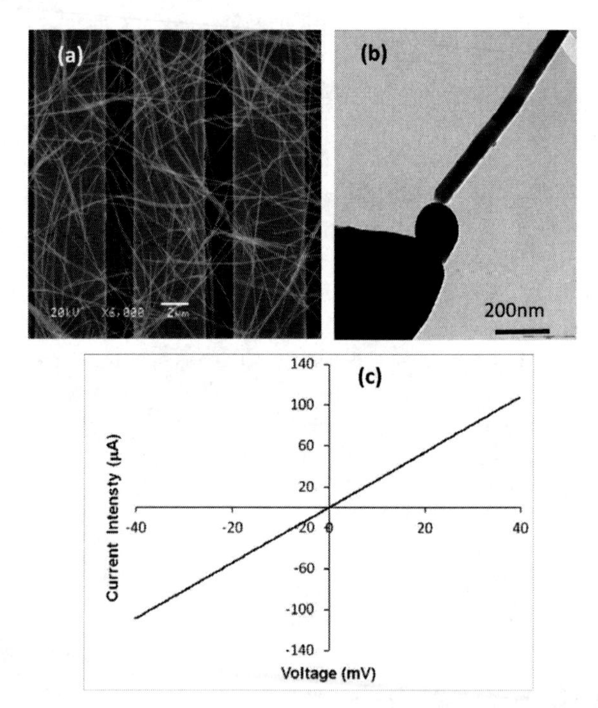

Figure 2.5 (a) A SEM image of a group of Pd NWs randomly distributed on Pt electrodes (gray contrast strips) [42]. (b) A TEM image of a tungsten tip in electrical contact with a Pd NW. The other end is in contact with a fixed gold electrode. (c) The *I*–*V* curve obtained by sweeping the voltage for the NW shown in Fig. 2.5b.

Studying individual NWs typically requires specialized devices. For sensors working under the resistive mode, for example, new in situ transmission electron microscopy (TEM) holders can be used to determine the electrical resistance and other properties of a NW. This is beyond simply measuring the total resistance of the NW. Rather, these TEM holders allow the application of selected voltages between different points of a NW. A typical procedure is to set the NW in the holder with one end in electrical contact with a conductive substrate. A moving conductive nanotip controlled by the TEM holder is used to touch the NW at the other end to close the circuit and to obtain current–voltage (*I*–*V*) curves for the nanostructure. The effects of the contact resistances can be minimized by repeating the measurements with the tip at different positions along the NW and calculating the slope of the resistance versus position graph. The resistivity of the NW can be easily determined from the data once the cross-sectional area of the NW is measured with the TEM. Figure 2.5b shows a Pd NW in electrical contact with a tungsten tip using a STM-TEM holder. Figure 2.5c shows the *I*–*V* curve obtained with such a configuration.

TEM holders for in situ STM, AFM, and indentation are available to characterize sensor nanostructures [43]. However, outside of an electron microscope, specialized microdevices have also been designed to characterize single NWs. In the case of temperature sensors, for example, the determination of the electrical and heat conductivities, and the Seebeck coefficient can be difficult when dealing with NWs. Figure 2.6a shows a chromium silicide ($CrSi_2$) nanowire deposited on a microdevice prepared at the University of Texas at Austin (UTA) [44]. This device has been used to determine the thermoelectric properties of carbon nanotubes [44,45] and $CrSi_2$ NWs [46], among others. As marked in Fig. 2.6a, the system consists of two Pt resistor thermometers (PRT) and four Pt electrodes deposited on free standing silicon nitride membranes to minimize heat and charge leaks. One PRT is also used as a heater to form the temperature gradient along the NW. With this device the thermoelectric parameters of a nanowire are easily determied.

A similar device has been used to measure the change in resistance of individual NWs due to gas exposure [42,47]. As an example, Fig. 2.6b shows a Pd NW deposited on the device and Fig. 2.6c shows the corresponding change in current intensity when exposed to a gas mixture of 90% Ar and 10% H_2. Its four-point-probe configuration can be used to minimize the confusing contributions of the contact resistances.

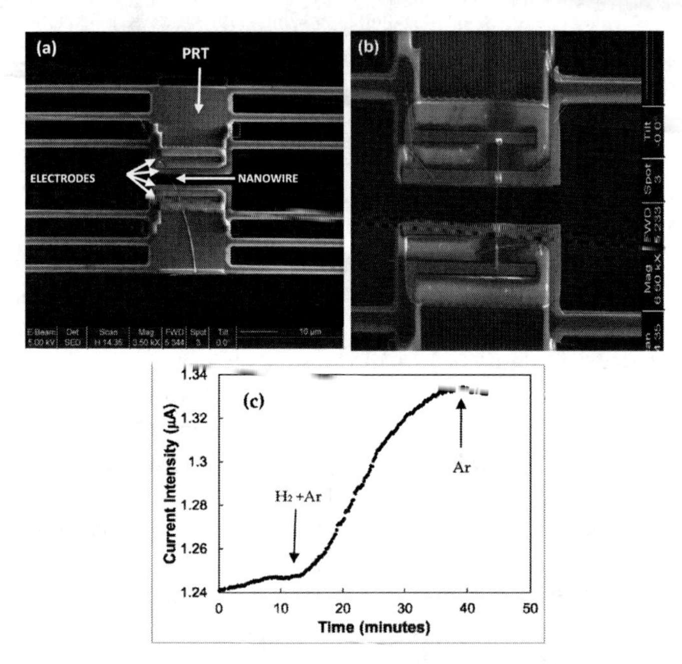

Figure 2.6 (a) SEM image of microdevice prepared to measure thermoelectrical properties of single NWs. The device has two separated sections: upper and lower. Each includes six self-supported Pt contact beams to which a PRT zigzag and two Pt electrodes are connected. A $CrSi_2$ NW is deposited between the upper and the lower sides. (b) SEM image of similar device with a Pd NW [42], (c) current intensity passing through the NW when exposed to a gas mixture of 10% H_2 and 90% Ar. The arrows show the period of time for the H_2 flux.

High-resolution and atomic-resolution TEMs and their related spectroscopic techniques with nanoscale lateral resolution; the combination of focused ion beam, electron beam, and manipulation and testing devices in the same vacuum chamber; new in situ TEM holders with STM, AFM, indentation tips, voltage bias electrodes, and environmental cells; and new specialized ex situ microdevices are part of the available methods that can be used to study new sensor nanostructures.

However, while these capabilities exist, these characterization structures and techniques present their own set of technical challenges. For example, for larger material systems, making a four-lead measurement does not have the same complexity as the system shown in Fig. 2.6. In contrast, the use of approaches such as that described in Fig. 2.6 may not provide all the necessary data. Further maturation of both mounting characterization techniques can accelerate the ability to understand the properties of nanostructure-based material systems, as well as the ability to properly characterize for future application implementation.

2.4 Sensing Mechanisms

Nanomaterials are recognized as a potentially superior form of metal oxide semiconducting material for reasons of size, surface area relative to depletion depth, stability, and sensitivity. At the extremes, very different nanostructures exist, either single-crystal or polycrystalline. The unknown defect density of single-crystal nanowires in comparison to the variable response of junction potentials of the polycrystalline nanofibers opens the question as to which morphology is best for sensing applications. Detailed comparisons between one-dimensional elements of single and polycrystalline morphology provide the best opportunity to answer this question and are an example of the types of investigation necessary in order to understand the sensing mechanisms associated with nanostructures. This section describes such an investigation and the considerations involved.

Different forms of one-dimensional morphology sensing elements require very different fabrication and integration processes for commercial sensing devices [13–14]. Thermal evaporation condensation (TEC) synthesized nanowires offer uniform crystal surfaces as well as resistance to sintering, and their synthesis may be done apart from the substrate. However, with higher crystalline perfection, potentially fewer chemisorption sites exist, resulting in lower sensitivity and dynamic range. Thermal evaporation condensation nanowires will require liquid phase deposition as a wash coat and perhaps an additional binder such as a sol-gel solution. Tin oxide can readily be produced by TEC.

In contrast, controlled oxidation offers a synthesis route for nanowires of materials not readily accessible via a TEC approach. Examples include refractory oxides such as iron oxide (Fe_2O_3), tungsten oxide (WO_3), titanium dioxide (TiO_2), and molybdenum oxide (MoO_3), etc. However, the method is extremely sensitive to both the nascent metal grain structure and process conditions, and in particular, the oxidizer concentration. Harvesting is required and purification necessary, with both steps plagued by the adhesion strength of the nanowires to the supporting (oxidized) metal substrate.

Another method of fabrication is electrospinning, which offers direct deposition, composition control, and potentially a very reactive surface reflecting the polycrystallinity of the material. Precursors are expensive, and calcination will involve the entire substrate. Electrospun nanofibers offer a dry fabrication process on the sensor chip different from the sol-gel plus polymer precursor solution. During deposition of electrospun nanofibers, the substrate temperature elevates, unless an approach involving collection followed by subsequent dispersal and deposition is applied (such as done in TEC). While the resulting individual particles may be single-crystalline, the overall film composed of particles will necessarily be polycrystalline. Fewer chemisorption sites and susceptibility to sintering may result.

Figure 2.7 summarizes the study of SnO_2 nanostructures fabricated with two different processing approaches: TEC and electrospinning. Figures 2.7a–c relate to a TEC SnO_2 nanowire coated with a Pd catalyst, while Figs. 2.7d–f relate to SnO_2 electrospun nanofibers both with and without a Pd catalyst. Each processing approach produces sensor materials with different properties. The effect of the addition of a palladium (Pd) catalyst is examined for both materials. The purpose of this comparison is to understand how processing and crystal structure can affect the sensor response. This allows both optimization of processing to provide improved sensors, but also begins to suggest some of the dominant sensing mechanisms that are present for materials of these one-dimensional nanomaterials. Figures 2.7c,f are HRTEM images of the different crystalline structures of the TEC produced nanowires and electrospun nanofibers whose sensing response is compared. Significantly different crystal structure is noted between the two processing techniques with the TEC produced nanowires being single crystal and the electrospun nanofibers being nanocrystalline.

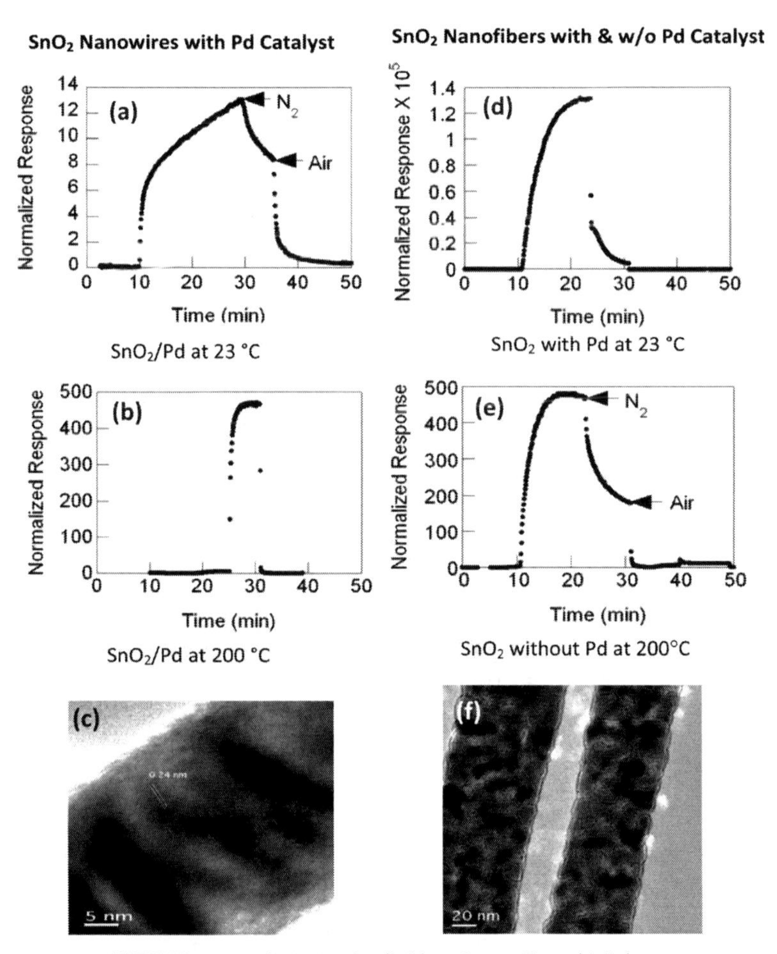

HRTEM Images of a nanowire (left) and nanofibers (right).

Figure 2.7 Response curves for SnO_2/Pd nanowires made by TEC as well as SnO_2 and SnO_2/Pd nanofibers made by electrospinning at the indicated temperatures to air, nitrogen (N_2), 0.5% H_2 in N_2, N_2, and air. As a further point of comparison, it is interesting to note that the nanofibers without Pd catalyst exhibit the same response magnitude as the nanowires with Pd catalyst at 200°C. The lower panels show HRTEM images of these materials, highlighting the differences between the single-crystal nanowires and polycrystalline nanofibers [14].

The nascent materials without catalyst exhibit divergent responses. As illustrated in Fig. 2.7, the TEC-produced nanowire (Fig. 2.7a) response is very low at ambient and a Pd catalyst is necessary to realize detectable response. A modest response is noted at the operating temperature of 200°C (Fig. 2.7b). In contrast, the bare electrospun nanofiber response without catalysts is high at 200°C with ~500-fold normalized response (Fig. 2.7d). This suggests that junction potentials present in the polycrystalline electrospun nanofibers are superior to a continuous surface depletion layer of the single crystal TEC nanowire as a transduction mechanism for sensing chemisorptions at a given temperature.

Overall, using a catalyst deposited upon the surface in the form of nanoparticles, yields dramatic gains in sensitivity for both nanostructured one-dimensional forms relative to their uncatalyzed counterparts at the same temperature (not all data shown in Fig. 2.7). The response magnitude and response rate for the TEC-produced nanowire with catalyst uniformly increase with increasing operating temperature (Figs. 2.7a,b). Such changes are interpreted in terms of accelerated surface diffusion processes, yielding greater access to chemisorbed oxygen species and faster dissociative chemisorption, respectively [48,49] (as observed by comparison of the rise and fall of the response curves in Figs. 2.7a,b).

In contrast, the normalized response of the nanofibers with catalyst decreases with increasing temperature, being the highest at ambient, 23°C (Fig. 2.7c). This decreasing response is interpreted as reflecting the open porosity created by the polycrystalline structure of the nanofiber in conjunction with its small radius. Adsorbates can access all exposed surfaces already at ambient temperature. Accessible surface area, as nominally governed by diffusional processes, does not increase with increasing temperature. Rather, with increasing temperature, chemisorbed oxygen species may be lost (desorbed) and/or transformed into more strongly chemisorbed species, thereby accounting for the decreasing response with increasing temperature. Nevertheless, the temporal response of the electrospun nanofibers improves with operating temperature, reflecting faster dissociation of adsorbing hydrogen.

Regardless of operating temperature, sensitivity of the nanofibers is a factor of 10 to 100 greater than that of nanowires with the same catalyst for the same test condition. In summary, nanostructure appears critical to governing the reactivity, as measured by electrical resistance of SnO_2 toward reducing gases. For both morphological forms, catalyst nanoparticles produce a high response amplitude (advantageous for ambient temperature response of the nanowires), but their effect is strongly moderated by the metal oxide nanostructure. Significantly, the Pd catalyst enables useful operation from the nanowires at ambient temperature. In concert with Pd catalyst, the polycrystalline nanostructure of the electrospinning-produced nanofibers for gas sensing is superior to the single-crystal TEC-produced nanowires. We note that this conclusion is based upon only one catalyst, Pd. Preliminary testing of SnO_2 nanowires with Pt as catalyst has shown either comparable or superior responses compared to the nanofibers with Pd catalyst. Such results suggest that both nanostructures of the metal oxide couple strongly with these noble metal catalysts.

In summary, this study highlights that for SnO_2 nanostructures of comparable dimensions, while a catalyst will improve response, the nascent and doped sensing properties of that material can be notably different. This work highlights these variations for a single material prepared within a relatively narrow set of parameters. Given that the properties of nanostructures for sensors are still at the relatively early stage of investigation, a range of studies such as this are necessary to begin to understand the sensing mechanisms of nanostructures and how they might be optimized for a given application.

2.5 Smart Nanosensor Systems

One possible implementation of nanotechnology in sensors is to more fully enable "smart sensor systems." The definition of a smart sensor may vary, but at a minimum a smart sensor is the combination of a sensing element with processing capabilities provided by a microprocessor [50]. One embodiment of a smart sensor system is illustrated in Fig. 2.8: a complete self-contained sensor system that includes the capabilities for logging, processing with a model of sensor response and other data, self-contained power, and an ability to transmit or display informative data to an outside user. The fundamental idea of a smart sensor is that the integration of silicon microprocessors with sensor technology can not only provide onboard data interpretation and customized outputs but also significantly improve sensor system performance and capabilities. Multiple sensors can be included in a single smart sensor system to detect multiple species and parameters, cover different concentration ranges, and allow cross-correlation of signals. The processed data becomes information, which can then be transmitted to external users, stored onboard, or used to provide caution and warning.

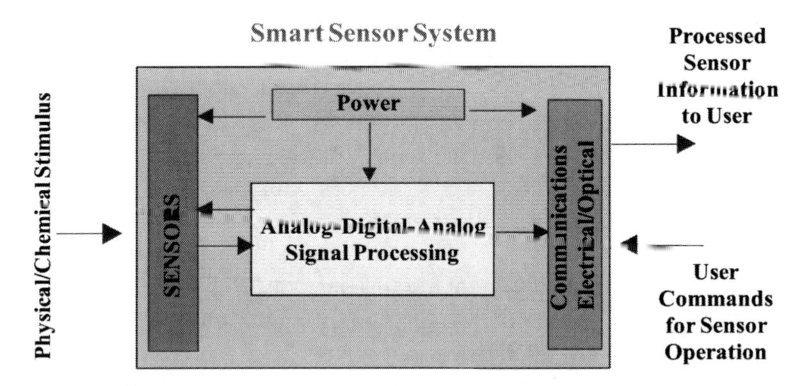

Figure 2.8 A Smart sensor system as presented herein. The core of a standalone smart sensor system includes sensors, power, communication, and signal processing [50].

Smart sensors can be networked through the integrated communication interface so that each sensor can be identified and its input correlated with the other sensors on the network. Given the onboard processing capabilities, remote or on-site reprogramming or recalibration of the smart sensor system can be performed as necessary. The output from a number of sensors within a given region can be correlated not only to verify the data from individual sensors but also to provide better situational awareness meeting the needs of the specific application. These types of capabilities will provide for a more reliable and robust system because they are capable of networking among themselves to provide the end user with coordinated data that is based on redundant sensory inputs. Thus, important data can be provided to the user with increased reliability and integrity. A driving goal in the development of smart sensor systems is the implementation of systems in a nonintrusive manner so that the information is provided to the user wherever and whenever needed and in the form tailored for the application. In effect, one objective of smart sensor research is the development of sensor systems that can be implemented into the environment without causing disruption and providing the users what they need to know in order to make sound decisions.

An example of a smart sensor systems is the "Lick and Stick" leak sensor system [18]. This is a multifunctional system with a microsensor array fabricated by MEMS based technology designed to detect hazardous conditions due to fuel leaks. Included within this system is the hydrogen sensor shown in Fig. 2.1. The complete system has three sensors, signal conditioning electronics, power, data storage, calibration tables, built-in self-test, telemetry, and an option for self-power in the surface area comparable to a postage stamp. The approach is to be able to place sensors in a vehicle, like postage stamps, where they are needed without rewiring or drawing power from the vehicle. The electronics can be programmed to provide the user with certain information required on a regular basis, but also much further diagnostic information when needed. A prototype model of the "Lick and Stick" leak detection sensor system is shown in Fig. 2.9a. The ability to have one "Lick and Stick" sensor system send data by telemetry, as well as have several "Lick and Stick" sensor systems sending data to a central processing hub, has been demonstrated [18]. Figure 2.9b shows the operation of the electronics with the three sensor system simultaneously with data sent telemetrically. Smart sensors systems using this "Lick and Stick" system as a core have been adapted to applications as broad as fire detection, breath monitoring, environmental monitoring, and operation on rocket engine test stands [16]. For example, one smart system is being developed to enable combined multiparameter environmental and fire detection, and another smart system is being developed for high temperature engine applications [51,52].

Smart sensor systems potentially represent a new generation of sensing capability core to enabling future intelligent systems [53]. Such systems can have a profound impact on a wide range of applications such as safety and hazard detection and warning, environmental monitoring, health monitoring and medical diagnostics [54], and industrial and aerospace applications. Smart sensor systems can enable intelligent vehicle systems, which can monitor themselves and respond to changing conditions optimizing safety and performance while decreasing fuel burn and emissions. The integration of sensors and algorithms is also needed in order to properly provide context to the data. The smart sensor system approach can achieve distributed sensor systems feeding information from multiple locations to improve the overall understanding of system conditions.

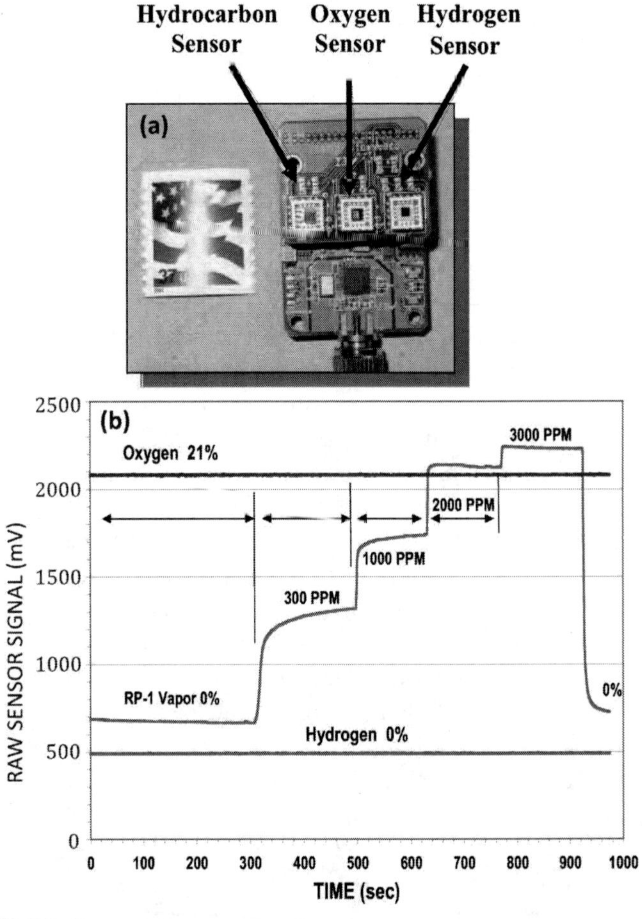

Figure 2.9 (a) A prototype version of a "Lick and Stick" leak sensor system with sensors combined with supporting electronics. (b) Response of three sensors to varying hydrocarbon concentrations in a constant background environment. The sensor signal is sent by telemetry and is the output from the signal conditioning electronics, which processes the measured sensor current [18].

However, in order to reach the promise of sensory systems, further advancements in both micro and nanotechnology, as well as associated smart sensor software algorithms, are necessary. Nanotechnology is proposed to be enabling for a new range of sensing capabilities. Control of nanotechnology can in principle allow sensors to operate in a wide range of conditions; have improved sensitivity and selectivity; significantly minimize size, weight, and power consumption; and have vastly improved reliability. The objective is that these nanotechnology-based sensors may truly enable "Lick and Stick" systems that can be placed wherever and whenever needed. This can be accomplished not only by the use of nanotechnology for sensor materials but also with advancements in power supplies, electronics, communication, packaging, and improved integration into vehicles. In the drive to make systems that are small and smart, nanotechnology can in principle drive sensors toward highly miniaturized, integrated, and self-contained units whose presence is unnoticed and whose implementation is as simple as dropping the sensor where you want it. A more detailed description of how nanotechnology, including smart sensor systems can revolutionize a range of applications has been published [3].

2.6 Summary and Future Possibilities

This chapter discusses sensor systems and how nanotechnology may provide the ability to monitor the environment, vehicle system, human health, or other applications in the future. The baseline is set by the existence of MEMS technology, and future implementation of nanotechnology must provide enhanced capabilities beyond those that are presently available. However, the stage of maturity of nanotechnology is not to the level of that of MEMS. Rather, the basic capabilities to make sensor systems, while predominantly in place for MEMS, are still under development for nanotechnology. Simply stated, this chapter discusses notable challenges associated with the basic capabilities to (1) repeatably fabricate sensors based on nanostructures, (2) characterize the resulting sensor nanostructures, and (3) determine the nano-structures sensing mechanism. Examples of each activity were provided as well as the challenges that still exist for future processing, characterization, and implementation of nanotechnology in sensor systems.

Nonetheless, the work in this chapter lays the foundation for further sensor development. A long-term vision for how nanotechnology can change the way sensor measurements are performed is the idea of designer chemical sensors. In effect, arrange the chemical sensor structure to "fit" the molecule in question and verify the presence of the molecule uniquely with the sensor designed specifically for that type of detection. A notable enabling feature of nanotechnology may not be in the ability to detect millions of molecules, but rather, the ability to enable atomic scale design of systems targeted for specific applications and meant to operate on this atomic level. For sensors, such an approach would significantly improve sensor selectivity by having unique identifiers for each species as well as drastically decreasing the number of sensors necessary for determination of a given species in a multispecies environment. In addition, such an atomic level manipulation approach would have notable impact on power, communication, and other applications. In the long-term, nanotechnology can fundamentally change the way sensing, communication, power generation, and a range of other capabilities are enabled, and in turn significantly affect a range of applications. However, tools to fully enable such a revolution are still at an early stage.

Acknowledgments

The authors would like to acknowledge the contributions of Professor C. C. Liu of Case Western Reserve University; Gus Fralick, D. Litt, Dr. L. Matus, and Dr. M. Zeller of the NASA Glenn Research Center (GRC); Dr. C. Chang of ASRC Aerospace/NASA GRC; Dr. L. Chen of the Ohio Aerospace Institute/NASA GRC; Dr. B. Ward and Dr. D. Makel of Makel Engineering, Inc.; Dr. J. Stetter of KWJ Engineering, Inc., Prof. P. Hesketh of Georgia Institute of Technology, J. Gonzalez, M. Artale, P. Lampard, C. Hampton of Sierra Lobo/NASA GRC. L. F. Fonseca acknowledges contribution from L. Valentín, J. Carpena, and Dr. D. Yang from University of Puerto Rico; Prof. L. Shi and Dr. M. Pettes from the University of Texas at Austin for providing thermoelectrical characterization microdevices, and the National Center for Electron Microscopy at the Lawrence Berkeley Lab who is supported by the U.S. Department of Energy under Contract # DE-AC02-05CH11231 for the use of their FIB facilities. L. F. Fonseca acknowledges support from NASA-URC grant #NNX08BA48A and NSF-grant #1002410.

References

1. Stetter, J. R., Hesketh, P. J., and Hunter, G. W. (2006). Sensors: Engineering Structures and Materials from Micro to Nano, *Interface* [the Electrochemical Society], **15**(1), 66–69.
2. See Chapter 1 in this book.
3. Nanotechnology-Enabled Sensing, Report on the National Nanotechnology Initiative Workshop, Arlington, Virginia, May 5–7 (2009). Available at http://www.nano.gov/.
4. Hunter, G. W. (2003). Morphing, Self-Repairing Engines: A Vision for the Intelligent Engine of the Future, *AIAA/ICAS International Air & Space Symposium, 100th anniversary of Flight, 14–17 July 2003* (Dayton, OH, AIAA 2003-3045).
5. Hunter, G. W., Xu, J. C., Liu, C. C., and Makel, D. B. (2006). Microfabricated Chemical Sensors for Aerospace Applications. In *The MEMS Handbook Second Edition: Design and Fabrication*, (Gad-El-Hak, M., ed.), Chapter 11, CRC Press, Baton Rouge.
6. Hunter, G. W., Xu, J. C., and Makel, D. B. (2008). Case Studies in Chemical Sensor Development. In *BioNanoFluidic MEMS.* (Hesketh, P. J., ed.), Chapter 8, Springer Science and Business Media, New York, pp. 197–231.
7. Ogawa, H., Nishikawa, M., and Abe, A. (1982). Hall Measurement Studies and An Electrical Conduction Model of Tin Oxide Ultrafine Particle Films, *J. Appl. Phys.*, **53**, 4448–4455.
8. Xu, C., Tamaki, J., Miura, N., and Yamazoe, N. (1991). Grain Size Effects on Gas Sensitivity of Porous SnO_2-Based Elements, *Sens. Actuators B*, **3**, 147–155.
9. Hunter, G. W., Xu, J. C., Evans, L. J., Vander Wal, R. L., Berger, G. M., Kulis, M. J., and Liu, C. C. (2006). Chemical Sensors Based on Metal Oxide Nanostructures, *ECS Trans.*, **3**(9), 199–209.
10. Yoo, D. J., Tamaki, J., Park, S. J., Miura, N., and Yamazoe, N. (1995). Effects of Thickness and Calcination Temperature on Tin Dioxide Sol-Derived Thin-Film Sensor, *J. Electrochem. Soc.*, **142**, L105–L107.
11. Vogel, R., Hoyer, P., and Weller, H. (1994). Quantum-Sized PbS, CdS, Ag_2S, Sb_2S_3, and Bi_2S_3 Particles as Sensitizers for Various Nanoporous Wide-Bandgap Semiconductors, *J. Phys. Chem.*, **98**, 3183–3188.
12. Hunter, G. W., Vander Wal, R. L., Xu, J. C., Evans, L. J., Berger, G. M., and Kulis, M. J. (2008). The Development of Metal Oxide Chemical Sensing Nanostructures, *ECS Trans.*, **16**(14), 73–84.

13. Vander Wal, R. L., Hunter G. W., Xu, J. C., Kulis, M. J., Berger, G. M., and Ticich, T. M. (2009). Metal-Oxide Nanostructure and Gas-Sensing Performance, *Sens. Actuators B*, **138(**1), 113–119.

14. Vander Wal, R. L., Berger, G. M., Kulis, M. J., Hunter, G. W., Xu, J. C., and Evans, L. E. (2009). Synthesis Methods, Microscopy Characterization and Device Integration of Nanoscale Metal Oxide Semiconductors for Gas Sensing, *Sensors,* **9**(10), 7866–7902.

15. Liu, C. C., Hesketh, P. K., and Hunter, G. W. (2004). Chemical Microsensors, *ECS Interface*, **13**, 22–29

16. Hunter, G. W., Xu, J. C., Evans, L., Biaggi-Labiosa, A., Ward, B. J., Rowe, S., Makel, D. B., Liu, C. C., Dutta, P., Berger, G. M., and Vander Wal, R. L. (2010). The Development of Micro/Nano Chemical Sensor Systems for Aerospace Applications, *SPIE Newsroom*, June 30 2010. Available at http://spie.org/x10819.xml?ArticleID=x40849.

17. Madou, M. (1997). *Fundamentals of Microfabrication*, CRC Press, Boca Raton.

18. Hunter, G. W., Xu, J., Neudeck, P. G., Makel, D. B., Ward, B., and Liu, C. C. (2006). Intelligent Chemical Sensor Systems For In-Space Safety Applications, *42nd AIAA/ASME/SAE/ASEE Joint Propulsion Conference & Exhibit, July 10–12, 2006,* (Sacramento, California, Tech. Rep AIAA-06-**58419**).

19. Comini, E., Faglia, G., Sberveglieri, G., Pan, Z., and Wang, Z. L. (2002). Stable and Highly Sensitive Gas Sensors Based on Semiconducting Oxide Nanobelts, *Appl. Phys. Lett.*, **81**(10), 1869–1871.

20. Fan, Z., and Lu, J. G. (2005). Zinc Oxide Nanostructures: Synthesis and Properties, *J. Nanosci. Nanotechnol.*, **5**(10), 1561–1573.

21. Ponzoni, A., Comini, A. E., Sberveglieri, G., Zhou, J., Deng, S. Z., Xu, N. S., Ding, Y., and Wang, Z. L. (2006). Ultrasensitive and Highly Selective Gas Sensors Using Three-Dimensional Tungsten Oxide Nanowire Networks, *Appl. Phys. Lett.*, **88**, 203101.

22. Subramanian, A., Vikramaditya, B., Nelson, B. J., Bell, D., and Dong, L. (2005). Dielectrophoretic Micro/Nanoassembly with Microtweezers and Nanoelectrodes, *Proceedings of 12th International Conference on Advanced Robotics*, pp. 208–215.

23. Hwang, I., Choi, Y., Park, J., Park, J., Kim, K., and Lee, J. (2006). Synthesis of SnO_2 Nanowires and Their Gas Sensing Characteristics, *J. Korean Phys. Soc.*, **49**, 1229–1233.

24. Mohney, S. E., Wang, Y., Cabassi, M. A., Lew, K. K., Dey, S., Redwing, J. M., and Mayer, T. S. (2005). Measuring the Specific Contact

Resistance of Contacts to Semiconductor Nanowires, *Solid-State Electron.*, **49**, 227–232.

25. Jin, S., Whang, D., McAlpine, M. C., Friedman, R. S., Wu, Y., and Lieber, C. M. (2004). Scalable Interconnection and Integration of Nanowire Devices with Registration, *Nano Lett.*, **4**, 915–919.

26. Whang, D., Jin, S., Wu, Y., and Lieber, C. M. (2003). Large-Scale Hierarchical Organization of Nanowire Arrays for Integrated Nanosystems, *Nano Lett.*, **3**, 1255–1259.

27. Tao, A., Kim, F., Hess, C., Goldberger, J., He, R., Sun, Y., Xia, Y., and Yang, P. (2003). Langmuir-Blodgett Silver Nanowire Monolayers for Molecular Sensing Using Surface-Enhanced Raman Spectroscopy, *Nano Lett.*, **3**, 1229–1233.

28. Fan, Z., Ho, J. C., Jacobson, Z. A., Yerushalmi, R., Alley, R. L., Razavi, H., and Javey, A. (2008). Wafer-Scale Assembly of Highly Ordered Semiconductors Nanowire Arrays by Contact Printing, *Nano Lett.*, **8**, 20–25.

29. McAlpine, M. C., Ahmad, H., Wang, D., and Heath, J. R. (2007). Highly Ordered Nanowire Arrays on Plastic Substrates for Ultrasensitive Flexible Chemical Sensors, *Nat. Mater.*, **6**, 379–384.

30. Tselev, A., Hatton, K., Fuhrer, M. S., Paranjape, M., and Barbara, P. (2004). A Photolithographic Process for Fabrication of Devices with Isolated Single-Walled Carbon Nanotubes, *Nanotechnology*, **15**, 1475–1478.

31. Huang, Y., Duan, X., Wei, Q., and Lieber, C. M. (2001). Directed Assembly of One-Dimensional Nanostructures into Functional Networks, *Science*, **291**, 630–633.

32. Smith, P. A., Nordquist, C. D., Jackson, T. N., Mayer, T. S., Martin, B. R., Mbindyo, J., and Mallouk, T. E. (2000). Electric-Field Assisted Assembly and Alignment of Metallic Nanowires, *Appl. Phys. Lett.*, **77**, 1399–1401.

33. Kumar, S., Rajaraman, S., Gerhardt, R. A., Wang, Z. L., and Hesketh, P. J. (2005). Tin Oxide Nanosensor Fabrication Using AC Dielectrophoretic Manipulation of Nanobelts, *Electrochim. Acta*, **51**, 943–951.

34. Morgan, H., and Green, N. G. (1997). Dielectrophoretic Manipulation of Rod-Shaped Viral Particles, *J. Electrostatics*, **42**, 279–293.

35. Kumar, S., Peng, Z., Shin, H., Wang, Z. L., and Hesketh, P. J. (2010). AC Dielectrophoresis of Tin Oxide Nanobelts Suspended in Ethanol: Manipulation and Visualization, *Anal. Chem.*, **82**, 2204–2212.

36. Chang, D. E., and Petit, N. (2005). Toward Controlling Dielectrophoresis, *Int. J. Robust Nonlinear Control.*, **15**, 769–784.

37. Diehl, M. R., Yaliraki, S. N., Beckman, R. A., and Heath, J. R. (2002). Self-Assembled, Deterministic Carbon Nanotube Wiring Networks, *Angew. Chem. Int. Ed.*, **41**, 353–356.

38. Krupke, R., Hennrich, F., Weber, H. B., Beckmann, D., Hampe, O., Malik, S., Kappes, M. M., and Lohneysen, H. V. (2003). Contacting Single Bundles of Carbon Nanotubes with Alternating Electric Fields, *Appl. Phys. A*, **76**, 397–400

39. Suehiro, J., Zhou, G., and Hara, M. (2003). Fabrication of a Carbon Nanotube-Based Gas Sensor Using Dielectrophoresis and Its Application for Ammonia Detection by Impedance Spectroscopy, *J. Phys. D*, **36**, L109–L114.

40. Hunter, G. W., Vander Wal, R. L., Evans, L. J., J. Xu, C., Berger, G. M., Kulis, M. J., and Biaggi-Labiosa A. (2012). Nanostructured Sensor Processing Using Microfabrication Techniques, *Sens. Rev.*, **32**, 106–117.

41. Evans, L. J., Hunter, G. W., Xu, J. C., Berger, G. M., and Vander Wal, R. L. (2010). Controlled Fabrication of Nanostructure Material Based Chemical Sensors, *MRS Proc.*, **1253**, 1253-K08-04.

42. Yang, D., Valentín, L., Carpena, J., Otaño, W., Resto, O., and Fonseca, L. F. (2012). Temperature- Activated Reverse Sensing Behavior of Pd Nanowire Hydrogen Sensors. *Small*, doi:10.1002/smll.201201639.

43. Ferreira, P. J., Mitsuishi, K., and Stach E. A. (2008). For a description of in situ TEM techniques and apparatus see for example: in situ Transmission Electron Microscopy, *MRS Bull.*, **33**(2) 83–90.

44. Shi, L., Li, D., Yu, C., Jang, W., Kim, D., Yao, Z., Kim, P., and Majumdar, A. (2003). Measuring Thermal and Thermoelectrical Properties of One-Dimensional Nanostructures Using a Microfabricated Device, *J. Heat Transfer*, **125**, 881–888.

45. Pettes, M. T., and Shi, L. (2009). Thermal and Structural Characterizations of Individual Single-, Double-, and Multi-Walled Carbon Nanotubes, *Adv. Funct. Mater.*, **19**, 3918–3925.

46. Zhou, F., Szczech, J., Pettes, M. T., Moore, A. L., Jin, S., and Shi, L. (2007). Determination of Transport Properties in Chromium Disilicide Nanowires via Combined Thermoelectric and Structural Characterizations, *Nano Lett.*, **7**, 1649–1654.

47. Shi, L., Yu, C., and Zhou, J. (2005). Thermal Characterization and Sensor Applications of One-Dimensional Nanostructures Employing Microelectromechanical Systems, *J. Phys. Chem. B*, **109**, 22102–2211.

48. Vander Wal, R. L., Hunter, G. W., Xu, J. C., Kulis, M. J., and Berger, G. M. (2009). Metal Oxide Nanostructure and Gas-Sensing Performance, *Sens. Actuators B*, **138**, 113–119.

49. Vander Wal, R. L., Hunter, G., Kulis, M. J., Xu, J. C., and Berger, G. M. (2009). Synthesis Methods, Microscopy Characterization and Device Integration of Nanoscale Metal Oxide Semiconductors for Gas Sensing in Aerospace Applications, *NASA TM 215607,* March 2009.

50. Hunter, G. W., Stetter, J. R., Hesketh, P. J., and Liu, C. C. (2011). Smart Sensor Systems, *ECS Interface*, **19**(4), Winter 2010, 29–34.

51. Hunter, G. W., Xu, J. C., Dungan, L., Ward, B., Dutta, P., Adeyemo, A. D., Liu C. C., and Gianettino, D. P. (2010) Smart Chemical Sensor Systems for Fire Detection and Environmental Monitoring in Spacecraft, *International Conference On Environmental Systems, Barcelona, Spain, AIAA766637.*

52. Hunter, G. W., Beheim, G. M., Ponchak, G. E., Scardelletti, M. C., Meredith, R. D., Dynys, F. W., Neudeck, P. G., Jordan, J. L., and Chen, L. Y. (2010). Development of High Temperature Wireless Sensor Technology Based on Silicon Carbide Electronics, *ECS Trans.*, **33**(8), 269–281.

53. Hunter, G. W., Oberle, L. G., Baakalini, G., Perotti, J., and Hong, T. (2011). *System Health Management: with Aerospace Applications,* (Johnson, S. B., Thomas Gormley, T., Kessler, S., Mott, C., Patterson-Hine, A., Reichard, K., Scandura, Jr., P., eds.), Chapter 25, John Wiley & Sons, Ltd, Chichester, West Sussex, UK.

54. Hunter, G. W., and Dweik, R. A. (2008). Applied Breath Analysis: An Overview of the Challenges and Opportunities in Developing and Testing Sensor Technology for Human Health Monitoring in Aerospace and Clinical Applications, *J. Breath Res.*, **2**, 037020.

Chapter 3

Nanoporous Materials in Atmosphere Revitalization

Arturo J. Hernández-Maldonado,[a,b] **Yasuyuki Ishikawa,**[a,b] **Raphael G. Raptis,**[a,b] **Bernadette Luna,**[c] **Lila Mulloth,**[c] **Christian Junaedi,**[d] **Subir Roychoudhury,**[d] **and Jay L. Perry**[e]

[a]*NASA-URC Center for Advanced Nanoscale Materials, University of Puerto Rico, Mayagüez Campus, Mayagüez, PR 00681, USA*
[b]*NASA-URC Center for Advanced Nanoscale Materials, University of Puerto Rico, Rio Piedras Campus, San Juan, PR 00931, USA*
[c]*NASA Ames Research Center, Moffett Field, CA 94035, USA*
[d]*Precision Combustion Inc., 410 Sackett Point Rd., North Haven, CT 06473, USA*
[e]*NASA George C. Marshall Space Flight Center, Huntsville, AL 35812, USA*

arturoj.hernandez@upr.edu

3.1 Introduction

Atmosphere revitalization (AR) is the term the National Aeronautics and Space Administration (NASA) uses to encompass the engineered systems that maintain a safe, breathable gaseous atmosphere inside a habitable space cabin. An AR subsystem is a key part of the Environmental Control and Life Support (ECLS) system for habitable space cabins. The ultimate goal for AR subsystem designers is to "close the loop," that is, to capture gaseous human metabolic products, specifically water vapor (H_2O) and carbon dioxide (CO_2), for maximal oxygen (O_2) recovery and to make other

Advanced Nanomaterials for Aerospace Applications
Edited by Carlos R. Cabrera and Félix A. Miranda
Copyright © 2014 Pan Stanford Publishing Pte. Ltd.
ISBN 978-981-4463-18-8 (Hardcover), 978-981-4463-19-5 (eBook)
www.panstanford.com

useful resources from those products. The AR subsystem also removes trace chemical contaminants from the cabin atmosphere to preserve cabin atmospheric quality, provides O_2, and may include instrumentation to monitor cabin atmospheric quality.

Long-duration crewed space exploration missions require advancements in AR process technologies in order to reduce power consumption and mass and to increase reliability compared to those used for shorter duration missions that are typically limited to low Earth orbit (LEO). For example, current AR subsystems include separate processors and process air flow loops for removing metabolic CO_2 and volatile organic trace contaminants (TCs). Physical adsorbents contained in fixed, packed beds are employed in these processors. Still, isolated pockets of high carbon dioxide have been suggested as a trigger for crew headaches [1,2], and concern persists about future cabin ammonia (NH_3) levels as compared with historical flights [3,4]. Developers are already focused on certain potential advancements. Environmental control and life support systems engineers envision improving the AR subsystem by combining the functions of TC control and CO_2 removal into a single regenerable process and moving toward structured sorbents-monoliths—instead of granular material [5]. Monoliths present a lower pressure drop and eliminate particle attrition problems that result from bed containment. New materials and configurations offer promise for lowering cabin levels of CO_2 and NH_3 as well as for reducing power requirements and increasing reliability. This chapter summarizes the challenges faced by ECLS system engineers and scientists in pursuing these goals, and the promising materials developments that may be a part of the technical solution for the challenges of crewed space exploration beyond LEO.

3.1.1 Development History of Atmosphere Revitalization Systems for Space

The early human space flight programs—Mercury, Gemini and Apollo—all utilized an expendable, granular lithium hydroxide (LiOH) canister for CO_2 control. Activated charcoal was located upstream of the LiOH in the same canister for odor control. The Skylab vehicle was much larger and missions lasted much longer. A swing-bed molecular sieve system operating on a 15 min half cycle was used for CO_2 control (see Fig. 3.1) [6,7]. Charcoal beds operated in parallel with 13X/5A zeolite beds to control odors. Flow rates in various legs of the system are shown in the figure. During desorption of water and CO_2 to space vacuum, flow was maintained in the charcoal and bypass legs of the desorbing side.

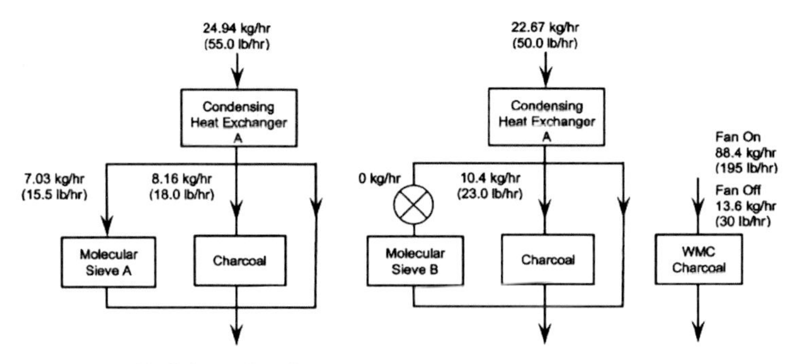

Figure 3.1 Skylab AR [6,7].

On the Shuttle, two radial flow beds containing 2.3 kg of granular LiOH (each) are used for CO_2 control. The LiOH beds produce water so the LiOH bed is followed by a condensing heat exchanger. Lastly, there is a radial-flow ambient temperature catalytic oxidizer (ATCO) containing 0.32 kg of Pt-on-charcoal catalyst (see Fig. 3.2).

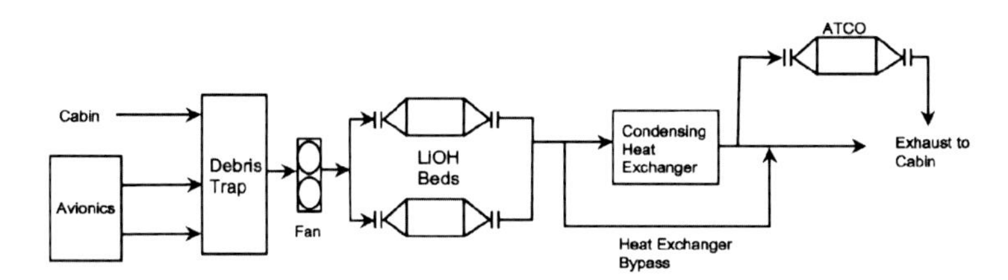

Figure 3.2 Shuttle AR. *Source*: NASA Shuttle Operational Data Book, Volume 1, NASA Johnson Space Center, October 1984.

On the U.S. Segment of International Space Station (ISS), CO_2 control is achieved using a molecular sieve swing-bed system consisting of two water-removal beds and two CO_2 removal beds. The equipment is called the Carbon Dioxide Removal Assembly (CDRA) [8], or the Four Bed Molecular Sieve (4BMS). Desiccant beds are separate from CO_2 removal beds to enable water recovery and avoid venting water to space. Water is a valuable resource on extended missions. International Space Station operates in a water-save mode; water removed from air to facilitate CO_2 adsorption is returned to the same air downstream for astronaut comfort.

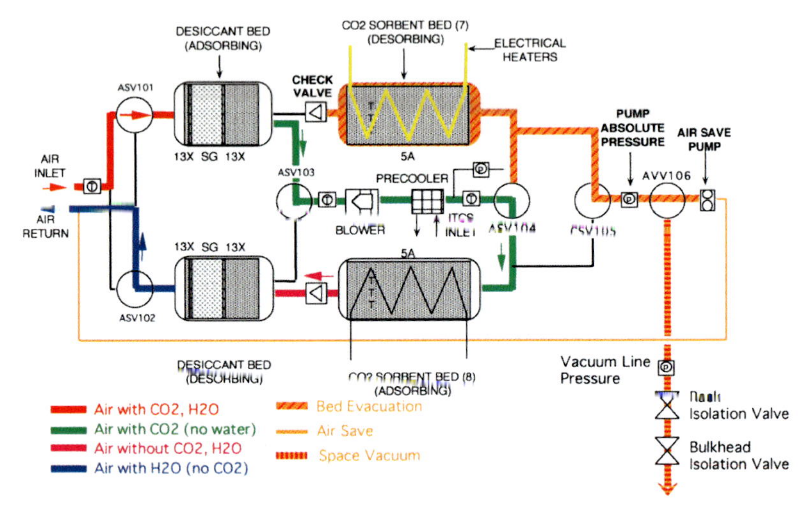

Figure 3.3 International space station (ISS) carbon dioxide removal assembly (CDRA) schematic [8].

3.1.2 Design Challenges and Considerations

Factors Influencing a Combined Trace Contaminant/Carbon Dioxide Removal System

On board the ISS U.S. On-Orbit Segment (USOS), TCs are removed using physical adsorption, chemical adsorption, and thermal catalytic oxidation. The TC control equipment employs a large packed bed of granular activated carbon (GAC) to remove high molecular weight volatile organic compounds (VOCs) from the cabin atmosphere. The GAC is treated with phosphoric acid (H_3PO_4) to remove NH_3. A packed bed containing thermal oxidation catalyst is located downstream of the GAC bed. The catalytic oxidation process removes light hydrocarbons such as formaldehyde (CH_2O) and methane (CH_4) as well as carbon monoxide (CO). A fixed bed of granular LiOH located downstream of the catalytic oxidation subassembly removes any acidic oxidation products from the process air stream. The carbon bed and LiOH bed are expendable and periodically refurbished. The system is shown schematically in Fig. 3.4.

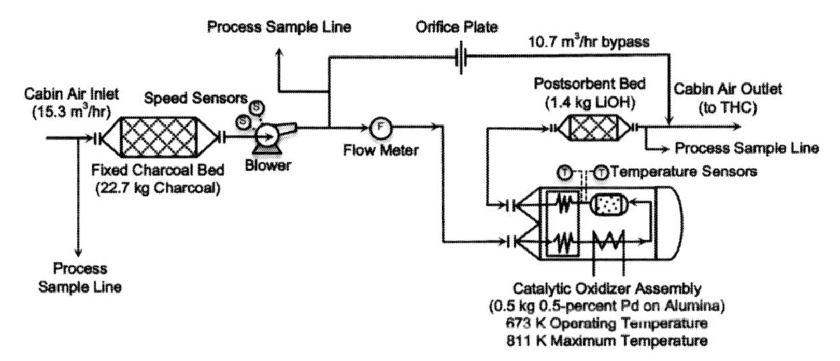

Figure 3.4 ISS TCCS [9].

In a parallel air purification process, CO_2 removal is accomplished via physical adsorption in a combined vacuum-swing/temperature-swing system that periodically vents CO_2 to space. As mentioned above, the ISS CDRA consists of four packed beds—two containing silica gel and zeolite 13X desiccant media and two containing zeolite 5A CO_2 sorbent media. Flow rates and bed sizes of the current TC and CO_2 removal systems are very different. These differences are driven by cabin air quality requirements such as maximum allowable concentration and contaminant generation rates, as well as the chemical and physical parameters of the system, such as surface area, porosity, and kinetics. Compounds produced at higher rates or consumed via inherently slow reactions require larger systems or higher throughput. Ammonia removal dictates the GAC bed size and flow rate while methane (CH_4) and carbon monoxide (CO) sets the catalytic oxidizer size and flow rate. The GAC bed for the ISS TC control equipment has a contact time of ~12 s and the thermal catalytic oxidation bed has a contact time of 0.67 s, reflecting the varying process flow rate and bed volumes. By comparison, process air flow through the CO_2 sorbent beds has resulted in a 1.6 s to 4 s contact time [10]. One amine-based CO_2 removal process under development employs a 0.25 s contact time. Historically, there has been a range of flow requirements needed to meet the cabin air quality specifications. Some overlapping middle ground exists such that one regenerable AR subsystem could incorporate both the TC and CO_2 removal functions in a new configuration, potentially incorporating shorter residence times and desorption cycles. This new vision necessitates an evaluation of newer and reversible sorbent materials.

Toxicology Considerations

Recently, for health and wellness reasons, NASA toxicology experts have lowered the spacecraft maximum allowable concentration (SMAC) for NH_3 to 7 ppm [11]. Carbon dioxide has also been subject to scrutiny as a contaminant in space cabins, concurrent with its examination as a terrestrial greenhouse gas. Pockets of high CO_2 concentration have been suggested as the trigger for crew-reported headaches. NASA has lowered the 180 day SMAC for CO_2 from 7000 to 5000 ppm. Lower levels of CO_2 are clearly better, more closely approximating the terrestrial environment and contributing to fewer physiological responses.

Desorption Dynamics and Vacuum Stability Considerations

The effect of vacuum exposure on sorbents must be understood. On board ISS, NH_3 is removed with Barnebey-Sutcliffe Type 3032 4 × 6 mesh GAC, which is periodically replaced. Type 3032 activated carbon was treated with H_3PO_4 and is no longer commercially available. It will likely be replaced with a similar material, and could be susceptible to acute vacuum exposures because some future crewed spacecraft design architectures have no airlock. Therefore an emergency extravehicular activity (EVA) would require cabin depressurization. For such spacecraft design architectures the sorbent material must be vacuum-stable. Similarly, in vacuum-venting systems such as the ISS CDRA, vacuum is necessary during desorption to remove CO_2 from the beds. Temperature is often used with vacuum desorption to enhance desorption kinetics, but most granular sorbent materials are inherently poor thermal conductors, making the process very inefficient. Thermal desorption from some new engineered materials, however, can be driven by directly-applied electrical current, offering a potential reduction in the power requirement and simplifying the thermal design challenge.

Opportunities for Spacecraft Resource Conservation

Beyond these challenges and considerations, there are opportunities for power savings and reliability improvements in the AR subsystem design. Across all categories of target contaminant, engineered structured sorbents—monoliths—offer the promise of lower pressure drop (and fan power), while minimizing the problem of particle attrition. Packing retention mechanisms typically use force (e.g., springs, clamps) to keep bed particles fixed within the flow circuit. The packing force, the flow itself, and reactant particle volume changes all contribute to particle attrition and the production of fines, which can be carried downstream and affect valve sealing surfaces and equipment such as fans and heat exchangers. Some sorbent substrates also allow for more efficient heating. NASA has pursued the development of monolithic adsorption systems for trace contaminants, CO_2, water vapor, and targeted TC catalytic oxidation. Those systems might appear in a three-tiered swing-bed system such as the one shown below in Fig. 3.5, NASA's concept for a Next Generation Atmosphere Revitalization system.

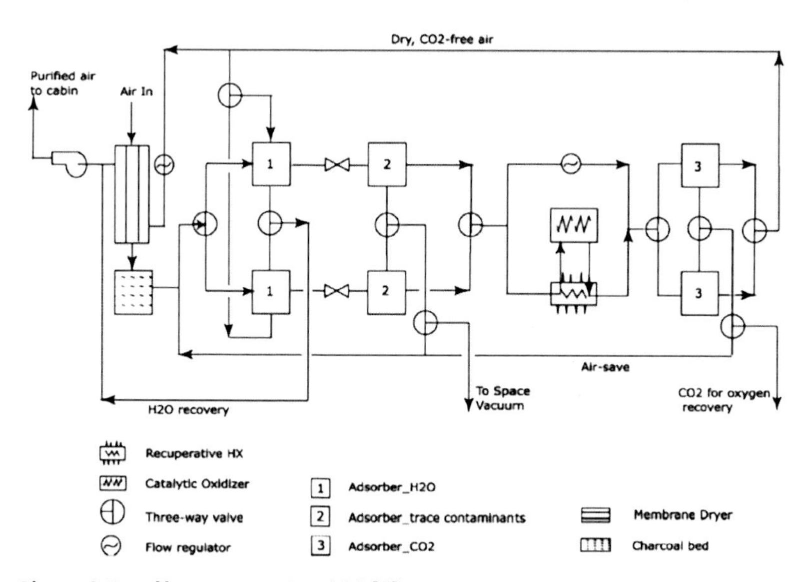

Figure 3.5 Next-generation ARS [5].

Lastly, in pursuit of loop closure, there are additional needs for specific gas separations. For example, one CH_4 processor produces a mixture of acetylene (C_2H_2) and hydrogen that would require separation and selective adsorption systems using nanoporous materials offer a promising means for accomplishing this.

The Center for Advanced Nanoscale Materials (CaNM) at the University of Puerto Rico and NASA have been working to address many of these challenges. The remainder of this chapter describes specific materials development efforts, experimental results and future plans.

3.2 Microporous Materials

Porous adsorbent materials are usually classified according to the International Union of Pure and Applied Chemistry (IUPAC) definitions. That is, microporous (<20 nm), mesoporous (20–50 nm) and macroporous (50 nm) materials. Among these, only the first two classifications are usually considered in fixed bed–type applications due to their inherent large specific surface area. The following sections will therefore focus in summarizing relevant properties of some microporous and mesoporous materials that are currently been used and/or could be considered for AR.

3.2.1 Zeolitic Adsorbents

Zeolites are inorganic porous materials with frameworks generally comprised of silicon and aluminum tetrahedra and with a multi-dimensional and interconnected pore system. The structure net charge is usually balanced with extra-framework cations that also serve as adsorption sites depending on their ultimate location. For example, cation locations in zeolite type-A (LTA) vary according to Fig. 3.6. The cations located in site S I are inaccessible to even molecules such as CO_2 given the small dimensions of the Sodalite cage windows surrounding these [12].

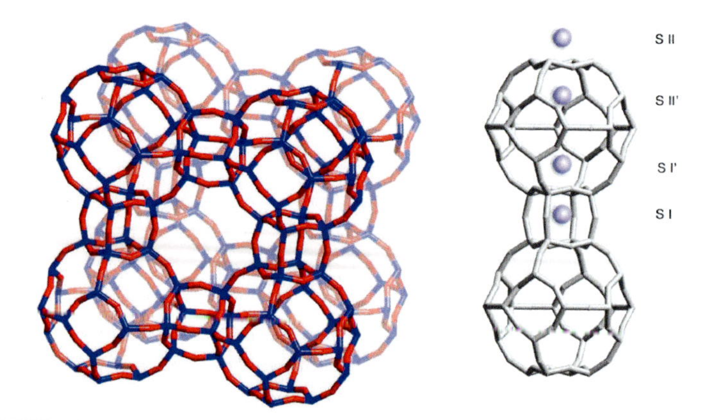

Figure 3.6 Zeolite type A (LTA) unit cell (left) and cation locations (right).

The average pore size of LTA zeolites depends on the nature of the extra-framework cation. For example, sodium-containing LTA (Na^+-LTA; 4A Zeolite), the average pore diameter is ca. 0.4 nm (4 Å) and the unit cell contains about 12 extra-framework cations. Only a fraction of these cations are available for interaction with adsorbates (i.e., cations occupy site S II). NASA currently employs Ca^{2+}-LTA (5A Zeolite) for CO_2 removal on board the ISS as part of the CDRA system (see Fig. 3.3), due to its excellent adsorption capacity at room temperature and ease of regeneration. A set of equilibrium isotherms and associated isosteric heat of adsorption profile are shown in Fig. 3.7.

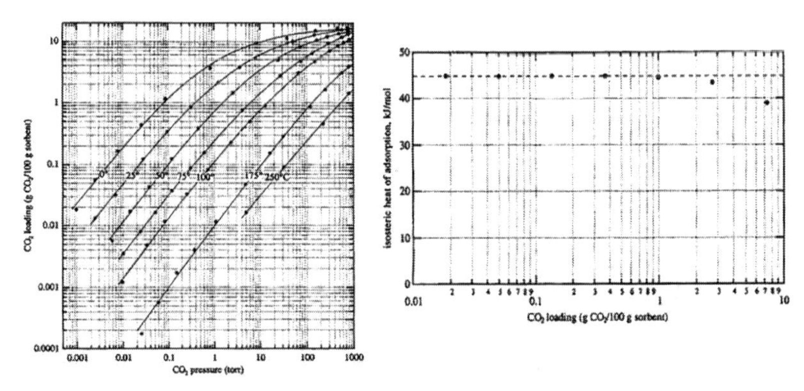

Figure 3.7 Adsorption isotherms of CO_2 on Ca^{2+}-LTA at different temperatures (left) and isosteric heat of adsorption (right) [13].

According to the isosteric heat profile, the surface-adsorbate interaction is at the physisorption level, which is suitable for onboard regeneration using vacuum and/or a moderate temperature swing. One of the main limitations of LTA materials, however, is their high degree of hydrophilicity due to coordination of multiple water molecules to cations. The typical water vapor loading at room temperature and about 50% relative humidity is ca. 22 wt% and complete removal usually requires temperatures greater than 350°C. In fact, one of the main challenges of the CDRA system used in the ISS (Fig. 3.3) is to avoid water vapor from reaching the CO_2 adsorption bed (i.e., Ca^{2+}-LTA). Furthermore, applicable SMACs for CO_2 will be lower for longer manned-space missions, and adsorption processes that utilize LTA zeolites will require more generation cycles to compensate for the decrease in working capacity, as evidenced by the lower CO_2 uptakes at lower partial pressures (see Fig. 3.7).

For several decades, we have seen a great effort in the field of crystal growth, leading to the discovery of hundreds of new topologies with frameworks whose composition and nature depart considerably from the one exhibited by zeolites. This offers an excellent window of opportunity to tailor-make adsorbents with properties more suitable for space applications, including lower hydrophilicity, and specific surface interactions to provide better CO_2 working capacities at lower partial pressures. Recently, Hernández-Maldonado and co-workers reported a strontium-based silicoaluminophosphate (Sr^{2+}-SAPO-34) designed with these challenges in mind [14–17].

SAPO-34 materials have a framework made of oxygen-bridged silicon, aluminum, and phosphorous centers all in tetrahedral coordination [18–21]. These at the same time form a 3-D interconnected pore network with windows of about 4 nm (4 Å) in diameter, similar to those found in Ca^{2+}-LTA zeolites. A typical SAPO-34 unit cell and associated extra-framework cation locations [22,23] are shown in Fig. 3.8. Loading of strontium via ion-exchange processes (liquid or solid phase) usually results in about one or two Sr^{2+} ions per unit cell. Due to charge balancing and repulsion criteria, these cations are found in S II and S II, positions, which is desired for interaction with adsorbates. Figure 3.9 shows CO_2 equilibrium isotherms at different temperatures and the associated isosteric heat of adsorption of Sr^{2+}-SAPO-34 adsorbents. When compared to the results corresponding to Ca^{2+}-LTA at 25°C (see Fig. 3.7) and a CO_2 partial pressure of ca. 1 Torr (or 1300 ppm), the Sr^{2+}-SAPO-34 adsorbents display a 60% increase in equilibrium adsorption capacity. Even more important is the fact that the observed average heat of adsorption is quite similar for both sorbents, which indicates that the SAPO-34–based material might be suitable for regeneration schemes such as the one employed in the CDRA system. Studies performed by Hernández-Maldonado and co-workers have found that the interaction between the CO_2 and Sr^{2+}-SAPO-34 arises from a chemical bond of ionic character and that charge donation leads to reduction of the positive charge on the metal cation [16], which results in a lower CO_2 adsorption energy (i.e., physisorption). Therefore, it should be experimentally possible to design an improved system for ultra-deep removal of

CO_2 by varying the silicoaluminophosphate composition in the Chabazite-like framework to control the CO_2 adsorption capacity.

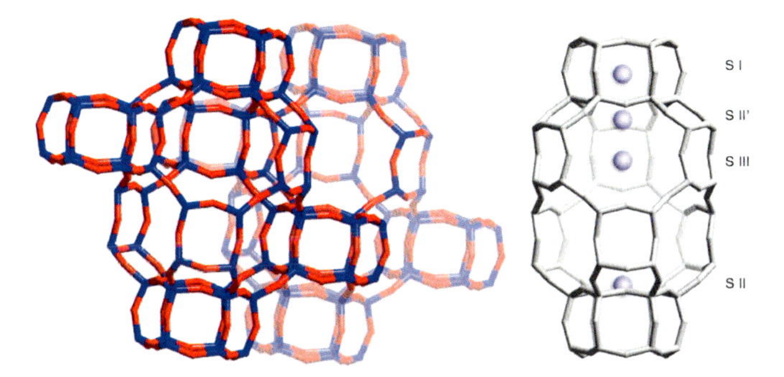

Figure 3.8 SAPO-34 unit cell (left) and cation locations (right).

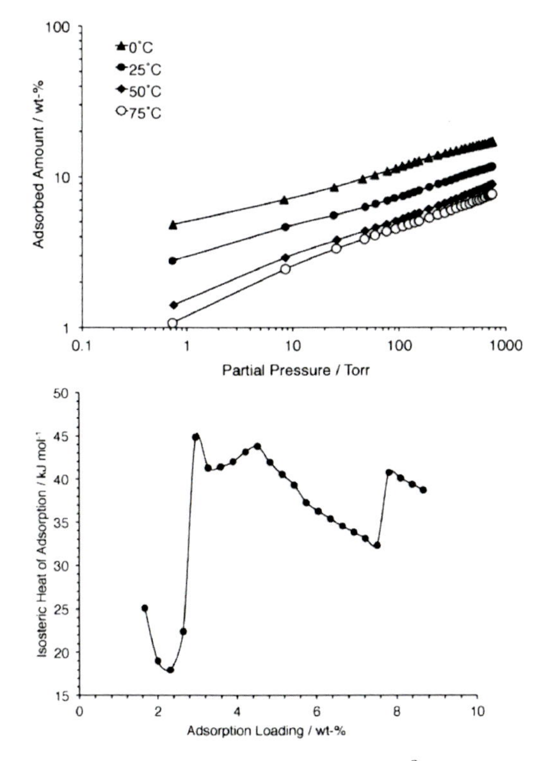

Figure 3.9 Adsorption isotherms of CO_2 on Sr^{2+}-SAPO-34 at different temperatures (up) and isosteric heat of adsorption (down). *Source*: Data reprinted with permission from [15]. Copyright 2008 American Chemical Society.

An important aspect to be considered when designing porous adsorbents for CO_2 removal in closed-volume applications is the kinetics governing the adsorption process. Tests performed by Hernández-Maldonado and co-workers for a step loading of 1000 ppm CO_2 concentration indicate a diffusion half time of ca. 0.2 s, which should be suitable for fixed bed–type processes without sacrificing much working capacity. In fact, tests performed using a small-scale fixed-bed CO_2 adsorption system at NASA Ames Research Center indicate that Sr^{2+}-SAPO-34 materials display a breakthrough time of ca. 200 min during treatment of a stream of nitrogen containing 1300 ppm CO_2 (see Fig. 3.10). For a detection limit of 2 ppm CO_2, the aforementioned data corresponds to a loading of ca. 2.6 wt%, which matches well with the equilibrium amount observed at similar conditions (see Fig. 3.9) and, therefore, evidences that there are no apparent resistance to diffusion of CO_2.

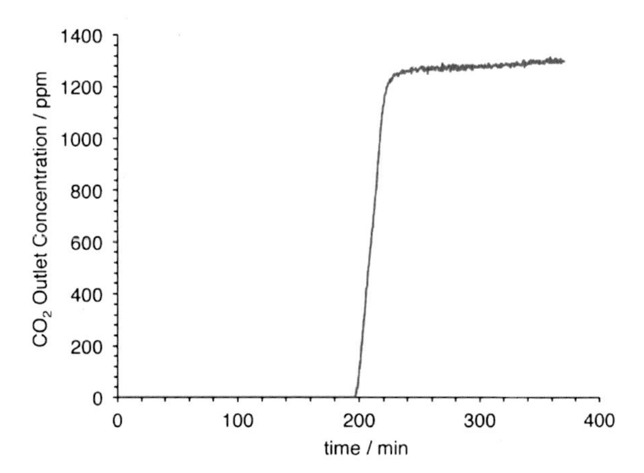

Figure 3.10 CO_2 breakthrough curve in Sr^{2+}-SAPO-34 at 25°C. Carrier gas was N_2 at 0.5 L/min.

Zeolites have garnered NASA interest in recent years with regard to their capacity to reversibly adsorb ammonia gas. Ammonia has traditionally been removed with phosphoric acid treated GAC, which has a capacity of between 2–4% by weight, and must be replaced periodically. NASA's ISS GAC is no longer available. Given that a new material must be identified and that NASA seeks to use regenerable swing-beds for combined CO_2 and trace contaminant control, transition metal ion-exchanged Y-zeolites offer promise. That conclusion is based on the data published by Liu and Aika [24], which documented capacities for reversible ammonia adsorption of 7.03 and 6.49 mmol/g on Co-Y and Cu-Y, respectively, at 323 K (~11% by weight). Luna and Hernandez-Maldonado prepared small quantities of these ion-exchange zeolites and pressed pellets ranging in size from 20 to 40 mesh. Figure 3.11 shows the ammonia weight gain in mg/g of a 1 g sample of each of these materials subjected to a 500 sccm, 50 ppm (NH_3 in nitrogen) mixture. The current ISS carbon is shown for comparison.

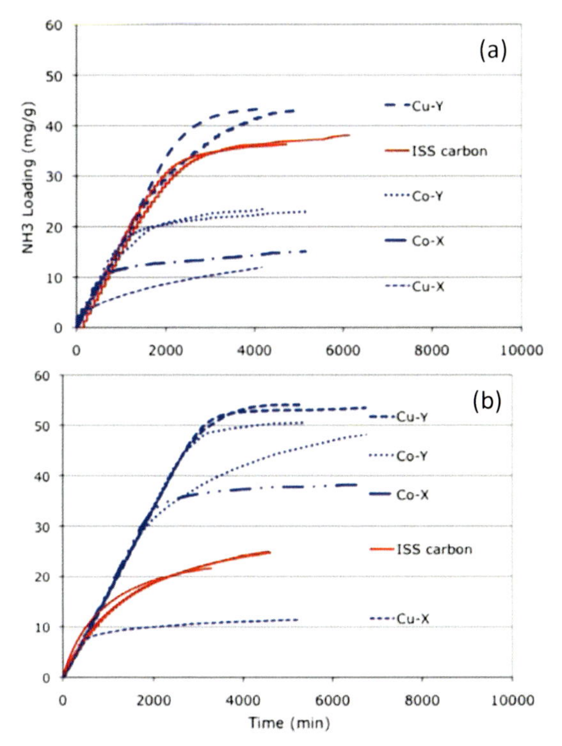

Figure 3.11 Ammonia loading on Cu- and Co-exchanged X and Y zeolites. Barneby-Sutcliffe 3032 granular activated carbon, used on ISS for ammonia removal, is shown for comparison. (a) with 40% relative humidity and (b) in dry nitrogen.

Although zeolites and zeolitic materials offer attractive CO_2 adsorption characteristics for purification processes necessary in long term missions, their powder nature gives rise to handling problems even when employed in combination with clay-based substrates. The alternative would be the development of zeolitic membranes synthesized by means of secondary growth techniques [25–32] or the use of metal-based frameworks/substrates. The latter will be thoroughly discussed later in this Chapter.

3.2.2 Amine-Based Adsorbents

The notion of an amine embedded adsorbent originated from amine scrubbing, which has been used to remove CO_2 from natural gas and hydrogen since 1930, when it was patented [33,34]. The process involves a reaction between the CO_2 with aqueous amine via a zwitterion mechanism to form carbamates [35]. Despite improvements in recent years, there are still many challenges related to scrubbing methods. These include considerable energy requirements for solvent regeneration and corrosion control. However, embedding of the amine onto the surface of a porous material makes it feasible for space related applications as discussed below.

Between the *Skylab* and Shuttle space flight programs, amine compounds gained a CO_2 removal role. Earliest concepts were steam-desorbed [36,37]. The current regenerable "solid" amine sorbent has undergone a lengthy development, and stems from the use of liquid amines for industrial CO_2 capture and for CO_2 control onboard submarines. Aerospace technologists sought a solution that would retain the basic chemistry of the liquid-phase amine-CO_2 system but eliminate the difficulties associated with corrosive solutions and with micro-gravity liquid management [38,39]. These sorbents are prepared by immersing high surface area plastic beads in an amine solution and subsequently drying. The amine bonds covalently with the plastic but weakly with CO_2 and water. Various post-treatments have been evaluated for improved performance.

The Shuttle program experimented with a vacuum-swing, amine-based CO_2 removal process via detailed test objectives (DTO) during several flights as part of the Extended Duration Orbiter efforts in the 1990s [39–47]. This process, the Regenerable CO_2 Removal System (RCRS) employed the Hamilton Sundstrand HSC and HSC$^+$ sorbent media. This media required the presence of atmospheric moisture to effectively remove CO_2. The units were not designed to control atmospheric moisture as the condensing heat exchanger performed that function. LiOH was used as the functional backup during the DTOs. The experience gained with amine-based

CO_2 removal during the Extended Duration Orbiter DTO activities gave rise to an improved sorbent that addressed various technical issues involving lifetime stability and functional redundancy issues [46–48]. An amine-based AR subsystem process architecture was selected for the Crew Exploration Vehicle (CEV), the flagship crew transport vehicle of the Constellation program [49].

The CEV system consists of thermally linked beds of a proprietary sorbent, designated SA9T, alternately exposed to cabin air for removal of CO_2 and water vapor, and to vacuum for desorption [50,51]. SA9T sorbent, small diameter amine-loaded polymethylmethacrylate (PMMA) beads, are poured into beds containing a brazed aluminum foam support structure. The aluminum foam serves as a conduction path for the heat of adsorption to move from the adsorbing bed toward the desorbing bed, thereby maintaining near-constant temperature and obviating the need for active temperature control. The CEV's CO_2 removal system is regenerable; that is, the sorbent material is not consumed. However, SA9T sorbent attracts both water and CO_2, and therefore both resources are lost upon vacuum desorption. Because the cost of this loss is unacceptable over a long period of time, amine-based systems are not a candidate for long-term missions. However, the same material has been proposed for use in the portable life support system of space suits [52,53], again with a loss of resources deemed acceptable to the mission.

These amine materials for CO_2 removal have been found to have some affinity for trace contaminants at the expense of CO_2 [54–56], and to lose CO_2 capacity and offgas NH_3 over time [45]. Nevertheless, solid amines are a proven means of CO_2 removal for short duration space missions where recovery of CO_2 and water is not a mission objective.

3.2.3 Metal Organic Frameworks

Metal organic frameworks (MOFs) allow the synthesis of tailor-made materials with predetermined, tunable porosity and designed chemical nature of pore surface [57–61]. Among the wide variety of porous MOFs described in the recent literature, there are some eye-catching examples of large pore sizes [62–65], flexible frameworks [66], and post-synthetic modified materials [67]. On the other hand, MOFs are not as robust as zeolites and zeolite-like materials and often collapse after a few cycles of pressure and/or temperature swings.

A wide variety of MOFs with pores larger than the kinetic diameter of CO_2 (3.3 Å) have been shown to sorb the latter efficiently. Zeolitic imidazolate frameworks (ZIFs) in particular outperform most other sorbents at ambient conditions [68]: The Zn-containing ZIF-78, with pore diameter of 4.4 Å and BET surface area of 950 $m^2 g^{-1}$, absorbs ~83 L CO_2 per 1 L of sorbent (~ 11.7 wt%) at 273 K and ambient pressure [69]. The CO_2 sorption capacity of MOFs can be improved by the incorporation of NH_2-functionalities; A Zn-aminotriazole-based MOF absorbs >15 wt% at 273 K, ~1 atm [70], while an ethylenediamine functionalized Cu-containing MOF has shown a large increase of its CO_2 sorption capacity at 298 K compared to a non functionalized analogue [71]. A cobalt(II)-adeninate, bio-MOF-11, with pore diameter of 5.8 Å, performs even better absorbing 20.9 wt% at 298 K and 1 bar [72].

The storage capacity of MOFs increases dramatically at high pressure: A Cu-based MOF uptakes 35.8 wt% at 298 K and 15 bar [73], while the Zn-containing MOF-177 adsorbs 59.6 wt% at 32 bar (BET surface area, 4500 $m^2 g^{-1}$) and MOF-200 and MOF-210, with BET surface areas of 4530 and 6240 $m^2 g^{-1}$, adsorb ~71.5 wt%, respectively, at 298 K and 50 bar (Fig. 3.12) [62,74].

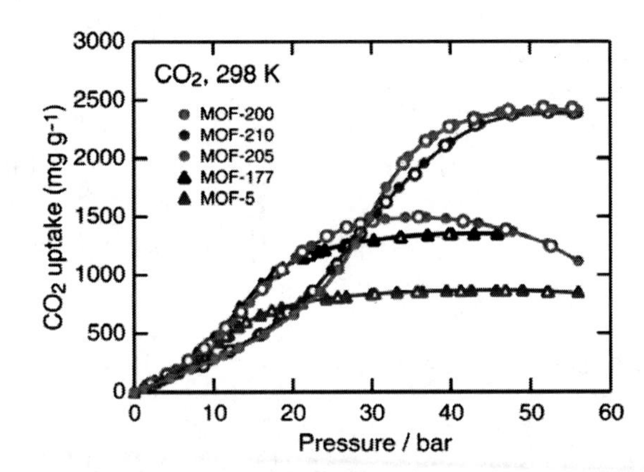

Figure 3.12 High-pressure CO_2 isotherms of MOF-5, –177, –200, –205, and –210 at 298 K. Source: Data reprinted with permission from [62]. Copyright 2010 *Science*.

An increase of CO_2 sorption capacity has been observed as the number of metal ions present per unit volume of the sorbent increases [75]. In addition, low-temperature X-ray crystallographic evidence shows that CO_2 is physisorbed on the pores of MOFs through end-on O···M interactions with coordinatively unsaturated metal cations, or side-on interactions between amine or hydroxyl groups and the carbon atom of CO_2 (Fig. 3.13) [70]. Theoretical work has also focused on interpreting the experimentally determined isosteric heats of CO_2 adsorption in terms of dipolar interactions [76].

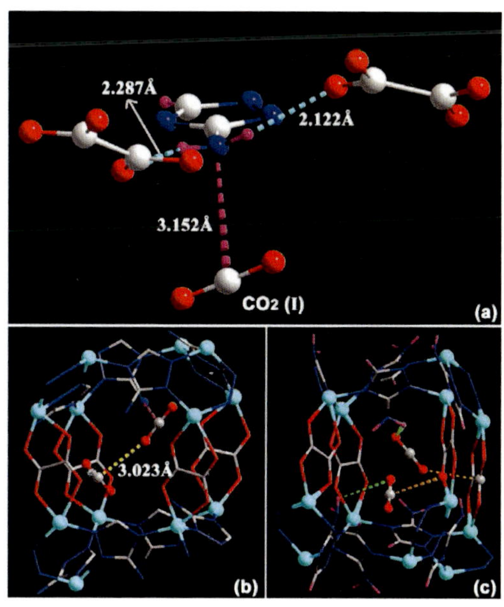

Figure 3.13 X-ray structure of CO_2 binding in MOF $Zn_2(Atz)_2(ox)\cdot(CO_2)_{1.3}$ (Atz, 3-amino-1,2,4-triazole; ox, oxalate) at 173 K. (a) The role of the amine group of Atz in binding CO_2-I is depicted. The H atoms of the amine group (located crystallographically) H-bond to oxalate O atoms, directing the N lone pair toward the C(δ+) atom of the CO_2 molecule. H-bond distances shown are for H-acceptor interactions. (b) Both crystallographically independent CO_2 molecules are shown trapped in a pore, showing the cooperative interaction between CO_2-I and CO_2-II molecules. The CO_2...NH_2 interaction is represented as a dotted purple bond, and the CO_2...CO_2 interaction is indicated as a dotted yellow bond. (c) This panel shows the other interactions present. The CO_2-I...Ox interactions are shown in orange, and the CO_2...NH_2 hydrogen bond interactions are shown in green. For clarity, H atoms are shown in purple. Source: Data reprinted with permission from [70]. Copyright 2010 *Science*.

Current experimental work at the University of Puerto Rico involving the systematic modification of pendant groups and lattice charge in Cu- and Ag-based MOFs, coupled with density functional theory (DFT) calculations of CO_2 sorption energies on their pore surfaces, is pursuing an in-depth understanding of the role of weak van der Waals interactions.

3.3 Microlith®-Based Structured Adsorbents

Regardless of sorbent material or target contaminant, containment of particles in packed beds presents challenges. Compressive forces and particle density changes contribute to particle attrition and the creation of dust. Dust can interfere with valve operation and increase system pressure drop. Implementation of engineered structured sorbents (ESS) is a potential alternative to the traditional packed bed system for the environmental control applications that addresses these challenges. This approach offers the inherent performance and safety attributes of molecular sieve, zeolites (e.g., aluminosilicate), and other sorbent materials capable of effectively removing CO_2, H_2O, and organic contaminants with greater structural integrity, regenerability, and process control. Precision Combustion, Inc. (PCI) and NASA's Marshall Space Flight Center (MSFC) have been developing one ESS approach based on the Microlith® technology [77] to meet the requirements of future, extended human spaceflight explorations [78–82]. The Microlith®-based ESS consist of metal substrates that provide structural integrity (i.e., less partition of sorbents compared to the pellet-based system) and enhanced thermal control during the adsorption/desorption process. This offers improved durability and efficiency over current, state-of-the-art, pellet-based systems. This particular ESS concept also offers a unique internal resistive heating capability that has shown potential for shorter regeneration times and reduced power requirement compared to conventional systems [79–81].

The Microlith® technology, patented and trademarked by PCI [77], consists of a series of ultra short-channel-length, low thermal mass metal meshes as shown in Fig. 3.14. It replaces the long channels of conventional monoliths with a series of short channel length substrates. Whereas in conventional honeycomb monoliths a fully developed boundary layer is present over a considerable length of the device, the very short channel length characteristic of the mesh-type substrate avoids boundary layer buildup. A Computational Fluid Dynamics (CFD) analysis (Fig. 3.14) illustrates the difference in boundary layer formation between a monolith and mesh-type elements. Since heat and mass transfer coefficients depend on the boundary layer thickness, minimizing boundary layer buildup enhances transport properties. Additionally, the mesh-type substrate can pack more active surface area into a small volume, providing increased adsorption area for a given pressure drop. The effectiveness of the Microlith® technology and the long-term durability of sorbent and catalyst coatings have been rigorously demonstrated in space station cabin air cleaning application [78–84] and other catalytic applications, including engine exhaust after-treatment [85], combustion, and fuel processing [86–88].

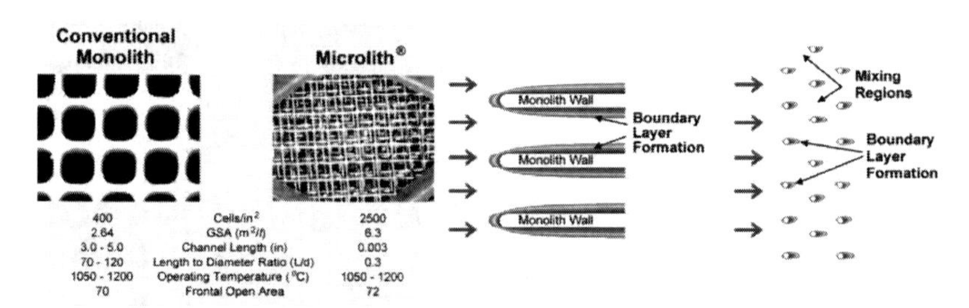

Figure 3.14 Physical characteristics of conventional monolith and short-channel-length Microlith® screens, and CFD analysis of boundary layer formation for a conventional monolith and a linear stack of three mesh-type Microlith® elements.

The development of zeolite and other sorbent coating protocols on the mesh-type substrate requires rigorous evaluation of washcoat formulations in order to produce an adherent, durable coating. The substrate geometry poses unique challenges in the development of sorbent coatings. The washcoats must be capable of being readily applied, and the resulting coatings must have a high degree of adhesion and cohesion and must be sufficiently abrasion resistant in order to withstand routine handling and multiple thermal cycles. At the same time, the formulation must retain the desired chemical and physical characteristics of the sorbents to achieve the expected adsorption capacity and removal efficiency. Washcoats that allow for rapid application of sorbent coatings for high-volume production have been developed. Scanning electron microscopy (SEM) micrograph of the sorbent-coated Microlith® substrates is shown in Fig. 3.14, indicating uniform coatings with a complete coverage around the Microlith® strands.

Several Microlith®-based regenerable adsorption modules have been examined at NASA-MSFC for performance evaluation and optimization. A one-person load CO_2/trace contaminant adsorber prototype was evaluated to demonstrate the potential of the mesh-type structured sorbent technology for cabin air-cleaning applications. In 2008, a complete one-person Microlith®-based ESS system, which consisted of a residual drier (i.e., moisture removal module), CO_2 removal module, and trace contaminants removal module was designed and developed. Figure 3.15a shows an example of the Microlith®-based ESS adsorber design and Fig. 3.15b shows one of the adsorber modules mounted on the test rig. Performance evaluation at NASA MSFC indicated the capability of removing the targeted 1 kg CO_2/day with an average power requirement of 40 W and a 20 min regeneration period [81]. The adsorption capacity and CO_2 removal efficiency were stable over the 20 h test and after subjecting the modules to at least 60 thermal cycles (i.e., adsorption–desorption cycles) [81]. Finally, a four-crew two-leg Microlith®-based ESS system was developed to study continuous CO_2 removal at a rate of 4 kg CO_2/day. These modules are being examined for performance demonstration, system optimization, and long-term durability.

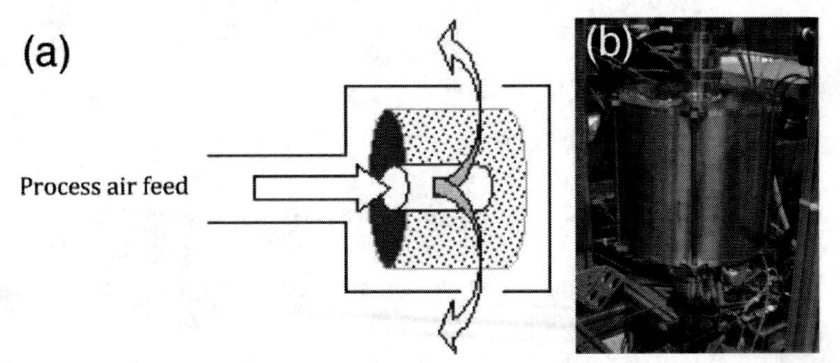

Figure 3.15 (a) Simplified Microlith®-based ESS adsorber design and (b) Adsorber module on the test rig.

3.4 Closing Remarks

It is quite evident that the technology and processes used to provide life support systems in space related endeavors have evolved considerably in response to the arrival of new SMACS and emerging space exploration mission challenges. As we proceed to plan for long-term missions (i.e., Mars exploration), it is imperative to accomplish highly efficient resource recovery for these future space exploration vehicle and habitat platforms. Highly efficient resource recovery that in effect "closes the mass loop" requires complex systems engineering that incorporates unique technical solutions. Fortunately, there has been considerably gain in the knowledge of bottom-up design of adsorbents to deal with the stringent atmospheric (i.e., CO_2 and NH_3) and processing requirements, including the design of third-generation porous frameworks that provide structural flexibility and surface interactions that can be controlled by an external variable. Furthermore, we have seen substantial advancement at the engineering level as well, such as the ESS approach that has the potential to significantly enhance the performance of current and future sorbent and catalyst materials that will enable AR subsystem architectures that will posses lower operational costs.

References

1. Law, J., Watkins, S., and Alexander, D. (2010). In-flight carbon dioxide exposures and related symptoms: association, susceptibility, and operational implications, NASA Document ID: TP-2010-216126.
2. Matty, C. M. (2010). Overview of carbon dioxide control issues during international space station/space shuttle joint docked operations, *International Conference on Environmenal Systems*, AIAA Paper 2010-6251.
3. Luna, B., Somi, G., Winchester, J. P., Grose, J., Mulloth, L. M., and Perry, J. L. (2010). Evaluation of commercial off-the-shelf sorbents and catalysts for control of ammonia and carbon monoxide, *International Conference on Environmental Systems*, AIAA Paper 2010-6062.

4. Luna, B., Ehresmann, D. J., Howard, D. J., Salas, L. J., Podolske, J. R., Mulloth, L., and Perry, J. L. (2008). Evaluation of commercial off-the-shelf ammonia sorbents and carbon monoxide catalysts, *International Conference on Environmental Systems*, SAE Paper 2008-01-2097.
5. Mulloth, L. M., Perry, J. L., and LeVan, M. D. (2004). Integrated system design for air revitalization in next generation crewed spacecraft, *International Conference on Environmental Systems*, SAE Paper 2004-01-2373.
6. Perry, J. L. (1998). Elements of spacecraft cabin air quality control, NASA Document ID: TP-1998-207978.
7. Ray, C. D. (1974). Skylab atmospheric contamination control, NASA Document ID: NASA/TM-X-64900.
8. Knox, J. C. (2000). International space station carbon dioxide removal assembly testing, *International Conference on Environmental Systems*, SAE Paper 2000-01-2345.
9. Perry, J. L., Curtis, R. E., Alexandre, K. L., Ruggiero, L. L., and Shtessel, N. (1998). Performance testing of a trace contaminant control subassembly for the international space station, *International Conference on Environmental Systems*, SAE Paper 981621.
10. Wieland, P. O. (1998). Living together in space: the design and operation of the life support systems on the international space station. vol. 1, NASA Document ID: TM-1998-206956.
11. James, J. T. (2008). Spacecraft maximum allowable concentrations for airborne contaminants, NASA Document ID: JSC-20584.
12. Breck, D. W. (1973) Zeolite Molecular Sieves (Wiley, New York).
13. Mulloth, L. M., and Finn, J. E. (1998). Carbon dioxide adsorption on a 5A zeolite designed for CO_2 removal in spacecraft cabins, NASA Document ID: NASA/TM-1998-208752.
14. Rivera-Ramos, M. E., and Hernandez-Maldonado, A. J. (2007). Adsorption of N_2 and CH_4 by ion-exchanged silicoaluminophosphate nanoporous sorbents: interaction with monovalent, divalent, and trivalent cations, *Ind. Eng. Chem. Res.*, **46**, 4991–5002.
15. Rivera-Ramos, M. E., Ruiz-Mercado, G. J., and Hernandez-Maldonado, A. J. (2008). Separation of CO_2 from light gas mixtures using ion-exchanged silicoaluminophosphate nanoporous sorbents, *Ind. Eng. Chem. Res.*, **47**, 5602–5610.
16. Arevalo-Hidalgo, A. G., Santana, J. A., Fu, R., Ishikawa, Y., and Hernandez-Maldonado, A. J. (2010). Separation of CO_2 from light gas mixtures using nanoporous silicoaluminophosphate sorbents: effect of multiple-step

ion exchange and adsorption mechanism via computational studies, *Micropor. Mesopor. Mater.*, **130**, 142–153.

17. Zhang, L., Primera-Pedrozo, J. N., and Hernandez-Maldonado, A. J. (2010). Thermal detemplation of Na-SAPO-34: Effect on Sr^{2+} ion exchange and CO_2 adsorption, *J. Phys. Chem. C*, **114**, 14755–14762.
18. Lok, B. M., Messina, C. A., Patton, R. L., Gajek, R. T., Cannan, T. R., and Flanigen, E. M. (1984). Crystalline silicoaluminophosphates. United States Patent U.S. Patent 4,440,871.
19. Lok, B. M., Vail, L. D., and Flanigen, E. M. (1988). Magnesium-aluminum-phosphorus-silicon-oxide molecular sieve compositions. United States Patent U.S. Patent 4,758,419.
20. Hartmann, M., and Kevan, L. (1999). Transition-metal ions in aluminophosphate and silicoaluminophosphate molecular sieves: location, interaction with adsorbates and catalytic properties, *Chem. Rev.*, **99**, 635–663.
21. Tan, J., Liu, Z. M., Bao, X. H., Liu, X. C., Han, X. W., He, C. Q., and Zhai, R. S. (2002). Crystallization and Si incorporation mechanisms of SAPO-34, *Micropor. Mesopor. Mater.*, **53**, 97–108.
22. Djieugoue, M. A., Prakash, A. M., and Kevan, L. (1998). Electron spin resonance and electron spin echo modulation studies on reducibility, location, and adsorbate interactions of N(I) in Ni(II)-exchanged SAPO-34, *J. Phys. Chem. B*, **102**, 4386–4391.
23. Mortier, W. J. (1982) Compilation of extra-framework sites in zeolites (Butterworth Scientific Limited, Guildford).
24. Liu, C. Y., and Aika, K. (2003). Ammonia adsorption on ion exchanged Y-zeolites as ammonia storage material, *J. Jpn. Pet. Inst.*, **46**, 301–307.
25. Jeong, H. K., Krohn, J., Sujaoti, K., and Tsapatsis, M. (2002). Oriented molecular sieve membranes by heteroepitaxial growth, *J. Am. Chem. Soc.*, **124**, 12966–12968.
26. Lai, Z. P., Tsapatsis, M., and Nicolich, J. R. (2004). Siliceous ZSM-5 membranes by secondary growth of b-oriented seed layers, *Adv. Funct. Mater.*, **14**, 716–729.
27. Mabande, G. T. P., Ghosh, S., Lai, Z. P., Schwieger, W., and Tsapatsis, M. (2005). Preparation of b-oriented MFI films on porous stainless steel substrates, *Ind. Eng. Chem. Res.*, **44**, 9086–9095.
28. Choi, J., Ghosh, S., King, L., and Tsapatsis, M. (2006). MFI zeolite membranes from *a*- and randomly oriented monolayers, *Adsorption*, **12**, 339–360.

29. Choi, J., Ghosh, S., Lai, Z. P., and Tsapatsis, M. (2006). Uniformly a-oriented MFI zeolite films by secondary growth, *Angew. Chem. Int. Edit.*, **45**, 1154–1158.

30. Choi, J. Y., Lai, Z. P., Ghosh, S., Beving, D. E., Yan, Y. S., and Tsapatsis, M. (2007). Layer-by-layer deposition of barrier and permselective c-oriented-MCM-22/silica composite films, *Ind. Eng. Chem. Res.*, **46**, 7096–7106.

31. Snyder, M. A., and Tsapatsis, M. (2007). Hierarchical nanomanufacturing: From shaped zeolite nanoparticles to high-performance separation membranes, *Angew. Chem. Int. Edit.*, **46**, 7560–7573.

32. Fan, W., Snyder, M. A., Kumar, S., Lee, P. S., Yoo, W. C., McCormick, A. V., Penn, R. L., Stein, A., and Tsapatsis, M. (2008). Hierarchical nanofabrication of microporous crystals with ordered mesoporosity, *Nat. Mater.*, **7**, 984–991.

33. Rochelle, G. T. (2009). Amine scrubbing for CO_2 capture, *Science*, **325**, 1652–1654.

34. Bottoms, R. R. (1930). Separating acid gases. US Patent 1,783,901.

35. D'Alessandro, D. M., and McDonald, T. (2011). Toward carbon dioxide capture using nanoporous materials, *Pure Appl. Chem.*, **83**, 57–66.

36. Wood, P., and Wydeven, T. (1987). Stability of IRA-45 solid amine resin as a function of carbon dioxide absorption and steam desorption cycling, *Intersociety Conference on Environmental Systems*, SAE Paper 871452.

37. Brennan, S. M., and Donovan, R. M. (1986). Space station benefits from ECLS: propulsion system synergism, AIAA/ASME/SAE/ASEE 22nd Joint Propulsion Conference, AIAA Paper 1986-1407.

38. Otsubo, K., Tanemura, T., Nitta, K., Oguchi, M., Nakabayasi, N., Kimura, S., and Kuroda, H. (1992). Evaluation of solid amine CO_2 adsorbing characteristics, *International Conference on Environmental Systems*, SAE Paper 211334.

39. Satyapal, S., Filburn, T., Trela, J., and Strange, J. (2001). Performance and properties of a solid amine sorbent for carbon dioxide removal in space life support applications, *Energy Fuels*, **15**, 250–255.

40. Ouellette, F. A., Winkler, H. E., and Smith, G. S. (1990). The extended duration orbiter regenerable CO_2 removal system, *Intersociety Conference on Environmental Systems*, SAE Paper 901292.

41. Jeng, F. F., Williamson, R. G., Ouellette, F. A., Edeen, M. A., and Lin, C. H. (1991). Adsorbent testing and mathematical modeling of a solid amine regenerative CO_2 and H_2O removal system, *International Conference on Environmental Systems*, SAE Paper 911364.

42. Graf, J. C., Dall-Bauman, L. A., and Jeng, F. F. (1993). Characterization of an improved solid amine for a regenerative CO_2 removal systems, *International Conference on Environmental Systems*, SAE Paper 932292.

43. Ouellette, F. A., Allen, G., Baker, G. S., and Woods, D. J. (1993). Development and flight status report on the extended duration orbiter regenerable carbon dioxide removal system, *International Conference on Environmental Systems*, SAE Paper 932294.

44. Kazemi, A. R., and Mitchell, S. M. (1993). Advanced testing and modeling of a modified solid amine regenerative CO_2 and H_2O removal system, *International Conference on Environmental Systems*, SAE Paper 932293.

45. Genovese, J., and Nalette, T. (1994). Life characterization of enhanced solid amine CO_2 sorbents, *International Conference on Environmental Systems*, SAE Paper 941395.

46. Graf, J., Filburn, T., Lantzakis, M., and Taddey, E. (1998). An orbiter upgrade demonstration test article for a fail-safe regenerative CO_2 removal system, *International Conference on Environmental Systems*, SAE Paper 981536.

47. Papale, B., and Dean, W. C. (2002). Development, testing, and packaging of a redundant regenerable carbon dioxide removal system (RRCRS), *International Conference on Environmental Systems*, SAE Paper 2002-01-2530.

48. Filburn, T., Nalette, T., and Graf, J. (2001). The design and testing of a fully redundant regenerative CO_2 removal system (RCRS) for the shuttle orbiter, *International Conference on Environmental Systems*, SAE Paper 2001-01-2420.

49. Trinh, E. H., and Joshi, J. (2005). Critical technologies for sustained human space exploration, AIAA Aerospace Sciences Meeting and Exchibit, AIAA Paper 2005-129.

50. Button, A., Sweterslitsch, J., and Cox, M. (2010). Space suit environment testing of the orion atmosphere revitalization technology, *International Conference on Environmental Systems*, AIAA Paper 2010-6155.

51. Button, A., and Sweterlitsch, J. (2010). 2009 Continued testing of the atmosphere revitalization technology, *International Conference on Environmental Systems*, AIAA Paper 2010-6163.
52. Paul, H., and Rivera, F. (2010). Portable life support rapid cycle amine repackaging and subscale test results, *International Conference on Environmental Systems*, AIAA Paper 2010-2066.
53. Conger, B., Chullen, C., Barnes, B., and Leavitt, G. (2010). Proposed schematic for an advanced development lunar portable life support system, *International Conference on Environmenal Systems*, AIAA Paper 2010-6038.
54. Button, A., Sweterlitsch, J., Broerman, C., and Campbell, M. (2010). Trace contaminant testing with the orion atmosphere revitalization technology, *International Conference on Environmental Systems*, AIAA Paper 2010 6299.
55. Monje, O., Brosnam, B., Flanagan, A., and Wheeler, R. (2010). Characterizing the adsorption capacity of SA9T using simulated spacecraft gas streams, *International Conference on Environmental Systems*, AIAA Paper 2010-6063.
56. Monje, O., Brosnan, B., and Wheeler, R. M. (2010). Characterizing the dynamic performance of SA9T, *International Conference on Environmenal Systems*, AIAA Paper 2010–6269.
57. Tranchemontagne, D. J., Mendoza-Cortes, J. L., O'Keeffe, M., and Yaghi, O. M. (2009). Secondary building units, nets and bonding in the chemistry of metal-organic frameworks, *Chem. Soc. Rev.*, **38**, 1257–1283.
58. Perry, J. J., Perman, J. A., and Zaworotko, M. J. (2009). Design and synthesis of metal-organic frameworks using metal-organic polyhedra as supermolecular building blocks, *Chem. Soc. Rev.*, **38**, 1400–1417.
59. Robson, R. (2008). Design and its limitations in the construction of bi- and poly-nuclear coordination complexes and coordination polymers (aka MOFs): a personal view, *Dalton Trans.*, 5113–5131.
60. Farha, O. K., and Hupp, J. T. (2010). Rational design, synthesis, purification, and activation of metal-organic framework materials, *Acc. Chem. Res.*, **43**, 1166–1175.
61. Zaworotko, M. J. (2008). Materials science: Designer pores made easy, *Nature*, **451**, 410–411.
62. Furukawa, H., Ko, N., Go, Y. B., Aratani, N., Choi, S. B., Choi, E., Yazaydin, A. O., Snurr, R. Q., O'Keeffe, M., Kim, J., and Yaghi, O. M. (2010). Ultrahigh porosity in metal-organic frameworks, *Science*, **329**, 424–428.

63. Farha, O. K., Yazaydin, A. O., Eryazici, I., Malliakas, C. D., Hauser, B. G., Kanatzidis, M. G., Nguyen, S. T., Snurr, R. Q., and Hupp, J. T. (2010). De novo synthesis of a metal-organic framework material featuring ultrahigh surface area and gas storage capacities, *Nat. Chem.*, **2**, 944–948.

64. Ferey, G., Mellot-Draznieks, C., Serre, C., and Millange, F. (2005). Crystallized frameworks with giant pores: are there limits to the possible?, *Acc. Chem. Res.*, **38**, 217–225.

65. Chen, B. L., Eddaoudi, M., Hyde, S. T., O'Keeffe, M., and Yaghi, O. M. (2001). Interwoven metal-organic framework on a periodic minimal surface with extra-large pores, *Science*, **291**, 1021–1023.

66. Horike, S., Shimomura, S., and Kitagawa, S. (2009). Soft porous crystals, *Nat. Chem.*, **1**, 695–704.

67. Tanabe, K. K., and Cohen, S. M. (2011). Postsynthetic modification of metal-organic frameworks: a progress report, *Chem. Soc. Rev.*, **40**, 498–519.

68. Wang, B., Cote, A. P., Furukawa, H., O'Keeffe, M., and Yaghi, O. M. (2008). Colossal cages in zeolitic imidazolate frameworks as selective carbon dioxide reservoirs, *Nature*, **453**, 207-U206.

69. Phan, A., Doonan, C. J., Uribe-Romo, F. J., Knobler, C. B., O'Keeffe, M., and Yaghi, O. M. (2010). Synthesis, structure, and carbon dioxide capture properties of zeolitic imidazolate frameworks, *Acc. Chem. Res.*, **43**, 58–67.

70. Vaidhyanathan, R., Iremonger, S. S., Shimizu, G. K. H., Boyd, P. G., Alavi, S., and Woo, T. K. (2010). Direct observation and quantification of CO_2 binding within an amine-functionalized nanoporous solid, *Science*, **330**, 650–653.

71. Demessence, A., D'Alessandro, D. M., Foo, M. L., and Long, J. R. (2009). Strong CO_2 binding in a water-stable, triazolate-bridged metal-organic framework functionalized with ethylenediamine, *J. Am. Chem. Soc.*, **131**, 8784–8786.

72. An, J., Geib, S. J., and Rosi, N. L. (2010). High and selective CO_2 uptake in a cobalt adeninate metal-organic framework exhibiting pyrimidine- and amino-decorated pores, *J. Am. Chem. Soc.*, **132**, 38–39.

73. Liang, Z. J., Marshall, M., and Chaffee, A. L. (2009). CO_2 adsorption-based separation by metal organic framework (Cu-BTC) versus zeolite (13X), *Energy Fuels*, **23**, 2785–2789.

74. Millward, A. R., and Yaghi, O. M. (2005). Metal-organic frameworks with exceptionally high capacity for storage of carbon dioxide at room temperature, *J. Am. Chem. Soc.*, **127**, 17998–17999.

75. Britt, D., Furukawa, H., Wang, B., Glover, T. G., and Yaghi, O. M. (2009). Highly efficient separation of carbon dioxide by a metal-organic framework replete with open metal sites, *Proc. Natl. Acad. Sci. USA*, **106**, 20637–20640.

76. Yazaydin, A. O., Snurr, R. Q., Park, T. H., Koh, K., Liu, J., LeVan, M. D., Benin, A. I., Jakubczak, P., Lanuza, M., Galloway, D. B., Low, J. J., and Willis, R. R. (2009). Screening of metal-organic frameworks for carbon dioxide capture from flue gas using a combined experimental and modeling approach, *J. Am. Chem. Soc.*, **131**, 18198–18199.

77. Pfefferle, W. C. (1991). Microlith catalytic reaction system. US Patent 5,051,241.

78. Roychoudhury, S., Walsh, D., and Perry, J. (2004). Microlith based sorber for removal of environmental contaminants, *International Conference on Environmental Systems*, SAE Paper 2004-01-2442.

79. Roychoudhury, S., Walsh, D., and Perry, J. (2005). Resistively-heated Microlith-based adsorber for carbon dioxide and trace contaminant removal, *International Conference on Environmental Systems*, SAE Paper 2005-01-2866.

80. Junaedi, C., Roychoudhury, S., Walsh, D., Knox, J. C., Perry, J. L., Howard, D. F., and Sullivan, P. D. (2008). Adsorption system based on Microlith® technology and its progress in fuel cell, spacecraft, and chem-bio warfare defense applications, Proc. AIChE Separations Div., AICHE, paper 131352.

81. Howard, D. F., Perry, J. L., Knox, J. C., and Junaedi, C. (2009). Engineered structured sorbents for the adsorption of carbon dioxide and water vapor from manned spacecraft atmospheres: applications and testing 2008/2009, *International Conference on Environmental Systems*, SAE Paper 2009-01-2444.

82. Roychoudhury, S., Perry, J., and Walsh, D. (2006). Regenerable adsorption system. US Patent 7,141,092 B1.

83. Carter, R., Bianchi, J., Pfefferle, W. C., Roychoudhury, S., and Perry, J. L. (1997). Unique metal monolith catalytic reactor for destruction of airborne trace contaminants, *International Conference on Environmental Systems*, SAE Paper 972432.

84. Perry, J. L., Carter, R. N., and Roychoudhury, S. (1999). Demonstration of an ultra-short channel metal monolith catalytic reactor for trace contaminant control applications, *International Conference on Environmental Systems*, SAE Paper 1999-01-2112.

85. Roychoudhury, S., Bianchi, J., Muench, G., and Pfefferle, W. C. (1997). Development and performance of Microlith™ light-off preconverters for LEV/ULEV, International Congress & Exposition, SAE Paper 971023.

86. Lyubovsky, M., Karim, H., Menacherry, P., Boorse, S., LaPierre, R., Pfefferle, W. C., and Roychoudhury, S. (2003). Complete and partial catalytic oxidation of methane over substrates with enhanced transport properties, *Catal. Today*, **83**, 183–197.

87. Roychoudhury, S., Castaldi, M., Lyubovsky, M., LaPierre, R., and Ahmed, S. (2005). Microlith catalytic reactors for reforming iso-octane-based fuels into hydrogen, *J. Power Sources*, **152**, 75–86.

88. Lyubovsky, M., Roychoudhury, S., and LaPierre, R. (2005). Catalytic partial "oxidation of methane to syngas" at elevated pressures, *Catal. Lett.*, **99**, 113–117.

Chapter 4

Nanotechnology in Advanced Life Support: Water Recycling

Michael Flynn,[a] Eduardo Nicolau,[b] and Carlos R. Cabrera[b]

[a]*NASA Ames Research Center, Moffett Field, CA 94035, USA*
[b]*NASA-URC Center for Advanced Nanoscale Materials, University of Puerto Rico, Rio Piedras Campus, San Juan, PR 00931, USA*

eduardo.nicolau@upr.edu

4.1 Introduction

Almost all chemical and separation processes ultimately function at the nanoscale. This is particularly true in the field of life support. Life support is the study of all aspects of technologies that are required to keep astronauts alive in space. These aspects include water recycling, air recycling, solid waste treatment and recycling, thermal control, and energy recovery systems, which are largely defined by molecular level flux, reaction kinetics, multi-phase equilibriums, and biological processes.

The study of nanotechnology is critical for understanding the function of life support systems. Even though most life support systems are relatively large and complicated processes, the fundamental physical effects that are the key to their function occur at the nanoscale level. In addition, these effects can be dependent on gravity complicating function in microgravity. The following

Advanced Nanomaterials for Aerospace Applications
Edited by Carlos R. Cabrera and Félix A. Miranda
Copyright © 2014 Pan Stanford Publishing Pte. Ltd.
ISBN 978-981-4463-18-8 (Hardcover), 978-981-4463-19-5 (eBook)
www.panstanford.com

sections describe some of the interactions between nanotechnology and life support technologies. The focus is primarily on water recycling technologies. It is provided to stimulate nanotechnology research that is relevant to life support applications.

4.2 Membranes

Membrane-based systems are typically used to perform separations of different molecules based on molecular size, charge, volatility, or polarity. Membranes are typically used in contactors, which provide a platform to expose a contaminated feed to the membrane. A driving force such as pressure, osmotic potential, or vapor pressure difference is then used to drive the separation. There are wide ranges of membrane types that are used to perform specific separations.

4.2.1 Pervaporation Membranes

Pervaporation membranes provide a method of separating liquids and gasses, solubilizing gasses in liquids, and distilling liquids. These membranes typically are composed of hydrophobic materials with nano-sized pores.[1] The pore size is limited by the optimization of surface tension forces that produce a meniscus. The meniscus restricts the flow of liquids through the pores and provides the bases for gas liquid separations. Figure 4.1 shows the orientation of the meniscus relative to the membrane pores. The resulting air gap within the media allows only water vapor to pass through it.

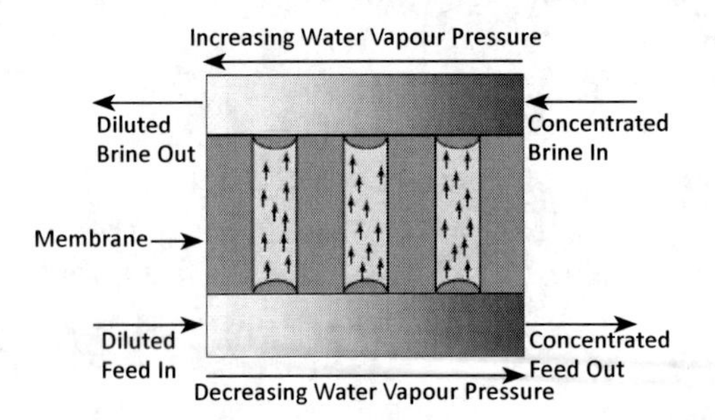

Figure 4.1 Cross section of pervaporation membrane showing meniscus at membrane pore and the flux of water vapor between the feed side meniscus and the product side meniscus. Copyright: Michaels, A. S. (1998). A new option: osmotic distillation. *Chem. Eng. Prog.*, 7, 49–61.

The key parameters in the structure of these membranes are total pore area, pore structure, and pore diameter. Pore area is related to the flux rate of the membrane as it defines the contact area for vapor-liquid equilibrium. The higher the total pore area, the more flux the membrane can produce. Typically, commercially available membranes have pore areas less than 50%. Pore structure helps to determine the rejection characteristics of the membrane. The more torturous the pore path, the higher chance there is for a given molecule to reflect back to the inlet of the pore. This can result in improved rejection characteristic in comparison with non-membrane evaporation processes.[2] Pore diameter is important in that it defines the maximum pressure the membrane can withstand prior to the meniscus collapsing. Pervaporation membranes with large pores cannot withstand high pressures. As the pore diameter increases, the meniscus area increases, and the force it must withstand increases. When the surface tension of water is exceeded, the meniscus fails, and liquid water enters the pore causing the membrane to leak.[3]

4.2.2 Reverse Osmosis Membranes

Reverse osmosis (RO) membranes are typically hydrophilic membranes that allow water to pass through while rejecting contaminates. They are widely used for separating liquids and total dissolved solids.[4] Reverse osmosis membranes are commonly composed of a thin dense layer deposited on a polymer matrix that provides a supporting substrate. The membranes are designed to allow only water to pass through the dense layer. The porous polymer matrix allows the membrane to withstand high pressures, i.e., up to 1200 psi for seawater. The three most common types of reverse osmosis membranes are cellulosic, aromatic polyamide, and thin-film composite.[5] These membrane materials can be cast in sheets or as hollow fibers.

Cellulosic membranes are composed of a thin dense surface layer of cellulose acetate and a porous substructure. The thin dense layer is responsible for solute rejection and the porous substructure provides strength. Aromatic polyamide membranes also are composed of a micro-thin dense skin and a porous substructure. They have more resistance to physical deterioration as a result of pH, hydrolysis, and degradation due to biological attack compared to cellulosic membranes.[6]

Thin-film composite membranes are made by forming a nanothin, dense, solute-rejecting surface film on top of a porous substructure. The thin surface layer typically has a thickness from 10 to 100 nm. Several types of thin-film composite membranes have been developed, including aromatic polyamide, alkyl-aryl poly urea/polyamide, and polyfurane cyanurate. The supporting porous sublayer is usually made of polysulfone.[7]

4.2.3 Forward Osmosis Membranes

Forward osmosis (FO) membranes are similar to RO membranes except that they are optimized for the significantly lower forces used in forward osmosis. Pure water has zero osmotic potential where as salt water has a high osmotic potential. The difference in osmotic potentials is the driving force for pure water to move from the feed water to a salt water solution. The most common type of membrane used for FO is cellulose tri-acetate.[8] A comparison between RO and FO membranes is provided in Fig. 4.2. The figure shows SEM images of the cross section of (a) cellulose acetate (CA) asymmetric

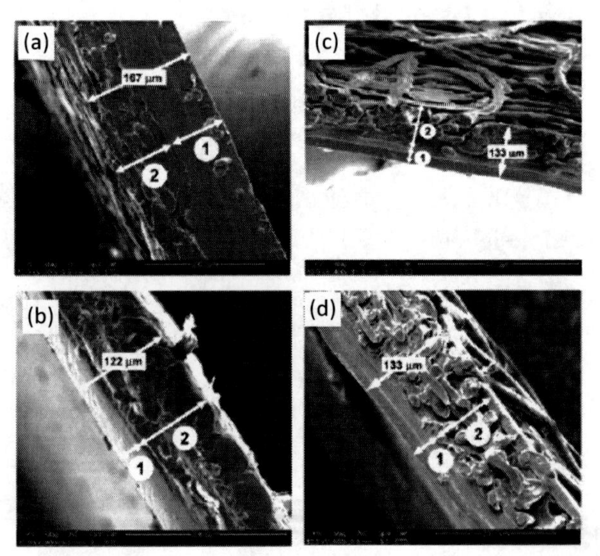

Figure 4.2 SEM images of the cross section of (a) CA membrane, (b) AD membrane, (c) FO membrane, and (d) FO membrane at higher resolution. (1) is the dense selective layer and (2) is the support layer. Copyright: *Environ. Sci. Technol.*, 2006, **40**(7), 2408–2413.

RO membrane (GE Osmonics, Minnetonka, MN), (b) polyamide composite (AD) RO membrane (GE Osmonics), (c) A cellulose triacetate FO membrane (Hydration Technology Innovations), and (d) the same FO membrane at higher resolution. The images show that the FO membrane has a much less dense backing material than the RO membranes. The FO membranes are not designed to withstand high pressures. Having a less dense more porous backing material reduces mechanical strength but also reduces the concentration polarization of ions at the membrane.

FO membranes are strongly impacted by concentration polarization (CP). Concentration polarization is the buildup of ionic and nonionic concentration gradients at the surfaces of the membranes and within the support matrix of the pores.[9] These contaminates can be ions and organics of the feed side and pure water internal to the membrane on the product side. Concentration polarization works to reduce the osmotic potential across the membrane and thus reduce the flux rate across it. This is particularly true on the product side of the membrane where the backing material produces a porous media that becomes saturated with pure water. The salts in the draw solution must diffuse through this backing material to maintain an osmotic potential across the membrane.[10]

4.2.4 Future Directions in Membrane Development

The design of separation membranes at the nanoscale level has resulted in a number of new formulations that offer advantages over the current state of the art technologies. Some of these include zeolitic ceramic membranes, catalytic nanoparticle membranes, mixed matrix membranes, carbon nanotube membranes, block copolymer membranes, and bio-inspired membranes such a lipid and aquaporin protein-based membranes.

4.2.4.1 Zeolite ceramic membranes

Zeolites are aluminosilicate minerals. They are characterized by sub-nanometer scale crystalline structures that are highly uniform and are composed of amorphous silicate, aluminosilicate or aluminophosphate crystalline structures.[11] Zeolites are porous structures composed of a dense framework. The pore size determines the ion selectivity, and the framework density largely determines water permeability. Open porous structures facilitate better transport.[12] The ability of zeolites to act as a molecular sieve is determined by channel widths; therefore, changing the atoms in the framework results in a change in the properties of the sieve.[13] Zeolite membranes provide the potential to engineer the pore structure to improve selected rejection rates.

4.2.4.2 Catalytic nanoparticle membranes

Catalytic nanoparticle coated membranes are used to both oxidize and separate contaminates. Reactive surfaces are typically applied as semiconductor-based titania, zinc oxide, or ferric oxide. These surfaces are activated by UV light to engage in redox reactions for the degradation of organic compounds.[14,15] Membranes are typically mixed matrix organic polymeric-based structures. These are typically inorganic molecular sieves within a polymer matrix.[16,17] These membranes can provide both separation of solids and organic removal. Membrane contactor design and packing is complex due to the need for UV light.

4.2.4.3 Mixed matrix membranes

Mixed matrix membranes are composed of additives that are used to provide preferential flow paths for target species to pass through. They use inorganic molecular sieves embedded within a polymer active layer.[18,19] They are often composed of zeolites and silicates. These additives provide a pathway for diffusion of the target molecules. Mixed matrix membranes present an opportunity for engineered membranes that provide increased selectivity, targeted functionalities, and improved thermal, chemical, and mechanical stability.

4.2.4.4 Nanoparticle membranes

A variation of the mixed matrix membrane is the use of nanoparticles. The addition of nanoparticles during membrane synthesis or by surface attachment produces thin-film nanoparticle membranes. These membranes offer the potential benefits of enhanced separation capabilities, resistance to fouling, and improved antimicrobial characteristics.[20]

4.2.4.4.1 *Carbon nanotube membranes*

One of the most developed of these nanoparticle mixed matrix membranes is the carbon nanotube–based membrane.[21] Carbon nanotubes have many unique properties[22] and exhibit mass transfer rates that are up to 2–3 times higher than the current state of the art membrane technologies.[23] Typically, nanotube membranes are produced using nonaligned nanotubes where the bulk characteristics of a nanotube mat are transferred to the membrane by embedding them in a polymeric support film. Aligned nanotubes could improve membrane performance, but are not widely available.[24]

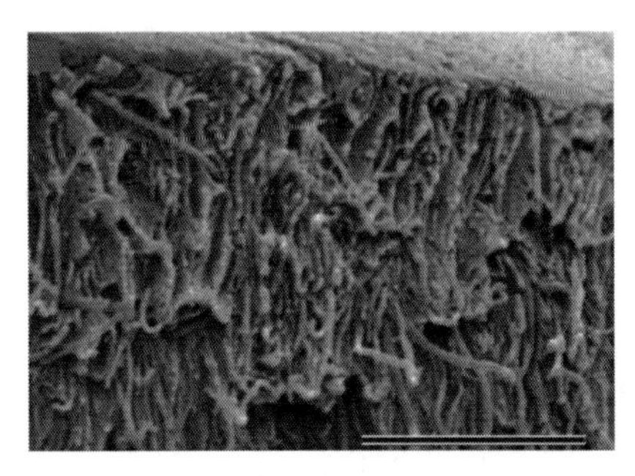

Figure 4.3 The edge of an aligned carbon nanotube membrane after exposure to OH plasma oxidation. Scale bar, 2.5 μm. Copyright *Science* 2 January 2004: **303** (5654) 62–65.

4.2.4.5 Block copolymer membranes

Block copolymers membranes are formed when one of the copolymers used to create the membrane is stripped from the membrane using a solvent. This produces a membrane with the ability to self-assemble into highly ordered structures based on copolymer molecular structures.[25] These membranes are formed with two copolymers such as linear polyethylene and polylactide. When a solvent is used to etch away the polyactide, a porous structure results. Block copolymer membranes provide a way of designing membranes with a wide range of molecular cutoffs.

4.2.4.6 Bio-inspired membranes

Bio-inspired membranes offer unique advantages over the current technologies. Lipid-based membranes are based on the natural repeating structure of lipid proteins. They offer advantages in improved flux rates, rejection characteristics, and the ability of self-repairing.[26] NASA has recently tested a proprietary lipid-based membrane that offers a 5 times improvement in flux over traditional membranes operating in FO.[27] The pores of these membranes are formed between lipid molecules in the mesostructure. Pores can also be formed by defects in the lipid structure at the macroscopic level and can be formed by entrapment of solvents during formation and subsequent removal of that solvent. One of the main benefits of these membranes is that they are potentially regenerable. By washing the membranes with the lipid molecules associated with defects will be replaced.

4.2.4.7 Aquaporin membranes

Aquaporins are the proteins that are found in nature and enable water flux across biological membranes. They provided passive transport of water across membranes using osmotic gradients.[28] Aquaporins are found in animal, plant, and microbial cells. Aquaporins are four-channel tetramers about 120 kDa in size. In animals, these channels are found in red blood cells, brain, and kidneys. The protein poses an hourglass shape with selective extracellular and intracellular vestibules at each end. This structure allows water molecules to pass rapidly in a single-file line through the proteins.[29] NASA testing of aquaporin membranes has shown a two to three times improvement in flux over current commercially available FO membranes and up to a 95% rejection of urea.[27]

4.3 Catalysts

Catalysts are materials that reduce the activation energy of chemical reactions. They can be formed from a wide range of materials and many mechanisms have been suggested for their function. For space flight life support applications, three broad categories of catalysts have been used. These categories include heterogeneous thermal catalysts, electrochemical catalysts, and photo catalysts.

Thermal catalysts are typically composed of high porosity granular or pelletized supports with catalytic materials deposited on the surface and in the pores. The catalytic material is typically deposited as molecules, and catalytic function is derived from the development of hydroxide structures or bridges between the catalyst and the support.[30] The functionality of these catalysts is limited by diffusion of the reactants to the surface, adsorption of the reactants to the support, movement of the reactant across the support to reactions sites, reaction kinetics, and desorption of the reactants from the catalyst.

Electrochemical catalysts are typically composed of porous structured electrodes with the catalytic material deposited on the electrodes. Catalysts are typically crystalline forms of noble metals. Typically a proton exchange membrane is used to separate the anode and cathode chambers.[31]

Photo catalytic catalysts are typically formed from semiconductor particles either fixed in a bed or suspended in a solution. Photon activation via UV light is used to create hydroxide radicals, which are responsible for oxidation. Catalyst packing density is an issue for fixed beds and catalyst shading is an issue for suspended beds.[32]

4.3.1 Catalytic Materials

4.3.1.1 Thermal

Thermal catalysts used in life support are typically oxidative and use noble metal catalysts. Platinum, ruthenium, and palladium are most often used although a wide range of materials have been researched.[33] Bi- and trimetallic catalysts are common and loading rates as high as 20 to 30% have been used.[34] The catalyst used on the International Space Station (ISS) Volatile Removal Assembly (VRA) is a bimetallic noble metal supported on alumina catalyst.[35] The metal loading rates are high, so deposition is completed in sequential steps with 1–2% deposition each time.

4.3.1.2 Electrocatalysts

Another area of recent interest for water recycling is electrochemical catalysis (electrocatalysis). In electrocatalysis a nano-sized catalyst is employed to promote the degradation or conversion of contaminants in water. This section is discussed in more detail in Sections 7.3.3.3.

4.3.1.3 Photocatalyst

The most common photo catalyst used is the anatase phase of titanium dioxide. When TiO_2 is exposed to UV light, electron holes are formed. The holes result in the formation of hydroxyl radicals from water, which are highly oxidative and react with organic molecules adsorbed to the surface of the catalyst. Oxygen in the feed stream scavenges the electrons (e-) to prevent electron-hole recombination and produces superoxide ions. Holes (h+) then react with water molecules to form hydroxyl radicals. The anatase form of the catalyst is desired due to its high surface area. UV light of 185–400 nm is typically used.[36]

4.3.2 Catalyst Supports

4.3.2.1 Thermal

Catalyst supports are most often materials such as alumina,[37] ceria,[38] and activated carbon.[39] Research has been performed to look at different support materials and the correlation between surface area and activity has been well established. For example, activated carbon has a surface area of about 500 m^2/g and alumina has a surface area of 200 m^2/g. In addition to surface area, pore shape and volume play a role in performance. Catalytic materials cannot be distributed in highly torturous pores and diffusion of reactants and products becomes limiting to these sites.

Mechanical stability is also important for thermal catalyst supports. Vibrations caused by hydrodynamic forces, pressure swings caused by cavitation, and depressurization can cause catastrophic catalyst failures.[40] Cavitation and depressurization occur because most catalysts operate above the boiling point of water. When pressure is lost in these systems, the water within them will flash resulting in cavitation, which can damage or destroy the catalyst support.

In addition, carbon-based catalyst supports are prone to oxidative damage. This occurs when oxygen in excess to stoichiometric requirements reacts with the carbon catalyst support causing a loss of catalyst. To mitigate this, protective coatings have been investigated[41] and the use of carbon nanotubes as support materials have been studied.[42]

4.3.3 Catalytic Processes Developed for Flight Applications

4.3.3.1 Volatile removal assembly

On the International Space Station, final treatment of recycled water is accomplished using a Volatile Removal Assembly (VRA). The VRA is a thermal catalytic system that is used to oxidize organics into organic acids and carbon dioxide. It also provides a final quick pasteurization of the product to ensure sterility. The organic acids are then removed through adsorption onto an adsorption bed and the CO_2 is removed using a phase separator.[34]

Figure 4.4 Volatile Removal Assembly flight experiment. *Copyright NASA.*

The VRA is a two-phase system where oxygen is bubbled through the reactor and then solubilized internally.[43] In this respect, it is similar to a conventional wet oxidation system. The VRA uses a heterogeneous bi-metallic catalyst that is proprietary to Hamilton Sundstrand Space Systems Inc. Operational experience on the International Space Station has demonstrated that in microgravity, the system performance is de-rated by 10 to 20% due to microgravity operation.[44] The exact reason for this issue has not been determined, but the current hypothesis is that bulk two-phase mixing or micro/nanoscale buoyancy-driven mixing is responsible for the reduction in performance caused by microgravity.

4.3.3.2 Aqueous phase catalytic oxidation

Aqueous phase catalytic oxidation (APCO) is similar to the VRA except that it is a single-phase system where oxygen is solubilized into the feed using a gasification membrane.[33] As a result, it is limited to the amount of organic material that can be treated in it. The APCO offers the benefits of a single-phase flow, which creates a less abrasive environment in the catalyst bed. This improves reactor life and prevents the formation of catalyst fines, which are particles that can impact downstream components such as pressure regulators and valves. APCO systems have only been developed for ground-test applications; therefore, microgravity flight data does not exist.

4.3.3.3 Electrochemical

Due to the high cost of delivering supplies to space, the recovery of potable water from spacecraft wastewater is critical for life support of crewmembers in long-term missions (i.e., 120–400 days).[45] Water is one of the most massive components aboard spacecraft making it imperative to recycle wastewater into useful resources. Thus, it is clear that water purification is of extreme importance for future missions to the moon, mars, and abroad, but it is also imperative to the recovery or recycling of unused contaminants (i.e., separated from the waste stream) into useful resources, a concept known as in situ *resource utilization.* In 2009 Cabrera et al. reported on the development of a bioelectrochemical device to degrade urea into ammonia as a strategy to introduce the concept of water purification while generating power in the same process. The results from this research revealed the feasibility of using a biological component (i.e., urease) to hydrolyze urea to ammonia while the latter was oxidized at the interface of an electrode by applying a certain voltage to an electrochemical cell. Accordingly, 20% urea conversion efficiency was achieved; meaning that one out of five urea molecules were directly converted into electrical power.[46] Later in that year, Botte et al. investigated the direct urea/urine electrolysis for the production of hydrogen by the use of an alkaline electrolytic cell.[47] In this work, the authors claimed that with the proposed construction the urea electrolysis at standard conditions would occur under a thermodynamic cell electrolytic potential of 0.37 V versus SHE with an inexpensive nickel catalyst at the anode and a platinum cathode. Meanwhile, the results of the investigation showed that after 22 h of electrolysis at an applied potential of 1.5 V, 13% of the urea was converted to hydrogen, nitrogen, and potassium carbonate. Finally, the authors demonstrated the feasibility of using urea for hydrogen production, which would require 30% less energy and might produce 36% cheaper hydrogen in comparison to the hydrogen production from water electrolysis. Likewise, another research group has recognized urea as a potential energy carrier due to its higher energy density (16.9 MJ L^{-1}) compared to compressed (5.6 MJ L^{-1}) or liquid hydrogen (10.1 MJ L^{-1}) while offering practical advantages such as easy storage and transportation. Hence, Lan et al. reported the development of a direct urea fuel cell to generate power from human waste.[48] The authors presented a direct urea fuel cell, and thus proving an alternative approach to obtain power in a single step from urea. In this research, the performances of various electrode catalytic materials were assessed for both anode and cathode separated by a simple alkaline membrane. Namely, for the anode side Pt/C and Ni/C (i.e., urea oxidation, Eq. 4.1) and Pt/C, Ag/C and MnO_2/C (i.e., oxygen reduction, Eq. 4.2) on the cathode were investigated with the membrane made of an anion exchange resin and polyvinyl alcohol.

Anode

$$CO(NH_2)_2 + 6OH^- \rightarrow N_2 + CO_2 + 5H_2O + 6e^- \quad (4.1)$$

Cathode

$$O_2 + 2H_2O + 4e^- \rightarrow 4OH^- \quad (4.2)$$

Overall

$$2CO(NH_2)_2 + 3O_2 \rightarrow 2N_2 + 2CO_2 + 4H_2O \quad (4.3)$$

The results of this investigation revealed that the combination of Ni/C on the anode and MnO_2/C (i.e., non-noble catalyst cell) on the cathode provided the highest power density with 1.7 mW/cm^2 at a cell temperature of 50°C and 1M urea in water. However, at room temperature the cell composed of Pt/C on both anode and cathode provided a power density of ~0.55 mW/cm_2 against a ~0.27 mW/cm_2 for the non-noble catalyst cell in 1 M urea in water. Hence, the use of non-noble metal catalyst in urea fuel cells truly opens the window for the use of inexpensive and reliable materials to generate power from alternative fuel sources other than the usual methanol and hydrogen, and represents a substantial contribution to the area of power generation from urea. Overall, there is room for improvement because impurities on the urea solution, urea cross-flow and slow urea hydrolysis are known to adversely affect the performance of these systems.

4.3.3.4 Bioelectrochemical cells

A rapidly growing area of interest for wastewater treatment is microbial fuel cells (MFCs), which are novel in that they not only provide rapid microbial transformation of waste organics, but they also produce an electrical current for a potential power offset. While conventional aerobic bacteria consume organics and produce biomass and CO_2, MFCs consume organics to produce electrons that when combined with specialized electrodes result in the generation of an electrical current rather than biomass. This metabolic strategy minimizes the accumulation of biomass as well as associated maintenance issues. The bacteria grow fixed to the electrodes and can be configured to treat organics, ammonia, and even fix CO_2 into valuable organic intermediates. Figure 4.5 shows the interrelationship between inputs and potential outputs.

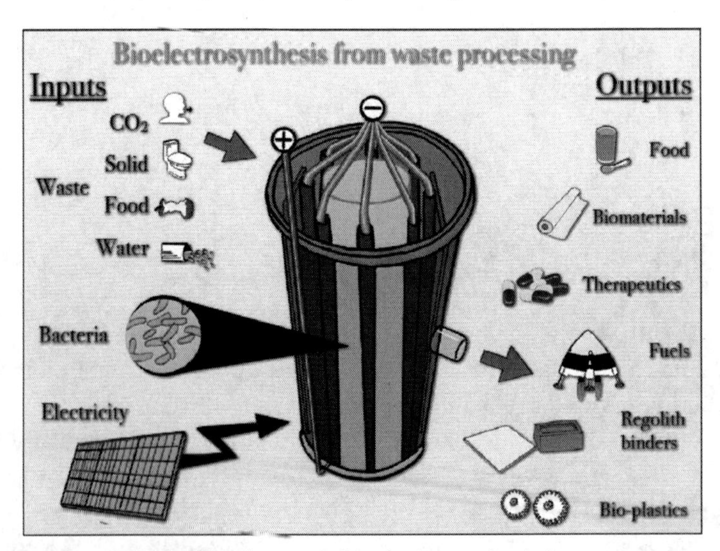

Figure 4.5 Relationship between bioelectrosynthesis inputs and outputs. Copyright NASA.

Current research topics include development of novel reactor design, design of electrodes, design of proton exchange membranes and application of synthetic biology techniques to create highly adapted organisms that excel within MFC growth conditions. Design objectives include maximizing power generation, water recovery, sequestration of carbon, and integration with dissolved solids removal technologies.

4.3.3.5 Adsorption process

Porous materials such as zeolites, metal oxides, mesoporous carbon, activated carbons and carbon molecular sieves are of importance in several areas of research and development, such as in adsorption applications.[49,50] Among these materials, activated carbons and carbon molecular sieves have been used in adsorption and separation processes.[51,52] However, most of the research with these materials has focused in the removal or adsorption of large molecules such as hormones, cyanotoxins, and oils.[53–55] More recently Cabrera and Flynn et al., investigated on the adsorption behavior and performance of these materials with regard to small molecules such as urea and urine components adsorption.[56]

4.3.3.5.1 *Performance of carbon-based adsorbents*

In Fig. 4.6 the amount of adsorbed urea as a function of time is presented for different carbon-based materials (oxidized-granulated activated carbon (6a), granulated activated carbon (6b), carbon molecular sieve-563 (6c), carbon molecular sieve-X (6d), carbon molecular sieve-Z (6d)). From this figure, it is noticeable that both the oxidized and non-oxidized activated carbons as well as the carbon molecular sieve-563 are able to adsorb higher concentrations of urea at contact times longer than 4 h, while the other carbon molecular sieves are able to adsorb more urea at contact time periods lower than 4 h. These results may suggest that both activated carbons and the Carbon molecular sieve-563 are able to attract multiple layers of urea molecules over time. Also, the Carbon molecular sieve-X and Carbon molecular sieve-Z reach adsorption equilibrium at a faster rate. At 4 h the urea adsorption trend is C-X > C-Z > GAC > C-563 > O-GAC. Other investigations have suggested that adsorption of molecules depends on both the pore structure and the surface chemistry.[57]

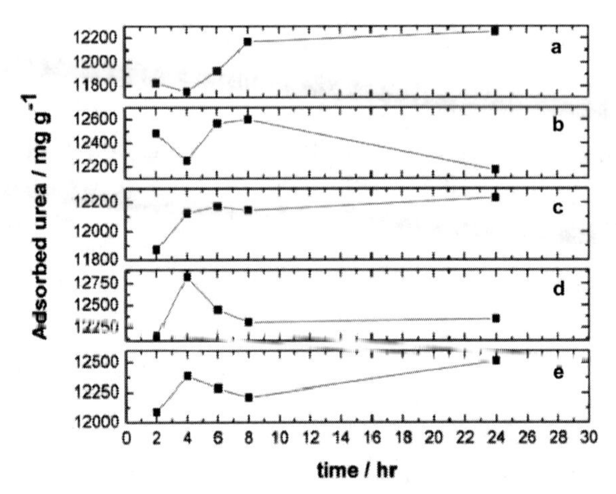

Figure 4.6 Contact time experiment of adsorbed urea, sodium chloride and potassium chloride ("urine mimic solution") in analytical quantities of 62,000 mg/L and 1,641 mg/L for both ionic salts, respectively, as a function of time for each material. O-GAC (a), GAC (b), C-563 (c), C-X (d), C-Z (e). Copyright: Nicolau, E.; Fonseca, J.; Cabrera, C. R.; Vu, C.; Tra-My Justine, R.; Flynn, M. *International Conference on Environmental Systems*, **2012**, 2012-3627, 1-8. NASA Rights.

These results suggest that the micropore distribution for the GAC is highly homogeneous while the O-GAC presents heterogeneity among its structure. Other researchers have observed a similar effect when the activated carbon is oxidized. In fact, is of general consensus that activated carbon oxidation leads to the production of oxygen-containing functional groups at the edge sites of the graphitic planes.[58] Moreover, the oxidation of this material is known also to alter the structural characteristics of the materials (i.e., BET surface area, pore volume and pore distribution).[59,60] A possible explanation for this behavior is given by Faust et al. that once oxidized, granulated activated carbon is an acidic oxygen surface.

A major portion of these oxygen groups are located presumably on the edges of the layer planes where they are not expected to interfere sterically with the adsorption of organic molecules on the basal plane. In Table 4.1, a summary for the amount of adsorbed urea and physical features of each material is provided. From this table, one can note that none of these physical features (i.e., micropore diameter and surface area) correlate with the adsorption behavior observed. Thus, this suggests that a combined effect of physical and chemical properties of each material provide the observed urea adsorption behavior. The chemical composition of each carbon molecular sieve is vendor's proprietary information, and therefore a further relationship of the chemical composition with the adsorption behavior of each material was not possible.

Table 4.1 Table of adsorbed urea (mg/g), surface area ($m^2\ g^{-1}$) and average micropore diameter (Å) for all materials tested

Adsorbent	Adsorbed urea ($mg\ g^{-1}$)	BET surface area ($m^2\ g^{-1}$)[a]	Average micropore diameter (Å)[a]
GAC	12,311	1,530[b]	—
O-GAC	11,390	1,260[b]	~0.2[c]
Carboxen 563	11,831	510	7–10
Carbopack X	12,040	240	100
Carbopack Z	12,859	220	255

[a]Specifications from vendor.

[b]Data obtained from Mochida et al.

[c]Data obtained from Szymanski et al.

4.4 Scale Formation and Precipitation

When recycling wastewater onboard a spacecraft, it is necessary to recover as much water as possible. Any water not recycled becomes waste and must be replaced with stored supplies. Typically, spacecraft water recycling systems are designed to achieve high water recovery ratios. For example, the International Space Station was originally designed to achieve an 85% water recovery ratio. However, the maximum water recovery ratio was reduced due to the formation of calcium-based scale solids.

In microgravity, the human body losses bone mass. This bone is solubilized in the body in excreted as calcium in the urine. An astronaut's urine has calcium concentrations as high as 200 mg/L.[61] This calcium can combine with bicarbonate or sulfate ions to form calcium carbonate or calcium sulfate when concentrated or heated. A variety of mitigation approaches are available to prevent this calcium scale from forming. Generally, these approaches prevent either the cause or the consequences of scale formation. The cause can be addressed by removing the problem ions through homogeneous nucleation. The consequences can be addressed by affecting the form of the precipitated solids through heterogeneous nucleation, which favors precipitation in bulk solution rather than on separation surfaces.

4.4.1 Molecular Scale Control

Molecular scale control is usually provided by the addition of anti-scale chemicals, which disperse the nanocrystals or nanosand of calcium carbonate, calcium phosphate, or calcium sulfate. Although there are a wide range of additives that are used for scale control, the most common are phosphonates. The phosphonates used most frequently for calcium carbonate scale control are amino-tri[methylene] phosphonic acid, 1-hydroxyethylidene 1,1-diphosphonic acid, and 2-phosphonobutane-1,2,4-tricarboxylic acid. An active dosage of 1.5 to 3 mg/L will increase the solubility of calcium carbonate by approximately a factor of 3.[62]

There are also a variety of specialized polymers that can inhibit various types of scale formation as well as disperse suspended solids. Homopolymers such as polyacrylate, polymethacrylate, and polymaleate are used to keep calcium carbonate in solution. Homopolymers and copolymers act as crystal modifiers. They distort calcium carbonate crystals such that they do not attach themselves to heat exchange surfaces but instead the crystals become suspended solids that can be removed through filtration. Usually dosages of 2 mg/L of active polymer will control calcium carbonate scale.[62]

4.4.2 Electrodialysis

In electrodialysis, water is desalinated by the transport of ions through ion exchange membranes under the driving force of an electric potential between a cathode and an anode. Cations are pulled from the feed water toward the cathode and anions are pulled toward the anode. Cations pass through cation selective membranes into the concentrate compartment and are held there by repulsion from the anion selective membrane. Anions similarly pass through anion exchange membranes and are held in the concentrate compartment by cation exchange membranes. The result is alternating compartments of purified water and a concentrate containing the removed ions. Figure 4.7 shows a fully assembled electrodialysis stack.

Figure 4.7 Electrodialysis Stack. Copyright NASA Ames Research Center.

The electrodialysis metathesis (EDM) is a hybrid of the electrodialysis process where ions of insoluble inorganic salts, such as $CaSO_4$, exchange bonds with soluble salts, such as NaCl, to prevent the precipitation of inorganic solids. In EDM, a $CaSO_4$-rich feed flows through one compartment of a cell stack and a solution of NaCl flows through an adjacent compartment. Due to an electric potential, the salts in the feed stream "change partners" and form concentrated solutions of mixed chloride salts (predominantly $CaCl_2$) as well as mixed sodium salts (predominantly Na_2SO_4). $CaCl_2$ and Na_2SO_4 have higher solubility's than $CaSO_4$ as shown in Fig. 4.8. As a result, the product of the EDM is a $CaSO_4$ depleted stream that can be concentrated to a higher level without producing a scale forming precipitate.

When the salt byproduct streams are removed from the EDM cell and are mixed in a separate mixing chamber under no electric potential, $CaSO_4$ re-forms and precipitates due to its lower solubility. This precipitation removes $CaSO_4$ from solution and a NaCl-rich supernatant is produced. The supernatant byproduct is then recycled back to the cell stack as the NaCl feed. The $CaSO_4$ is removed from the feed as a solid. No aqueous salt byproduct is produced only a solid $CaSO_4$. This approach is applicable for calcium carbonate scale as well.

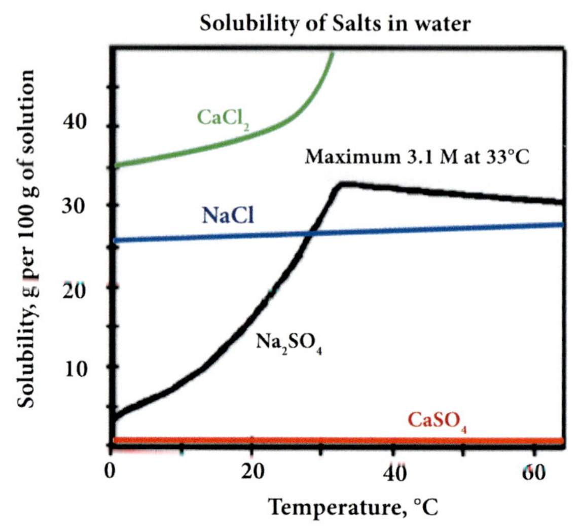

Figure 4.8 Solubility of mixed chloride and sodium salts. Image copyright to Thomas A. Davis. Director, Center for Inland Desalination Systems, University of Texas at El Paso, USA.

4.4.3 Nucleation Devices

Nucleation device media prevent the formation of scale by accelerating the transformation of the calcium minerals nanoparticles. The nucleation media acts as a catalyst and pulls the hardness minerals of calcium out of the solution, and then transforms these minerals into inactive nanocrystal particles. These nanoscopic particles make their way through plumbing systems without attaching on to pipes, fixtures, valves, or heating elements. Under the proper conditions, this approach has been shown to provide up to 99% scale prevention.

4.5 Effects of Microgravity on the Nanoscale

4.5.1 Concentration Polarization

A key aspect of controlling concentration polarization in FO separation membranes is nanoscale mixing at the pores of the membrane and in the backing material. Mixing at the nanoscale can occur due to Brownian motion derived diffusion or buoyancy-driven mixing. Buoyancy-driven mixing is typically the most important. Buoyancy-driven mixing occurs because of temperature differences across the membrane. These temperature differences can occur from bulk differences in the temperature of the feed and osmotic draw solution, and at the nanoscale, from energy released from the potential energy of mixing across osmotic gradients. In conditions of gravity environment (1 g), buoyancy drive mixing is of an order of magnitude larger than Brownian motion. In microgravity there is little to no buoyancy derived mixing, depending on the level of microgravity. Recent tests conducted on the Space Shuttle showed that the lack of mixing resulted in a 50% reduction in flux across the osmotic membrane.[63] Microgravity flight test was completed in 2011 on the STS-135 Atlantis space shuttle mission. The objective of this flight test was to evaluate the performance of Forward Osmosis bags X-Pack from Hydration Technologies (see Fig. 4.9). The X-Pack is currently used for disaster relief, backpackers, and military personnel in Iraq and Afghanistan to purify water (http://www.htiwater.com/divisions/military_regulatory/index.html). It can be filled with water from almost any contaminated stream, ditch, or puddle and is a passive device that produces a fortified drink or food product rather than pure water. NASA has a patent on the use of this technology to recycle urine in space [pat # 7,655,145,B1].

Figure 4.9 X-Pack Drink Bag. Copyright NASA.

Figure 4.10 shows the flight FO bag being tested in microgravity. Results of this flight experiment demonstrated that the FO process works in microgravity. However, flux rates for static feeds were reduced in microgravity by up to 50%. The data had a significant scatter with an error of ±20%.

Figure 4.10 STS-135 Atlantis FO bag microgravity flight test. Copyright NASA.

Results of this flight experiment demonstrated that the FO process works in microgravity. However, flux rates for static feeds were reduced in microgravity by up to 50%. The data had a significant scatter with an error of ±20%. Ground-based testing indicates this error is primarily due to the materials of construction of the FO bag. Although there was clearly a significant reduction in flux in microgravity, the exact amount remains to be determined. The florescent dye used was also determined to be subject to photo bleaching in the ground tests. This also contributed to the observed scatter in the data.

4.5.2 Electrochemical

Experimentation in microgravity imposes a buoyancy-free and thus macroconvection-free environment. Due to this quiescent condition, researchers have shown interest to study under such environment several processes within the fields of biology,[64] crystal growth,[65] human health,[66] nanotechnology,[67] and electrochemistry.[68] Within the field of electrochemistry, its impact on energy production is of particular importance. In an effort to find alternative means to generate electrical energy, fuel cells have been proposed with the use of different fuels such as ethanol, methanol, ammonia, and others.

Water electrolysis has been proposed as a way to generate hydrogen to be used as fuel in polymer electrolyte membrane fuel cells (PEMFC), and it has been investigated under microgravity conditions. These investigations have shown that the three-phase zone between the electrode, gas bubbles and electrolyte plays a key role in microconvection.[69,70] Kaneko et al. investigated water electrolysis in microgravity by parabolic flight and found that gas bubbles become larger and current densities decreased in microgravity. This suggests that the mass transfer of water molecules to the electrode controls water electrolysis. However, under normal gravity conditions water electrolysis is kinetically limited.[71] When methanol is employed as fuel in a fuel cell under microgravity, a similar effect is observed as when water electrolysis is performed, this is enlargement of gas bubbles and a decrease in overall cell performance.[72] H. Matsushima et al. indicated that in alkaline solutions, current density is much smaller under microgravity than under terrestrial gravity, and that ohmic resistance increases in microgravity due to the formation of a gas bubble froth layer and as well as an increase in electrode surface blockage by bubbles.[69,73] Iwasaki et al. found that the electrolysis current on the ground reached a steady state faster than that under microgravity conditions. Since the diffusional mass transfer was predominant over the process in microgravity, the characteristic time was longer than that of the process on the ground where macroconvection is predominant.[74]

Recently, Cabrera et al. investigated the effects of microgravity in the electrochemical oxidation of aqueous ammonia and demonstrated a decrease in performance of 20–65% over ground-based controls for the electrochemical oxidation of ammonia. This decrease in current is theorized to be due to the lack of buoyancy-driven mixing that occurs in microgravity. In microgravity only diffusional mixing occurs and therefore molecules may be unable to reach the nanocatalyst interface at the same rate as on the ground as can be observed in Fig. 4.11. These findings indicate that significant reductions in electrode performance for ammonia electrooxidation may occur if catalyst design measures are not taken in consideration to mitigate these microgravity effects. This conclusion is applicable specifically to electrochemical processes, but also may be relevant to all process constrained by nanoscale mixing of diffusion limited structures such as thermal catalysts, separation membranes, microbial growth and human cellular functions.

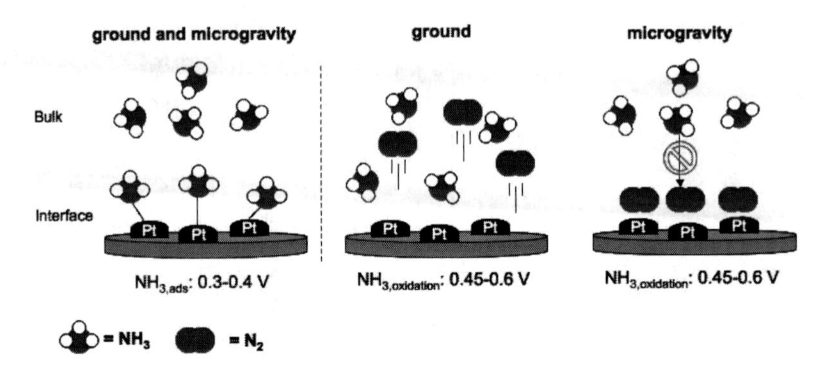

Figure 4.11 Schematic representation of the ammonia electrochemical oxidation effect in microgravity and ground.

Currently, investigations in microgravity (parabolic flights) are conducted by Cabrera et al. to find more suitable and selective catalysts that help to mitigate the effects of microgravity at the nanoscale level. The outcomes of this research are to be employed in the next-generation of energy harvesting devices such as fuel cells. In fact, Cabrera et al. is currently developing a prototype of an electrochemical ammonia removal system (EAR) that is of utmost importance to control the levels of ammonia in wastewater while generating electrical energy and thus achieving resource recovery (relates to Section 4.6).

4.5.3 Biology—Gas Diffusion

In microgravity, where buoyancy-driven mixing is minimal, the buildup of CO_2 from metabolism in solutions containing microbes can influence growth rates, final microbial biomass, and mask more subtle effects. Similarly, the delivery of O_2 and removal of metabolic byproducts in microgravity relies entirely on gas diffusion. The majority of investigations to date have indicated that microgravity has an influence on microbial growth and behavior. The mechanisms responsible for the observed biological responses, however, are not yet fully understood. It is believed that the major influence of a lack of gravity is a direct effect on the environment surrounding the cell simply due to the fact that there is no convection only diffusion. This suggests that in microgravity the observed changes in microbial growth and behavior are at least partially due to changes in the environment surrounding the cell.[75]

4.6 Current Integrated Strategies for Water Reclamation

The current strategy that integrates biochemical and physicochemical approaches to water reclamation is the forward osmosis secondary treatment (FOST) system. In general the FOST system is composed of an FO module and an RO module, as shown in Fig. 4.12. In this system, wastewater is first passed through a filter and then recirculated through a FO module where it is contacted with an OA (osmotic agent) across a semi-permeable membrane. Water then passes from the feed across the membrane into the higher osmotic potential OA. Concurrently, the OA is recirculated through a RO system water that passed through the FO membrane is removed from the OA. A UV light is used to control biological growth in the OA. An electrochemical ammonia removal (EAR) system is used to control ammonia levels in the OA. The feed filter will be a 100 micron knife-edge self-cleaning filter that discharges back into the bioreactor or into the brine tank. Most of the energy invested in the FO process is consumed by the RO subsystem when it reconcentrates the OA and harvests the product water from the OA. In order to reduce energy consumption an energy recovery RO system is used to minimize power consumption.

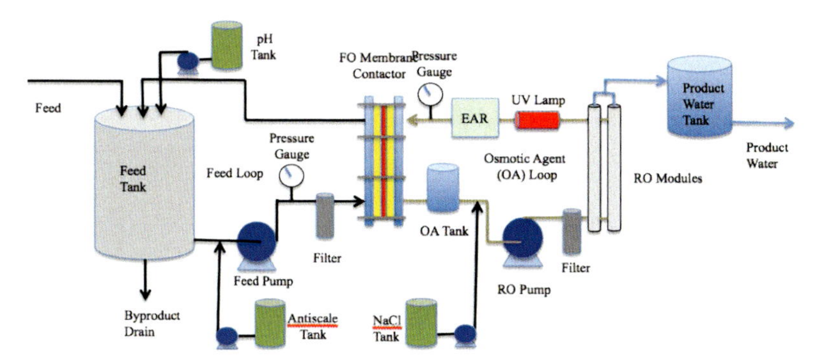

Figure 4.12 The FOST process flow diagram.

References

1. Feng, X., Huang, R. Y. *Industrial and Engineering Chemistry Research* 1997, **36**, 1048–1066.

2. Hogan, P. A., Canning, R. P., Peterson, P. A., Johnson, R. A., Michaels, A. S. In *Chemical Engineering Progress* 1998, **7**, 49–61.
3. Kunz, W., Benhabiles, A., Ben-Aïm, R. *Journal of Membrane Science* 1996, **121**, 25–36.
4. Lee, C. H. *Journal of Applied Polymer Science* 1975, **19**, 83–95.
5. Fritzmann, C., Löwenberg, J., Wintgens, T., Melin, T. *Desalination* 2007, **216**, 1–76.
6. Kwak, S.-Y., Jung, S. G., Yoon, Y. S., Ihm, D. W. *Journal of Polymer Science Part B: Polymer Physics* 1999, **37**, 1429–1440.
7. Cadotte, J. E., Petersen, R. J., Larson, R. E., Erickson, E. E. *Desalination* 1980, **32**, 25–31.
8. Cath, T. Y., Childress, A. E., Elimelech, M. *Journal of Membrane Science* 2006, **281**, 70–87
9. McCutcheon, J. R., Elimelech, M. *Journal of Membrane Science* 2006, **284**, 237–247.
10. Gray, G. T., McCutcheon, J. R., Elimelech, M. *Desalination* 2006, **197**, 1–8.
11. Kazemimoghadam, M. *Desalination*, **251**, 176–180.
12. Lobo, R. F. *Handbook of Zeolite Science and Technology*; Marcel Decker, Inc.: New York, 2003.
13. Dyer, A. An *Introduction to Zeolite Molecular Sieves*; John Wiley & Sons Ltd.: California, 1988.
14. Hoffmann, M. R., Martin, S. T., Choi, W. Y., Bahnemann, D. W. In *Chemical Reviews* 1995, **95**, 69–96.
15. Bahnemann, D. *Solar Energy* 2004, **77**, 445–459.
16. Moosemiller, M. D., Hill, C. G., Anderson, M. A. *Separation Science and Technology* 1989, **24**, 641–657.
17. Molinari, R., Mungari, M., Drioli, E., Di Paola, A., Loddo, V., Palmisano, L., Schiavello, M. *Catalysis Today* 2000, **55**, 71–78.
18. Mahajan, R., Burns, R., Schaeffer, M., Koros, W. J. *Journal of Applied Polymer Science* 2002, **86**, 881–890.
19. Jia, M., Peinemann, K.-V., Behling, R.-D. *Journal of Membrane Science* 1991, **57**, 289–292.
20. He, J., Kanjanaboos, P., Frazer, N. L., Weis, A., Lin, X. M., Jaeger, H. M. *Small*, **6**, 1449–1456.
21. Choi, J.-H., Jegal, J., Kim, W.-N. *Journal of Membrane Science* 2006, **284**, 406–415.

22. Ebbesen, T. W. *Annual Review of Materials Science* 1994, **24**, 235–264.
23. Brunet, L., Lyon, D. Y., Zodrow, K., Rouch, J. C., Caussat, B., Serp, P., Remigy, J. C., Wiesner, M. R., Alvarez, P. J. J. *Environmental Engineering Science* 2008, **25**, 565–575.
24. Kim, S., Jinschek, J. R., Chen, H., Sholl, D. S., Marand, E. *Nano Letters* 2007, **7**, 2806–2811.
25. Peinemann, K.-V., Abetz, V., Simon, P. F. W. *Nature Materials* 2007, **6**, 992–996.
26. Yee, C. K., Amweg, M. L., Parikh, A. N. *Journal of the American Chemical Society* 2004, **126**, 13962–13972.
27. Flynn, M., NASA Ames Research Center.Internal Proprietary Data. 2008.
28. Maurel, C. *Annual Review of Plant Physiology and Plant Molecular Biology* 1997, **48**, 399–429.
29. Kumar, M., Grzelakowski, M., Zilles, J., Clark, M., Meier, W. *Proceedings of the National Academy of Sciences* 2007, **104**, 20719–20724.
30. Weitamp, J., Puppe, L. *Catalysis and Zeolites: Fundamentals and Applications*; Springer, 1999.
31. Tsiplakides, D., Balomenou, S. Electrochemical promoted catalysis: Towards practical utilization, Chemical Industry and Chemical Engineering Quarterly 2008, Volume **14**, Issue 2, pp. 97–105.
32. Matthews, R. W. *Pure and Applied Chemistry* 1992, **64**, 1285–1290.
33. Akse, J. R., *Catalytic methods using molecular oxygen for treatment of PMMS & ECLSS waste streams final report.* Marshall Space Flight Center, National Aeronautics and Space Administration; Huntsville, Al.:, 1992.
34. Holder, D. W., Parker, D. *Volatile Removal Assembly Flight Experiment and KC-135 Packed Bed Experiment: Results and Lessons Learned*, Marshall Space Flight Center, 2000. International Conference on Environmental Systems, 2000-01-2251, 2000.
35. Kindt, L. M., Mullins, M. E., Hand, D. W., Kline, A. A., Carter, D. L., Garr, J. D. *International Conference on Environmental System* 1995, 951630.
36. Pozzo, R. L., Baltanás, M. A., Cassano, A. E. *Catalysis Today* 1997, **39**, 219–231.
37. Trimm, D. L., Stanislaus, A. *Applied Catalysis* 1986, **21**, 215–238.
38. Martínez-Arias, A., Fernández-García, M., Salamanca, L. N., Valenzuela, R. X., Conesa, J. C., Soria, J. *The Journal of Physical Chemistry B* 2000, **104**, 4038–4046.

39. Aksoylu, A. E., Madalena, M., Freitas, A., Pereira, M. F. R., Figueiredo, J. L. *Carbon* 2001, **39**, 175–185.

40. Dadyburjor, D. B. *AIChE Journal* 1988, **34**, 174–175.

41. Jolly, C., Flynn, M. *International Conference on Environmental System* 1997, 972271.

42. Frank, B., Rinaldi, A., Blume, R., Schlögl, R., Su, D. S. *Chemistry of Materials*, **22**, 4462–4470.

43. Guo, B.; Holder, D. W.; Tester, J. T., Two-Phase Oxidizing Flow in a Volatile Removal Assembly Reactor under Microgravity Conditions. *AIAA Journal* 2005, **43**(12), 2586–2592.

44. Carter, D. L., NASA MSFC internal proprietary data. 2009.

45. Dionysiou, D. D., Antoniou, M. G. *Catalysis Today* 2007, **124**, 215–223.

46. Nicolau, E., González-González, I., Griobenow, K., Flynn, M., Cabrera, C. R. *Advances in Space Research* 2009, **44**, 965–970.

47. Boggs, B. K., King, R. L., Botte, G. G. In *Chemical Communications*, RSC Publishing: 2009, 4859–4861.

48. Lan, R., Tao, S., Irvine, J. T. S. In *Energy & Environmental Science*, RSC Publishing: 2010, 438–441.

49. Punyapalakul, P., Takizawa, S. *Water Research* 2006, **40**, 3177–3184.

50. Fujita, H., Izumi, J., Sagehashi, M., Fujii, T., Sakoda, A. *Water Research* 2004, **38**, 159–165.

51. Werner, D., Hale, S. E., Tomaszewski, J. E., Luthy, R. G. *Water Research* 2009, **43**, 4336–4346.

52. Hartmann, M., Vinu, A., Chandrasekar, G. *Chemistry of Materials* 2005, **17**, 829–833.

53. Fukuhara, T., Iwasaki, S., Kawashima, M., Shinohara, O., Abe, I. *Water Research* 2006, **40**, 241–248.

54. Ho, L., Lambling, P., Bustamante, H., Duker, P., Newcombe, G. *Water Research* 2011, **45**, 2954–2964.

55. Uchiyama, M. *Water Research* 1978, **12**, 299–301.

56. Nicolau, E., Fonseca, J., Cabrera, C. R., Vu, C., Tra-My Justine, R., Flynn, M. *International Conference on Environmental Systems* 2012, 2012–3627, 1–8.

57. Bandosz, T. J. E.-S. Y. *Langmuir* 2005, **21**, 1282–1289.

58. Szymanski, G., Biniak, S., Siedlewski, J., Swiatkowski, A. *Carbon* 1997, **35**, 1799–1810.

59. Lopez, F., Medina, F., Prodanov, M., Guel, C. *Journal of Colloid and Interface Science* 2003, **257**, 173–178.

60. Mochida, I., Qiao, W., Korai, Y., Hori, Y., Maeda, T. *Carbon* 2002, **40**, 351–358.

61. Whitson, P. A., Pietrzyk, R. A., Morukov, B. V., Sams, C. F. *Nephron* 2001, **89**, 264–270.

62. Gill, J. *Corrosion* 1996, **96**, 24–29.

63. Flynn, M., Soler, M., Shull, S., Broyan, J., Chambliss, J., Howe, A. S., Gormly, S. *International Conference on Environmental System* 2012, 3599.

64. Mognato, M., Celotti, L. *Mutation Research* 2005, **578**, 417–429.

65. Lorber, B., Giege, R. *Journal of Crystal Growth* 2001, **231**, 252–261.

66. Borchers, A. T., Keen, C. L., Gershwin, M. E. *Nutrition* 2002, **18**, 889–898.

67. Hartmann, E., Marquardt, P., Ditterich, H., Steinberger, H. *Advances in Colloid and Interface Science* 1993, **46**, 221–262.

68. Nishikawa, K., Fukunaka, Y., Chassaing, E., Rosso, M. *Journal of Physics: Conference Series* 2011, **327**, 012045.

69. Matsushima, H., Nishida, T., Konishi, Y., Fukunaka, Y., Ito, Y., Kuribashayi, K. *Electrochimica Acta* 2003, **48**, 4119–4125.

70. Matsushima, H., Kiuchi, B., Fukunaka, Y., Kuribayashi, K. *Electrochemistry Communications* 2009, **11**, 1721–1723.

71. Kaneko, H., Tanaka, K., Iwasaki, A., Abe, Y., Negishi, A., Kamimoto, M. *Electrochimica Acta* 1993, **38**, 729–733.

72. Guo, H., Zhao, J. F., Ye, F., Wu, F., Lv, C. P., Ma, C. F. *Microgravity Science and Technology* 2008, **20**, 265–269.

73. Kiuchi, D., Matsushima, H., Fukunaka, Y., Kuribayashi, K. *Journal of the Electrochemical Society* 2006, **153**, E138–E143.

74. Iwasaki, A., kaneko, H., Abe, Y., Kamimoto, M. *Electrochimica Acta* 1998, **43**, 509–514.

75. Horneck, G., Klaus, D. M., Mancinelli, R. L. *Microbiology and Molecular Biology Reviews*, **74**, 121–156.

Chapter 5

Nanomaterials for Advanced Lithium-Ion Battery Anodes

Richard S. Baldwin, James J. Wu, and William R. Bennett

NASA John H. Glenn Research Center at Lewis Field,
21000 Brookpark Road, Cleveland, OH 44135, USA

william.r.bennett@nasa.gov

5.1 Introduction

The proliferation of compact, portable electronic devices has catalyzed continuous improvement in rechargeable battery technologies, driving the development of lithium-ion cells that combine high energy storage with smaller weight (high specific energy) and smaller volume (high energy density). In aerospace applications, improvements in specific energy and energy density are critical for achieving new missions.[1] Early commercial lithium-ion cells achieved specific energies of approximately 130 Wh/kg, using graphitic anode materials.[2] Ongoing development, driven by consumer electronics requirements, has resulted in commercial cells that achieve specific energies of 230 Wh/kg.[3] Recently, Panasonic announced plans to produce 18650-size cells with silicon-based

Advanced Nanomaterials for Aerospace Applications
Edited by Carlos R. Cabrera and Félix A. Miranda
Copyright © 2014 Pan Stanford Publishing Pte. Ltd.
ISBN 978-981-4463-18-8 (Hardcover), 978-981-4463-19-5 (eBook)
www.panstanford.com

anode materials, which reportedly will achieve specific energies of 250 Wh/kg.[4] The development of silicon-based anode materials has been a topic of considerable effort in recent years and its release in a commercial product marks a milestone in lithium-ion technology.[5]

Of course, the actual performance that is achieved by a given electrode technology is strongly influenced by the charge/discharge conditions (discharge rate, ambient temperature, voltage limits, etc.). The results presented here have not been measured under standardized conditions, and one must be careful when comparing performance projections from different sources.

With respect to anode materials for lithium-ion cell chemistries, nanotechnology research and development over the past decade has progressively led to significant performance enhancements and benefits for lithium-ion battery technology, albeit there are significant technical challenges to successfully overcome before the full potential of such can be practically realized. Compared to the behavior of bulk micron-size or larger electrode materials, nanosizing the same material (i.e., to <100 nm in at least one dimension) fundamentally allows enhanced access of lithium ions to preferred lattice or storage sites within the electrode material and also affects the reversibility of such a mechanism.

For intercalation nanomaterials, such as graphitic carbons, pathways for ionic diffusion are significantly reduced, and due to an increased area-to-volume ratio, the intercalation process mimics a surface reaction mechanism with very rapid kinetics. Complemented by a reduction in lattice flaws, electrode charge and discharge reactions can occur very rapidly. First-cycle coulombic efficiency and irreversible capacity losses due to solid electrolyte interphase (SEI) formation can be associated with structural and lattice imperfections. Smaller crystallite dimensions possess a lower susceptibility to phase imperfections, resulting in a lower degree of irreversibly trapped lithium, which leads to a larger useful cell capacity. This benefit can be partially offset by the increased surface area of active material and concomitant increase in irreversible capacity loss.

In addition to performance enhancements observed for intercalation or ion-insertion type materials as a result of nanosizing, alternative candidate anode materials, such as lithium alloy-forming elements and binary metal compounds, also exhibit improvements in performance at the nanoscale. Most of the materials under investigation and development, which have theoretically larger lithium storage capacities than carbonaceous materials, do not perform practically with micro-size particle dimensions, as crystal structure and diffusion channels are adversely affected and large irreversible capacity losses are exhibited.

Additional cell performance enhancements can be attributed to nanoscale electrode morphology. Improved cell cycle life can be attributed to the enhanced structural stability of nanoscale electrode materials, as they are more tolerant to mechanical stresses and crack proliferation caused by an expanding and contracting crystal lattice that occurs during lithiation and de-lithiation of intercalation-type electrode materials. The nanocrystal structure preservation is also enhanced by a more uniform lithium ion distribution, as well as a higher degree of surface reactions. The large surface-to-volume ratio of electrode nanomaterials also enhances low-temperature electrochemical performance, capacity, and rate capability, as the surface reaction kinetics are scarcely affected by temperature in comparison to bulk diffusion kinetics.

As the cell-level electrochemical performance benefits of nanoscale electrode materials continue to be identified and optimized in the laboratory, non-trivial technical and engineering challenges remain in order to successfully address practicality and safety issues for ultimate consumer relevance. The cost-effective scale-up of nanopowders with desired tolerances of purity and physical properties, such as particle size distribution, is an engineering challenge. The low densities associated with nanopowders have to be weighed against higher gravimetric energy density of a candidate material. Uniform and reproducible electrode substrate fabrication is a technical challenge with high-surface-area nanopowders, and processing problems need to be practically addressed, especially when thicker electrode coatings and useful loadings are required for specific applications, e.g., for advanced high-energy lithium-ion batteries for future aerospace mission applications. Although a greater electrode–electrolyte contact area is desirable, the high reactivity of the high-surface-area nanomaterials toward cell electrolytes can introduce issues of undesirably thick SEI layers and side reactions, and uniformly coated nanomaterials are being investigated to mitigate this issue. Also, and equally important, both the environmental and the health risks associated with the reactivity and physiological properties of nanomaterials are not well understood.

In summary, nanostructured electroactive materials for the negative electrode for lithium-ion batteries can potentially afford the following performance advantages compared to microstructured materials:

(1) Rapid charge–discharge performance characteristics as a result of high surface areas and the diminishing of slow solid-state lithium ion diffusion mechanisms necessary within bulk particles. Faster rates are more probable due to faster reactions near the particle surface.
(2) Improved cycle life and coulombic efficiency due to reduced stress-induced deformation and crystal lattice breakdown.
(3) Increased specific capacity due to the large interfacial area for lithiation and de-lithiation and improved capacity retention.
(4) Faster mass transport and reduced concentration polarization at higher charge/discharge rates due to shorter diffusion path lengths, which results in higher power capability.
(5) Improved low-temperature cell performance with respect to the cell electrode component.

Various nanomaterial morphologies and/or fabricated electrode structures are actively being explored and optimized to enhance anode and cell level performance with respect to energy density, cycle life, and safety. In addition to coated, uncoated, or immobilized nanoparticle and nanoalloy forms, nanotube structures, and nanostructured films and arrays possessing wire, rod, whisker, or columnar architectures are under development, and examples of such materials will be elucidated in the sections that follow.

5.2 Promising Nanomaterials as Alternative Anodes

5.2.1 Carbon-Based Anodes

In the history of the commercial lithium-ion cell, carbon has been the dominant anode material due to its advantages of high reversible specific capacity, low cost, favorable negative potential, satisfactory rate capability, and cycle life. Excellent review articles on the material exist and only the key points will be discussed here.[6,7] Graphitic carbons (ordered or soft carbon) have a theoretical capacity of 372 mAh/g, corresponding to the composition $Li_{x=1}C_6$. Non-graphitic carbons (disordered or hard carbons) can achieve higher specific capacities, exceeding 500 mAh/g due to the existence of additional lithium insertion sites. This corresponds to compositions of approximately $Li_{x>1.3}C_6$.[8] In Dahn's "house-of-cards" model for the nanostructure of disordered carbon, lithium is accommodated on both sides of individual graphene sheets, producing compositions of $Li_{x=2}C_6$ with theoretical capacities of 744 mAh/g.

Carbon nanotubes (CNTs) have received considerable attention as prospective anode materials for lithium-ion batteries and numerous geometries have been explored and reported in the literature. Since the 1990s, electrochemical studies of single-wall carbon nanotubes (SWNTs) and multiwall carbon nanotubes (MWNTs) have been performed for numerous electrode structures, including aligned and non-aligned structures. Gao et al. studied the effects of ball-milling on SWNTs, demonstrating up to 1000 mAh/g.[9] Vertically aligned MWNTs have demonstrated reversible cycling capacities approaching 800 mAh/g.[10] Jian Zhang et al. prepared carbon nanofibers (CNF) encapsulated in CNTs and achieved 410 mAh/g over 120 cycles.[11] Katar et al. studied thin films of bamboo-like CNTs that produced a reversible capacity equivalent to 205 $\mu Ah/cm^2$, per micron of film thickness.[12] This compares favorably with graphite. Assuming a pore-free film of graphite at 2.2 g/cm^3 and 372 mAh/g: a 1 micron film would have a theoretical capacity of 82 $\mu Ah/cm^2$. Therefore, Katar's thin-layer electrode appeared to achieve a capacity corresponding to 932 mAh/g, corresponding to the composition $Li_{x=2.5}C_6$.

The challenges of high irreversible capacity and voltage hysteresis have hindered the adoption of nanostructured carbon anodes in practical cells. There has been a great effort to seek new anode alternatives and the following sections highlight some of the promising nanomaterials being developed to replace graphite anodes in Li-ion batteries.

5.2.2 Silicon-Based Anodes

Metal alloys with lithium offer substantial theoretical increases in specific capacity relative to graphite. Investigation of binary alloys of lithium with group IV elements silicon, germanium, tin, and lead reveals that silicon is particularly attractive as an alternative anode material for lithium-ion cells.[13] Silicon is abundant and demonstrates a room temperature–specific capacity of 3579 mAh/g, corresponding to a composition of $Li_{15}Si_4$.[14] This represents an order-of-magnitude increase relative to graphite and is nearly equal to the specific capacity of metallic lithium (3863 mAh/g). However, the great capacity for lithium insertion in silicon is accompanied by an enormous volume change in cycling, with a volume increase of 370% for $Li_{15}Si_4$.[15] The associated strain leads to physical damage to the electrode material, loss of interparticle and particle–substrate contact, and eventual fading of capacity.[16] There are several approaches to reduce such mechanical strain induced by volume changes.[17] In one approach addition of conductive additives such as graphite flakes and/or nanoscale carbons into the micro-Si anode[18] is used to improve interparticle electronic contact and suppress the large Si volume change. Another approach is to narrow the cycling voltage window to decrease Li insertion/de-insertion levels which also lowers the stress.[18] The most widely used approach is to reduce the Si particle size to nanoscale[19,20,21,22,23] or change to various morphologies (for example: nanowires[24], nanorods,[25,26] and nanowisker[27] to release the stress and improve the performance.

Silicon-based anodes often exhibit significant first-cycle irreversible capacity loss and voltage hysteresis, similar to CNTs, but develop a useful capacity at a more practical voltage cutoff of ~1 V. Novel in situ stress measurements performed on thin silicon films suggest that a portion of the voltage hysteresis is due to mechanical work consumed by the volume change of the anode material.[16]

Methods for overcoming this practical issue have attracted considerable attention. An excellent review by Kasavajjula et al. presents the broad spectrum of silicon electrode research and development as of 2006, contrasting results for numerous morphologies and compositions.[5] Continued work in this area has expanded the knowledge of the limitations associated with silicon-based anode materials and some ingenious morphologies and compositions have been examined.

5.2.2.1 Si nanoparticles

The smaller particle size of Si reduces both the path of Li-ion transport/diffusion and the stress caused by the volume changes. For a nano-Si anode cycling between 0.0 V and 0.8 V at 0.1 mA/cm[2,27], a charge capacity of 2775 mAh/g and a discharge capacity close to 2097 mAh/g were obtained during the first cycle, which gave 76% coulombic efficiency. Its reversible capacity on the 10th cycle was 1729 mAh/g. The result showed that the capacity fade was much lower than that of a bulk Si because of reduced volume expansion of reduced particle size. Although the particle size reduction can reduce the volume change to a certain degree, the capacity fade cannot be eliminated completely.

5.2.2.2 Si nanowires

Si nanowires were grown directly on stainless steel by using a vapor–liquid–solid process and thus each nanowire was electrically connected to metallic current collector.[24] The anodes with the Si nanowires were able to accommodate large strain without cracking/pulverization, and provided good electronic contact and short Li conduction distances.[24] A theoretical charge capacity for Si anode was achieved, and discharge capacity close to 75% of the maximum was maintained with little fading of capacity at 0.05C discharge rate. The diameter of the nanowires (averaging ~89 nm) was observed to increase on cycling (averaging ~141 nm), which allowed for the large volume change without cracking. The nanowires also showed a drastic change in their atomic structure, gradually transforming from initially crystalline Si, to amorphous Li_xSi. Adequate space between adjacent nanowires provided by nanowire array, and improved contact between individual nanowires and the current collector, appear to help release the stress and improve the Si cycling performance.

5.2.2.3 Si nanorods

The Si nanorods were formed by oblique-angle deposition (OAD) with substrate rotation.[25,26] The OAD method allows fabricating Si nanorods with very large aspect ratio, controllable porosity, shape, and symmetry in a convenient, scalable and inexpensive way. The Si nanorod array accommodates the volume change during

Li cycling. The anodes with Si nanorods showed capacities of ~3600 mAh/g and good capacity retention (~83%) after 70 charge–discharge cycles.[29] The Si nanorods deposited by OAD on Cu foils also showed a stable capacity with ~1600 mAh/g, which is more than four times greater than graphite electrodes. During the first few cycles, there is an initial loss in the charge and discharge capacities, which indicates probable wettability issues between the electrode and electrolytes.

5.2.2.4 Silicon nanowhiskers

A novel concept of growing of Si nanowhisker on nanocarbon fibers with 1:1 weight ratio by vapor–liquid–solid process was demonstrated.[27] This composite anode is high in free volume, free of polycrystalline domains (not achievable for silicon anode by CVD), 50% or higher loading of silicon, a supporting matrix to act as an electronically conductive framework, and is can be produced using established procedures and equipment. The anodes showed 1600–1700 mAh/g discharge capacity with good columbic efficiency (80–90%).

5.2.2.5 Nanoscale silicon films

Some remarkable performance has been reported for thin silicon films. Takamura et al. studied 500 Å-thick films of vacuum-deposited, phosphorous-doped (n-type) silicon and achieved a stable cycling capacity of 3600 mAh/g for 200 cycles at 2C rate. A nine times thicker 4400 Å-thick film of (n-type) silicon achieved a stable cycling capacity of 2000 mAh/g after 200 cycles at a 1 C rate. Such films may have microbattery applications but are not practical for conventional lithium-ion batteries, due to the low areal loading issues. The calculated loading of a fully dense, 4400 Å, silicon film at 2000 mAh/g is ~0.2 mAh/cm^2, which is well below the level needed in practical cells (see Fig. 5.2). More practical loadings were addressed by Yin et al. who demonstrated stable cycling with 6 micron-thick films of silicon with a loading of 1.8 mAh/cm^2 for 250 cycles. This was achieved by limiting the cycling capacity to 45% of the reversible capacity.

5.2.2.6 Silicon/carbon composite

To improve the conductivity and reduce the volume expansion of Si anodes, carbon has been used as an active matrix because of its good conductivity, softness and compliance, small volume expansion, relatively low mass, and reasonable Li-insertion ability. Silicon dispersed onto a carbon matrix (Si/C) composite was extensively investigated. There are numerous approaches for Si/C anode preparation, which can be classified into the following types[5]: pyrolysis (or chemical/thermal vapor deposition), ball milling (or mechanical milling), pyrolysis in combination with mechanical milling, and chemical reaction of gels. All of the above Si/C composite anodes were prepared at high temperature between 600 and 900°C except ball milling. The Si/C composite anodes improve capacity retention and cycle life, however, the cycle life for these anodes is still insufficient for applications requiring lifetimes greater than 1000 cycles.

Composite anode materials of powdered silicon and carbon have produced promising results. A composite of nano-Si in SiO_x and C demonstrated 620 mAh/g after 200 cycles (0.01–1.5 V window), equivalent to 88% capacity retention.[29] Yushin's group present results for C–Si composites based on a hierarchal "bottom-up" approach, which demonstrated ~1400 mAh/g after 100 cycles at a 1 h rate (0.01–1.1 V window).[30]

5.2.3 Tin-Based Anodes

Both tin and tin oxides are interesting and attractive anode materials for the Li-ion batteries because of their high theoretical capacity (Sn: 994 mAh/g, SnO 875 mAh/g and SnO_2: 781 mAh/g), which is more than twice that of graphite. However, significant capacity fading with cycling is a problem due to volume changes during the Li insertion/de-insertion process. The volume expansion for tin can be as high as 259%.[31] It was demonstrated that the as-prepared SnO nanoflowers could be utilized as good anode materials for lithium ion rechargeable batteries with a high capacity of around 800 mAh/g, close to the theoretical value (875 mAh/g).[32] A hybrid structure of Sn nanoclusters covered with tin oxide (SnO_2) nanowires exhibited a high capacity of >800 mAh/g over 100 cycles with a low capacity fade of less than 1% per cycle.[33] The observed and enhanced stability with capacity retention is explained with the following factors: (a) the spacing between Sn nanoclusters on SnO_2 nanowires allowed the volume expansion; (b) high available surface area of Sn nanoclusters for Li-ion insertion/de-insertion;

and (c) the presence of Sn nanoclusters on SnO_2 allowed reversible reaction between Sn and Li_2O to produce Sn and SnO phases.

5.2.4 Transition Metal Oxide–Based Anodes

There is intensive interest in the development of nanomaterials of transition metal oxides, such as Mn_3O_4, Co_3O_4, and Fe_3O_4 as anode materials with higher specific capacity than graphite. The theoretical calculated capacities of Co_3O_4 and Mn_3O_4 are 890 mAh/g[34] and 936 mAh/g,[35] respectively. It was reported that three distinct Co_3O_4 nanostructures of nanoparticle, nanocubes, and hierarchical pompon-like microspheres were prepared, and the results showed that pompon-like microspheres Co_3O_4 exhibit the best electrochemical characteristics, revealed by the higher discharge capacity and excellent cycling retention.[34] Mn_3O_4 nanoparticles on graphene oxide as a hybrid anode showed that an unprecedented high capacity of ~900 mAh/g, near the theoretical capacity (~936 mAh/g), based on the mass of Mn_2O_3 (~810 mAh/g based on the total mass of the hybrid), with good rate capability and cycling stability.[35] The graphene-wrapped Fe_3O_4 composite showed a reversible specific capacity approaching 1026 mAh/g after 30 cycles at 35 mA/g and 580 mAh/g after 100 cycles at 700 mA/g as well as improved cyclic stability and excellent rate capability.[36] Considerable attention has been focused on the development of Fe_3O_4 as an anode material, due to its high specific capacity, low cost, ecofriendliness, and natural abundance.

Generally, the transition metal oxide anodes operate at higher potential than conventional, graphitic anodes. This introduces a potential safety benefit in terms of reduced likelihood for lithium plating during charge. It also introduces a disadvantage in terms of reduced cell voltage and lower specific energy.

5.3 Practical Considerations

5.3.1 Value of Anode Specific Capacity

In practical cell designs with graphite electrodes, the positive electrode occupies approximately 40% of the cell mass; therefore, improvements to cathode specific capacity have the greatest benefit in terms of reducing cell mass and increasing the specific energy.[37] Negative electrode materials based on graphitic carbon, have specific capacity that is approximately two times that of the positive electrodes, and occupy approximately 16% of the cell mass (see Fig. 5.1).

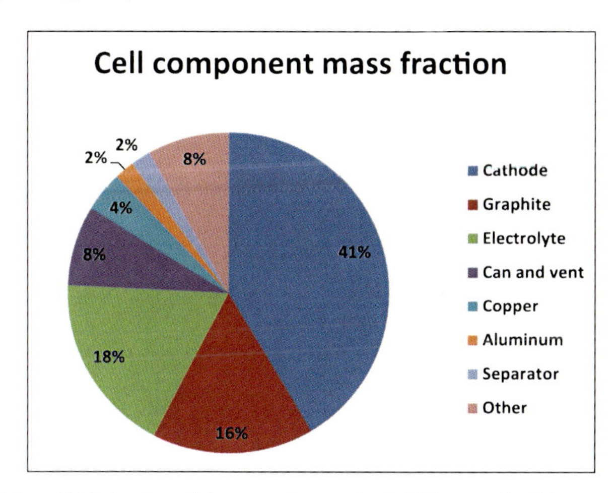

Figure 5.1 Weight breakdown of practical Li-ion cell. Values reported by Moshtev and Johnson.[2]

As improvements to cathode materials and cell designs are implemented, the fraction of mass occupied by the anode increases and the benefits of anode materials with greater specific capacity become more significant.

5.3.2 Advantages and Disadvantages of High-Surface-Area Electrodes

Nanostructured electrodes benefit from short diffusion distances and large exchange surface areas for mass transport. This generally enables faster kinetics and improved rate capability and power production in practical batteries. However, large specific area poses a burden in terms of irreversible capacity during the initial insertion of lithium into the carbon host. During the initial charge of a lithium-ion cell, the anode material eventually reaches a potential where electrolyte decomposition occurs. The decomposition products form the (SEI), which covers the anode surface and promotes stable performance in subsequent cycles.[38] Nanostructured electrodes present substantially greater surface area leading to a proportionately greater irreversible capacity. This has serious consequences in practical lithium-ion cells where capacity matching between the positive and negative electrodes must account for the total capacity of the electrodes. These cells are built in the discharged state and the positive electrode is the sole source of Li in the cell. Inordinately large negative irreversible capacity requires excess positive electrode material, which adds to the cell mass. In most cases, the additional cathode weight offsets any benefits due to increased reversible specific capacity at the anode. Possibilities exist for pre-lithiation of the anode, independently of the positive electrode capacity. A short communication by Jarvis et al. discusses this possibility.[39]

5.3.3 Importance of Electrode Loading

Electrode loading refers to the quantity of electrode material that is deposited per unit area of current collector foil. Practical lithium-ion cells, designed for high specific energy, require careful optimization of the electrode loading. Increased loading levels introduce excessive resistance within the electrode layer and diminish the utilization of active material at useful rates. Reduced loading levels will have enhanced rate capability but at the expense of cell mass and specific energy. Thinly loaded electrodes require increased electrode area for a given capacity, which in turn dictates a larger amount of metal foil current collector material. Thinner metal foils may be substituted but there exist practical limits imposed by the need to handle the foil in commercial cell assembly processes. A novel approach for supporting thin metal layers on tough polymer substrates has been described by Munshi.[40] Such an approach could help offset mass penalties in cell designs with low electrode loadings.

Moshtev and Johnson[2] presented a physical analysis of commercial lithium-ion cells (ca.1999) which can be used to calculate the electrode loading by dividing the cell capacity by the measured electrode active area. Calculations performed using their data show loadings ranging from 2.4 to 3.4 mAh/cm^2. In a separate work, the authors of this chapter calculated the sensitivity of impact of cell specific energy to electrode loading, with anode specific capacity as a parameter. For a hypothetical cell with "ordinary" cathode material and standard cell construction, the specific energy is seen to fall dramatically as loading is reduced below ~2.5 mAh/cm^2 (see Fig. 5.2). Note that the benefits of increased anode specific capacity are quickly lost for loadings less than 1 mAh/cm^2.

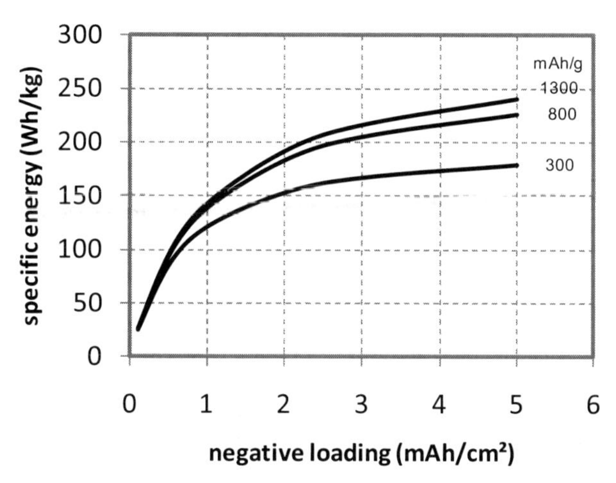

Figure 5.2 Effect of electrode loading in a hypothetical cell with anode specific capacity as a parameter.

The deleterious effect of reduced loading on cell level specific energy presents a challenge to the practical application of some nanostructured electrodes in commercial Li-ion cells. Many of the materials described in the literature are developed for use in thin film microbatteries. For example, the thin film anode described in reference 12 was stated to be 0.5 micron–thick, which corresponds to a loading of 0.1 mAh/cm^2. Note that this electrode was specifically prepared for microbattery applications. Frequently, loading information is not presented in the literature, which makes it difficult to assess if an electrode material can be applied in practical cells. Some nanostructured electrode materials will exhibit large specific capacity as thin films but cannot be scaled to commercial loading levels. Other suggestions for preparing experimental electrode structures with practical physical characteristics have been published by the group at Dahn's laboratory.[41]

5.3.4 Importance of Electrode Potential

Graphitic carbon is attractive, in part, because it maintains a low electrode potential over the range of its useful capacity, with little hysteresis between charge and discharge. This translates into relatively flat discharge voltage profiles in full cells and excellent energy efficiency in cycling. Disordered carbons and CNTs generally exhibit considerable hysteresis in cycling, which is a characteristic that is partly attributed to the structure of the carbon surface.[42] Often, optimistic values for specific capacity are reported for these materials without regard for the excessively broad voltage window used for cycling. This is illustrated in Fig. 5.3, which contrasts laboratory data for a MCMB carbon anode material with a hypothetical anode with large hysteresis. Here, the MCMB anode develops 330 mAh/g with a 1 V cutoff. The hypothetical anode develops significantly higher capacity 1000 mAh/g but at a 3.5 V cutoff. In a practical Li-ion cell, the matched cathode achieves its capacity at ~3 V and the cell would need to be discharged to 0 V in order to achieve the full capacity of the hypothetical anode. A 1 V cutoff potential represents a practical limit for the anode in such cells, assuming a 2 V cell discharge limit. The hysteresis exhibited by the hypothetical anode limits specific capacity to approximately 100 mAh/g, at the 1 V cutoff.

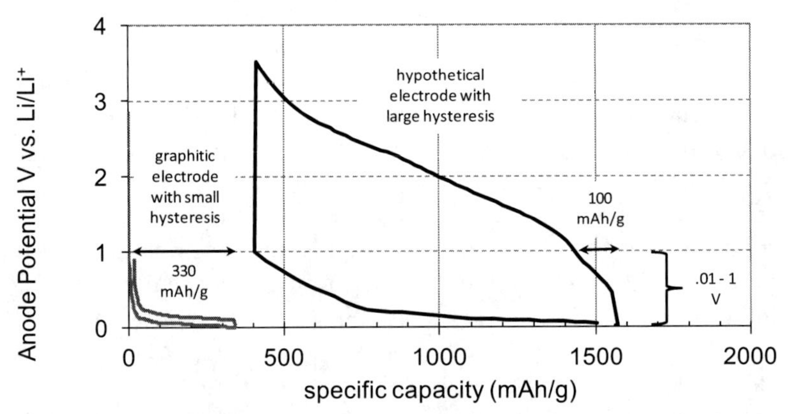

Figure 5.3 Comparison of conventional graphite anode with a hypothetical anode having large hysteresis.

The voltage/capacity performance for the electrode with large hysteresis is entirely hypothetical, drawn to illustrate the impact of voltage performance on practical capacity. Much of the data reported in the literature for disordered carbon and CNTs exhibit similar profiles with capacity reported to 3 V versus Li. A review of the performance of commercially available SWNTs using a more practical upper voltage limit of 1 V reveals specific capacities that are less than 200 mAh/g in limited cycling.[43]

References

1. Mercer, C. R., National Aeronautics and Space Administration, Battery and Fuel Cell Development Goals for the Lunar Surface and Lander, Presented at the Space Power Workshop, April 23, 2008, Huntington Beach, California.
2. Moshtev, R., Johnson, B. (2000), *J. Power Sources*, **91**, 86–91.
3. Brianne T. Scheidegger, Glenn Research Center, Cleveland, Ohio, Performance Characterization of High Energy Commercial Lithium-Ion Cells, NASA/TM—2010-216926.
4. Nikkei Electronics Asia—March 2010, http://techon.nikkeibp.co.jp/article/HONSHI/20100223/180545/.
5. Kasavajjula, U., et al. (2007), *J. Power Sources*, **163**, 1003–1039.
6. Flandrois, S., Simon, B. (1999), *Carbon*, **37**, 165–180.
7. Kaskhedikar, N. A., Maier, J. (2009), *Adv. Mater.*, **21**, 2664–2680.
8. Sun, H., et al. (2007), *Electrochim. Acta,* **52**, 4312–4316.
9. Gao, B., et al. (2000), *Chem. Phys. Lett.,* **327**, 69–75.
10. Welna, D. T., et al. (2011), *J. Power Sources*, **196**, 1455–1460.
11. Zhang, J., et al. (2008), *Adv. Mater.*, **20**, 1450–1455.
12. Katar et al. (2008), *J. Electrochem. Soc.*, **155**(2) A125–A128.
13. Obravac, M. N., et al. (2007), *J. Electrochem. Soc.*, **154**(9) A849–A855.
14. Obravac, M. N., et al. (2004), *Electrochem. Solid-State Lett.*, **7**(5) A93–A96.
15. Sethuraman, V. A., et al. (2010), *J. Power Sources*, **195**, 5062–5066.
16. Winter, M., Besenhard, J. O. (1999), *Electrochim. Acta*, **45**, 31.
17. Kasavajjula, U., Wang, C. S., Appleby, A. J. (2007), *J. Power Sources*, **163**, 1003–1039.
18. Liu, W. R., Guo, Z. Z., Young, W. S., Sheih, D. Z., Wu, H. C., Yang, M. H., Wu, N. L. (2005), *J. Power Sources*, **140**, 139.
19. Guo, Z. P., Wang, J. Z., Liu, H. K., Dou, S. X. (2005), *J. Power Sources*, **146**, 448.
20. Kim, H., Seo, M., Park, M.-H., and Cho, J. (2010), *Angew. Chem. Int. Ed.,* **49**, 2146–2149.

21. Yang, J., Winter, M., Besenhard, J. P. (1996), *J. Power Sources*, **90**, 281.
22. Kwon, Y., Kim, H., Doo, S. G., Cho, J. (2007), *Chem. Matter.*, **19**, 982.
23. Kwon, Y., Cho, J. (2008), *Chem. Commun.*, 1109.
24. Chan, C. K., Peng, H., Lin, G., Mcllwrath, K., Zhang, X. F., Huggins, R. A., Cui, Y. (2008), *Nat. Nanotechnol.*, **3**, 31.
25. Teki, R., Karabacak, T., Lu, T. M., Koratkar, N. (2006), *Appl. Phys. Lett.*, **89**, 193116.
26. Li, C., Wang, Z., Wang, P. L., Peles, Y., Koratkar, N., Peterson, G. P. (2008), *Small*, **4**, 1084.
27. Ma, J. Q., Newman, A., Lennhoff, J., Lang, C., Elliott, A., Constantine, K. (2009), NASA Aerospace Battery Workshop, Huntsville, AL.
28. Fleischauer, M. D., Li, J., Brett, M. J. (2009), *J. Electrochem. Soc.*, **156**, A33.
29. Morita, T., Takami, N. (2006), *J. Electrochem. Soc.*, **153**, A425.
30. Yushin et al., NASA Aerospace Battery Workshop, NASA Contract # NNX09CD29P (2009), *Nat. Mater.*, DOI: 10.1038/NMAT2725, Advance Online Publication (2010).
31. Boukamp, B. A, Lesh, G. C., Huggins, R. A. (1981), *J. Electrochem. Soc.*, **128**, 725.
32. Ning, J., Dai, Q., Jiang, T., Men, K., Liu, D., Xiao, N., Li, C., Li, D., Liu, B., Zou, B., Zhou, G., Yu, W. W. (2009), *Langmuir*, **25**.
33. Meduri, P., Pendyala, C., Kumar, V., Sumanasekera, G. U., Sunkara, M. K. (2009) *Nano Lett.*, **9**(2), 612–616.
34. Guo, B., Li, C., Yuan, Z. Y. (2010), *J. Phys. Chem. C*, **114**, 12805.
35. Wang, H., Cui, L. F., Yang, Y., Casalongue, H. S., Robinson, J. T., Liang, Y., Cui, Y., Dai, H. (2010), *J. Am. Chem. Soc.*, **132**, 13978.
36. Zhou, G., Wang, D. W., Li, F., Zhang, L., Li, N., Wu, Z. S., Wen, L., Lu, G. Q., Cheng, H. M. (2010), *Chem. Mater.*, **22**, 5306–5313.
37. Gaines, L., Cuenca, R. (2000), Costs of Lithium-Ion Batteries for Vehicles, ANL/ESD-42.
38. Balbuena, P. B., Wang, Y. (eds.) (2004), *Lithium-Ion Batteries: Solid Electrolyte Interphase,* Imperial College Press, London.
39. Jarvis, C. R., et al. (2006), *J. Power Sources*, **162**, 800–802.
40. Munshi, M. Z. A. (ed.) (1995), *Handbook of Solid State Batteries and Capacitors*, World Scientific Publishing Co. Pte. Ltd., River Edge, NJ, Chapter 17.

41. Dahn, J. R., et al. (2011), *J. Electrochem. Soc.*, **158**(1) A51–A57.
42. Frackowiak, E., et al. (1999), *Carbon*, **37**, 61–69.
43. Britton, D. L., NASA/TM—2007-214809.

Chapter 6

Advances in Designing High-Energy Cathode Materials for Li-Ion Rechargeable Batteries

Ram S. Katiyar and Gurpreet Singh

NASA-URC Center for Advanced Nanoscale Materials, University of Puerto Rico, Rio Piedras Campus, San Juan, PR 00931, USA

rkatiyar@hpcf.upr.edu

Materials with the reversible intercalation properties, such as $LiCoO_2$, $LiMn_2O_4$, $LiFePO_4$, and $LiNi_{1/3}Mn_{1/3}Co_{1/3}O_2$, have been explored as viable cathode materials for the lithium-ion batteries in the past two decades along with the extensive research on Li_2MnO_3 and its derivatives, also known as lithium rich solid solution. Intensive structural and electrochemical research has led to the lithium-ion technology to a stage where it is commercialized in almost every crucial electronic device of modern society. In order to enhance the energy density of the material and make it work for the longer time span, requirements such as high specific capacity, high operating voltage, and structural stability should be met. Researchers across the world have also made an effort to reduce the amount of costly and toxic materials ingredients so as to minimize the environmental impact without compromising

Advanced Nanomaterials for Aerospace Applications
Edited by Carlos R. Cabrera and Félix A. Miranda
Copyright © 2014 Pan Stanford Publishing Pte. Ltd.
ISBN 978-981-4463-18-8 (Hardcover), 978-981-4463-19-5 (eBook)
www.panstanford.com

the performance. This chapter reviews various experimental and computational approaches that have been utilized for obtaining the materials' best electrochemical properties. We have tried to cover all such crucial breakthroughs in the cathode materials' research related to layered, spinel, and olivine structures.

6.1 Introduction

In the past two decades, much emphasis has been made on the materials advancement for lithium-ion batteries. These batteries hold the future for the next-generation energy storage devices. Efforts have led to their successful application in most of the portable devices of today's life, such as camera, laptops, and next-generation mobile phones. In the vision of using these batteries as an alternative to petrol or diesel in cars, these batteries have been improved on at both materials as well as engineering level. Some of the major drawbacks of using these batteries in cars include their cost and the use of organic electrolytes. These factors make lithium-ion batteries less safe than other battery technologies. In the early development of these batteries, graphite in the form of MCMB (Meso carbon micro beads) and $LiCoO_2$ were used as anode and cathode material, respectively. Electrolytes composed of salts such as $LiAsF_6$ or $LiPF_6$, etc., dissolved in organic solvents such as DEC (di-ethyl carbonate), DMC (di-methyl carbonate), and EC (ethlyene carbonate), etc., have been used. It has been observed that the major contributor to the cost of these batteries is the cathode material, which comprise materials that can reversibly intercalate lithium ions in the structure. $LiCoO_2$ is a state-of-the-art cathode material in lithium-ion batteries due to its excellent performance from capacity, reversibility, and rate capability as proposed in the early 1980s [1]. However, this material is thermally unstable and also contains toxic cobalt, which is not an environment-friendly element. In an effort to identify an alternative to $LiCoO_2$, several materials based on other transition metals have been proposed. However, the layered materials' development without cobalt has not moved fast enough due to a number of requirements that battery materials should meet before their actual use in the market.

$LiMn_2O_4$ is considered an alternative material to $LiCoO_2$. The basic crystal structure of $LiMn_2O_4$, which is a spinel, is entirely different from $LiCoO_2$ [2]. $LiMn_2O_4$ meets almost all of the requirements for a battery cathode such as good thermal stability, reasonably good capacity, and capability to charge and discharge at high current rates. Mn is less toxic and inexpensive compared to Co. Therefore, this material has drawn much attention toward its use in the electric vehicles. Spinel-based $LiMn_2O_4$ possesses low charge/discharge capacity compared to $LiCoO_2$ and have some disadvantages regarding the stability of the structure, which lead to failure of the cell after certain number of cycles. Mn oxidation state in case of $LiMn_2O_4$ is +3.5, which implies that Mn^{3+} and Mn^{4+} are in present in 1:1 in ideal stoichiometry. Mn^{3+} is known as Jahn–Teller active ion, which undergoes distortion at low voltages. Therefore, the lower cutoff voltage in case of the spinel electrodes plays important role in the stability of the electrode material. It has also been pointed out that even though Jahn–Teller distortion is highly dependent on the voltage cutoff, but some structural fading also occur at high voltages because of the dissolution of the Mn ions in the electrolyte. Distortion in the structure leads to decrease in capacity with the increase in number of cycles. In order to improve Jahn–Teller distortion dopants such as Cr and Ni have been tried. An improved version of $LiMn_2O_4$ has been proposed by substituting some of Mn with Ni. Micron sized Ni doped $LiMn_2O_4$ has shown very good performance at high C-rates. Capacity as high as 130 mAh/g has been achieved at 1C rate of charging/discharging [3]. Li_2MnO_3 have also reported another manganese-based composition, that belongs to $C_{2/m}$ space group symmetry and it is electrochemically inactive [4]. However, it has been shown that it can be activated by chemically or electrochemically removing Li_2O from the host structure and leaving MnO_2 cage structure. Many articles have been published to explain the reaction mechanisms that occur at high potentials however the cause behind slow transformation of the layered to the spinel structure is still under research and is thorny issue in the lithium ion battery cathode material development. Another interesting aspect of this material is that it can be structurally integrated into other well-known layered materials to have composite structure with enhanced properties. $LiMnO_2$ is another compound that lies in the category of the manganese-based oxides for lithium-ion batteries. Therefore, cathode materials based on manganese have drawn much attention in the past and still a more advanced research is being carried out in this area and has been discussed in this review. $LiMnO_2$ shows very high capacity, however, due to the presence of Mn in its +3 oxidation state it undergoes Jahn–Teller distortion and it transforms to spinel structure. This is a major issue that needs to overcome because such Jahn–Teller distortion can lead to the formation of the lithiated spinel cathode and performance of the cell deteriorates as per the following reaction:

$$Li_2Mn_2O_4 \rightarrow Li_2MnO_3 + MnO$$

Mn^{2+} is soluble in the electrolytic solution and hence gives rise to poor cycling stability. In order to avoid the Jahn–Teller distortion a part of Mn was replaced by Ni and Co. $LiNi_{1/3}Mn_{1/3}Co_{1/3}O_2$ has been reported as a promising cathode compound with high energy density for its use in lithium-ion batteries [5].

A new composition based on iron phosphates was reported in the form of $LiFePO_4$, which is known to have olivine structure [6,7]. This material initially had very low electronic conductivity. However, this drawback has been overcome by its doping or coating with carbon. A breakthrough in research based on this material made by MIT researchers led to improvements in its rate capability [8]. It has been shown that the synthesis of off-stoichiometric phase $LiFe_{1-2y}P_{1-y}O_{4-\delta}$ gives rise to a material that is capable of charge and discharge at very high C-rates, because such kind of off-stoichiometry synthesis leads to the formation and coating of glassy phases that are highly conductive in nature. This communications has been criticized by John Goodenough and coworkers on the technical grounds in one of the letter to the editor of Journal of Power Sources.

6.2 High Voltage Spinel Electrode Materials

In the recent past, much focus has been made on the high energy density and high power density materials for lithium-ion batteries, which can be utilized for EV and HEV. The materials need to be low cost and nontoxic in nature. The thermal stability of these materials is another factor that should be taken into account. One of the promising candidates in this category is $LiMn_2O_4$. This material is quite inexpensive, less toxic and environment friendly. It is thermally stable compared to the layered $LiCoO_2$. However, its practical capacity is limited to ~120 mAh/g. The working voltage lies in the 4.2 to 3.0 V range and the voltage profile is flat compared to the layered $LiCoO_2$. Some of the reports related to thin-film batteries have also implied $LiMn_2O_4$ as a cathode material and checked its electrochemical properties as thin-film electrodes [9,10].

Efforts have been made in order to increase the average operating voltage by doping $LiMn_2O_4$ with the transition metals having higher redox potential, such as Ni and Cr. Ni doped $LiMn_2O_4$ with the composition "$LiNi_{0.5}Mn_{1.5}O_4$" can perform well in the 5 V region, which makes it an attractive candidate for the cathode material. This is termed as the 5 V cathode, with a plateau around 4.8 V.

In case of the $LiNi_{0.5}Mn_{1.5}O_4$, lithium can be extracted up to theoretical limit of 146.6 mAh/g. The performance was further improved by coating the material with ZnO or Li_3PO_4. Cr doping in $LiNi_{0.5}Mn_{1.5}O_4$ is shown to further increase the operating voltage and improve its electrochemical performance. The synthesis of $LiNi_{0.5}Mn_{1.5}O_4$ has been reported via many different procedures, viz. solid state, co-precipitation, sol-gel, emulsion drying, spray pyrolysis. Doping of $LiNi_{0.5}Mn_{1.5}O_4$ with rare earth elements has also shown some marginal improvement in the electrochemical performance. Here we are summarizing broad research progress relevant to $LiNi_{0.5}Mn_{1.5}O_4$.

$LiNi_{0.5}Mn_{1.5}O_4$ has been synthesized via sol-gel synthesis route and effect of synthesis temperature has been discussed in detail [11]. Samples prepared below 600°C in air were found to be oxygen deficient, with average Mn oxidation state less than +4. Above 650°C the disproportionation of the material occurs into $LiNi_xMn_{2-x}O_4$ ($x < 0.5$) and $Li_zNi_{1-z}O_4$ ($z \sim 0.2$). Therefore, the low temperature of ~600°C is sufficient to prepare pure $LiNi_{0.5}Mn_{1.5}O_4$ followed by slow cooling. $LiNi_{0.5}Mn_{1.5}O_4$ made at 600°C shows reversible capacity ~100 mAh/g, with operating voltage of 4.7 V, for tens of cycles.

Synthesis of $LiNi_{0.5}Mn_{1.5}O_4$ has also been reported via combustion synthesis route [12]. The samples synthesized at 500 and 800°C show good electrochemical behavior at 4.7 V with discharge capacity ~132 mAh/g. The capacity decreases with the increase in C-rate. At 5C rate, the discharge capacity has been observed to be ~85 mAh/g. Myung et al. have reported the synthesis of

$LiNi_{0.5}Mn_{1.5}O_4$ using emulsion drying method [13]. The synthesis temperature and time have been optimized to be 750°C and 24 h, respectively. The materials synthesized at 750°C show a flat discharge profile at ~4.6 V with a capacity ~110 mAh/g at cycling rate of 0.5C. It sustains a capacity ~100 mAh/g after 50 cycles at the cycling rate of 0.5C. This material has been shown to be oxygen deficient with average Mn oxidation state +3.84. Effect of oxygen stoichiometry on the electrochemical performance of $LiNi_{0.5}Mn_{1.5}O_4$ has also been studied [14]. Material has been synthesized via two different routes, viz. solid state and sol-gel. The oxygen contents of the samples (sol–gel method and annealed at a high P_{O2}) increased in comparison with the samples prepared by the solid-state method and is found to be closer to stoichiometric amount. The material was galvanostatically cycled between 5.0 and 3.6 V. In case of the sample synthesized with the help of the solid-state method, two voltage plateaus have been observed at ~4.1 and 4.7–5.0 V. The 4.1 V plateau is found to be absent in samples synthesized via sol-gel route and annealed at a high P_{O2}. $LiMn_{1.5}Ni_{0.5}O_4$ (sol-gel method) shows especially good cycle performance in the 5.0 V region. This can be correlated to the better particle size distribution and high surface area of the particles in case of sol-gel method. The galvanostatic charge–discharge curves of $LiNi_{0.5}Mn_{1.5}O_4$ at 25°C are shown in Fig. 6.1.

Comparative analysis of thin layers of $LiNi_{0.5}Mn_{1.5}O_4$ material comprising micro- or nano-sized particles on the Li^+ transport characteristics and electrochemical performance has also been reported in the literature [15]. Thin-layer electrodes comprising nano-sized particles develop faster electrochemical kinetics and better reversibility compared to electrodes with micron-sized particles. Particle size and morphology also plays a crucial role in the electrochemical performance of the material in lithium-ion batteries. The effect of particle size on the electrochemical performance of the material has been reported [16]. At room temperature after 100 cycles, 97% of initial discharge capacity was retained in case of micro-sized $LiNi_{0.5}Mn_{1.5}O_4$ powders sintered at 900°C. At −10°C and 1C rate nano-sized $LiNi_{0.5}Mn_{1.5}O_4$ sintered at 700°C deliver a capacity of 110 mAh/g. This shows that the material can perform well in the low temperature as well and hence it gives an evidence of the stability of the crystal structure. Cycling Performance of nano- and micro-$LiNi_{0.5}Mn_{1.5}O_4$ is shown in Fig. 6.2.

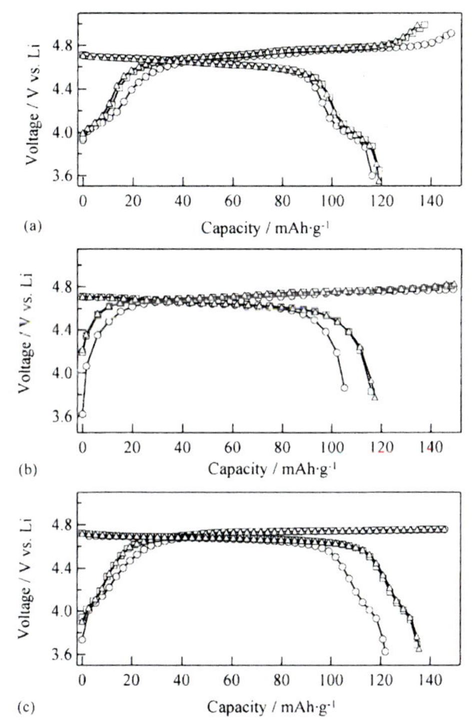

Figure 6.1 Galvanostatic charge-discharge curves of $LiNi_{0.5}Mn_{1.5}O_4$ at 25°C and 0.2 mA cm^{-2}. (a) Solid-state method, (b) 500°C, 60 h, P_{O2} = 2.02 MPa annealed, (c) sol-gel method. (○) first cycle, (□) second cycle, (Δ) third cycle [14].

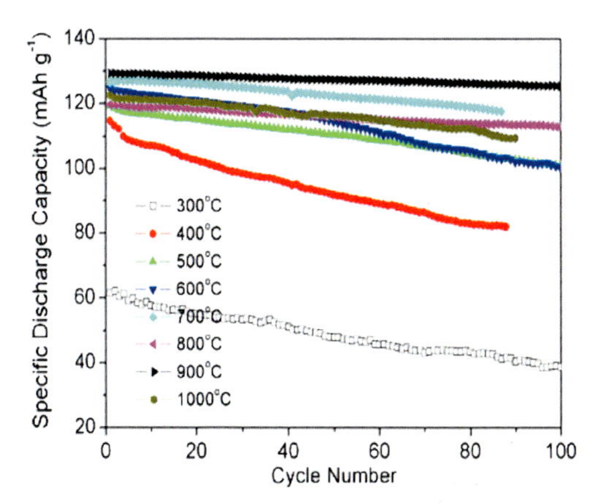

Figure 6.2 Cycling performance of nano- and micro-$LiNi_{0.5}Mn_{1.5}O_4$ at 1C [16].

In order to further improve the working voltage range of $LiNi_{0.5}Mn_{1.5}O_4$, Cr has been proposed as effective dopant because of its higher redox potential and variable oxidation state. It has been reported that Cr doping provides a flatter profile by suppressing the oxidation/reduction plateau corresponding to Mn^{3+} [17]. Also, the capacity retention increases with the increase in Cr content because Cr has greater electron affinity to oxygen and hence it makes the structure more stable, which has been demonstrated by the lesser loss of material weight in TG curves when the Cr concentration increases.

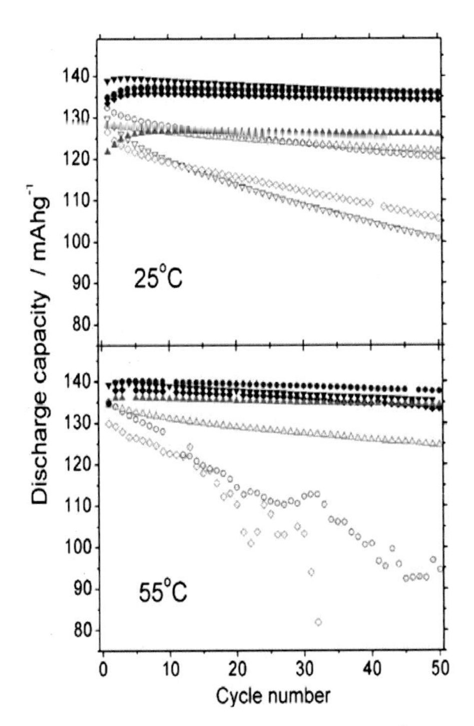

Figure 6.3 Evolution of discharge capacity vs. cycle number at 25 and at 55°C for $LiCr_{0.2}Ni_{0.4}Mn_{1.4}O_4$ samples heated at 700°C (▽), 750°C (◇), 800°C (○), 850°C (△), 900°C ▼), 950°C(♦), 1000°C (●, and 1100°C (▲). Potential range explored [18].

In order to check the performance of the Cr doped spinel cathode materials, electrochemical properties of $LiCr_{0.2}Ni_{0.3}Mn_{1.5}O_4$ at room and higher temperatures for 1C rate have been reported [18]. The discharge capacity in both the cases (at 25°C and at 55°C) was found to be the same (135 mAh/g) in spite of the difference in the particle size. However, cyclic performance is found to be quite dependent on the particle size. For particle size >500 nm a better cycling stability was achieved at 25 and 55°C than those with the particle size <500 nm. The spinels synthesized at high temperatures (1000 or 1100°C) showed good electrochemical performance with the operating voltage ~4.7 V and the capacity of 135 mAh/g for 250 cycles. Figure 6.3 shows the evolution of the discharge capacity versus cycle number at 25 and 55°C for $LiCr_{0.2}Ni_{0.4}Mn_{1.4}O_4$ samples heated at various temperatures.

In another report it has been shown that $LiMn_{1.4}Ni_{0.4}Cr_{0.2}O_4$ synthesized at 850°C gives good electrochemical performance showing initial discharge capacity of 128 mAh/g with 90% of capacity retention after 230 cycles when operated between 3.5 and 4.98 V [19]. It has also been demonstrated that $LiMn_{1.4}Ni_{0.4}Cr_{0.2}O_4$ and $LiNi_{0.35}Cr_{0.35}Mn_{1.3}O_4$ hold capacity at 4.9 V. The increase in the voltage limit of electrochemical cycling is due to the higher oxidation/reduction potential of the Cr^{3+}/Cr^{4+}.

It has also been clearly pointed out that $LiMn_{1.5}Ni_{0.4}Cr_{0.1}O_4$ can work in the high voltage region with voltage plateau at 4.87 V, which corresponds to the oxidation of Cr^{3+}/Cr^{4+} [20]. The rate performance of the material was also checked. At 0.2C and 1C, $LiMn_{1.5}Ni_{0.4}Cr_{0.1}O_4$ had shown initial discharge capacities of 126 mAh/g and 100 mAh/g. The discharge capacity retention was found to be 98.5% and 96% for 0.2 and 1C rates, respectively. Co has also been tried as a dopant in $LiNi_{0.5}Mn_{1.5}O_4$ in order to improve its electrochemical performance. Effect of Co on the electrochemical performance of $LiNi_{0.5}Mn_{1.5}O_4$ has been reported [21]. Doping the material with Co reduces the initial discharge capacity; however, the cyclability and the C-rate increase with Co doping. Co doped spinels, i.e., $LiNi_{0.45}Co_{0.05}Mn_{1.5}O_4$ showed discharge capacity of ~110 mAh/g for 10C rate in the voltage range of 5.0 to 3.5 V. Doping of $LiNi_{0.5}Mn_{1.5}O$ with Rh has also shown some improvement in the electrochemical performance. It has been shown that 4 % of Rh doping is sufficient to improve the electrochemical performance [22]. Improvement in the capacity retention to 93.5% after 50 cycles has been reported. It has been shown that the doping of $LiNi_{0.5}Mn_{1.5}O_4$ with Rh leads to the improvement in the structural stability, and hence the lattice parameter was found to be 8.217 and 8.202 Å before and after cycling the material for 50 cycles. Cyclic performance of the Rh doped $LiNi_{0.5}Mn_{1.5}O_4$ is shown in Fig. 6.4. Nd has also been tried as dopant in pristine $LiMn_2O_4$, and is found to improve the initial discharge capacity along with the improvement in the structural stability [23]. Initial discharge capacity of $LiMn_{1.99}Nd_{0.01}O_4$ was found to be 149 mAh/g, with the capacity retention of about 91% after 25 charge–discharge cycles. However, the explanation for such high capacity values was not given. These reports by the authors have shown that some of the rare earth element can be helpful in the improvement of the structural stability of the host structure.

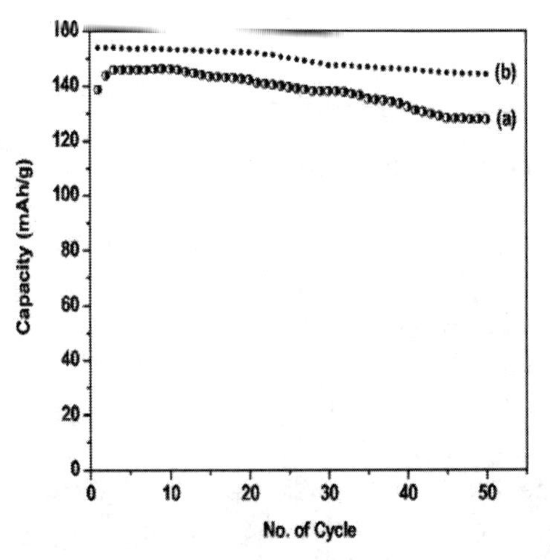

Figure 6.4 Specific capacity vs. number of cycle of (a) $LiMn_{1.5}Ni_{0.5}O_4$/(EC + DMC)/Li coin cell and (b) $LiMn_{1.5}Ni_{0.46}Rh_{0.04}O_4$/(EC + DMC)/Li coin cell [22].

As stated earlier, in order to improve the cycle life of "$LiNi_{0.5}Mn_{1.5}O_4$" at various temperatures, coating $LiNi_{0.5}Mn_{1.5}O_4$ with ZnO and Li_3PO_4 has been proposed by some of the research groups. Improvement in the cyclic stability of the material at elevated temperatures has been reported via coating $LiNi_{0.5}Mn_{1.5}O_4$ with ZnO [24]. At 55°C, ZnO-coated $LiNi_{0.5}Mn_{1.5}O_4$ showed discharge capacity of 137 mAh/g after 50 cycles whereas pristine $LiNi_{0.5}Mn_{1.5}O_4$ was able to retain only 10% of the initial capacity after 30 cycles. Charge–discharge curves for coated and uncoated samples of $LiNi_{0.5}Mn_{1.5}O_4$ at 55°C are shown in Fig. 6.5. In has also been demonstrated that the formation of HF is less in case of ZnO-coated $LiNi_{0.5}Mn_{1.5}O_4$ compared to the pristine $LiNi_{0.5}Mn_{1.5}O_4$ and it is known to be the reason behind the better cyclability of $LiNi_{0.5}Mn_{1.5}O_4$ [25]. Another report on the ZnO-coated $LiNi_{0.5}Mn_{1.5}O_4$ shows an improvement in the electrochemical performance of the material [26]. The initial discharge capacity of 146 mAh/g has been observed for ZnO-coated $LiMn_{1.5}Ni_{0.5}O_4$. The discharge capacity retention of 97% was observed after 50. Theoretical capacity of $LiNi_{0.5}Mn_{1.5}O_4$ is calculated out to be 146 mAh/g by assuming the complete removal of lithium ions from the structure.

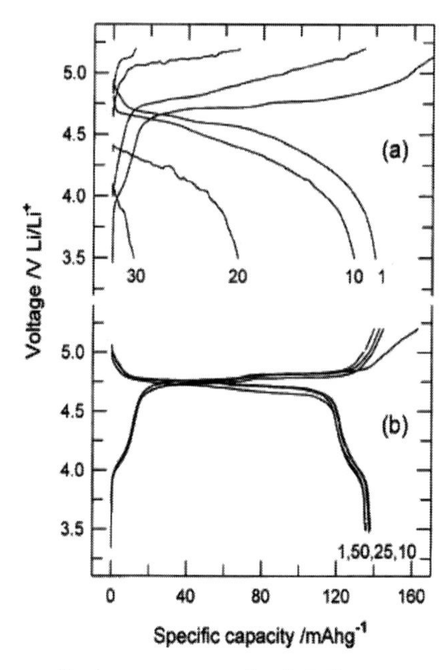

Figure 6.5 Charge–discharge curves for (a) the uncoated $LiNi_{0.5}Mn_{1.5}O_4$ and (b) ZnO-coated $LiNi_{0.5}Mn_{1.5}O_4$ electrodes at 55°C [24].

Since the electrolyte oxidation remains a big issue on high voltages near 5.0 V, which is close to the operating voltage of $LiNi_{0.5}Mn_{1.5}O_4$, therefore the compatibility of the spinel with the electrolyte should be strongly taken into account. Li_3PO_4/ $LiNi_{0.5}Mn_{1.5}O_4$| solid polymer electrolyte| Li and Li_3PO_4/ $LiNi_{0.5}Mn_{1.5}O_4$| $LiPF_6$ (EC:DMC :: 1:1)| Li battery has been demonstrated for high voltage region [27]. A layer of Li_3PO_4 on $LiNi_{0.5}Mn_{1.5}O_4$ has been applied using dip and dry process. Polyethylene oxide-based solid polymer electrolyte [ethylene oxide *co*-2-(2-methoxyethoxy)ethyl ether] was dissolved with organic peroxide and lithium tetrafluoroborate ($LiBF_4$) in dehydrated acetonitrile, and then cast directly onto the positive electrode, dried and cross-linked at 373 K for 3 h in an Ar atmosphere. Electrochemical performance of the cell in both cases, i.e., with liquid and solid electrolyte was found to be the same. It has been demonstrated that Li_3PO_4-coated $LiNi_{0.5}Mn_{1.5}O_4$ can be used at high potential both with liquid and solid electrolyte. Figure 6.6 shows the discharge voltage profiles of composite, polymer, and liquid-type cells.

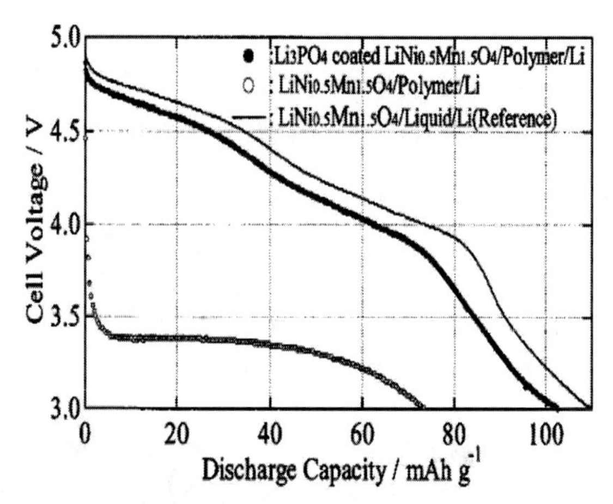

Figure 6.6 Discharge voltage profiles of composite, polymer, and liquid-type cells. (●) composite [Li_3PO_4-coated $LiNi_{0.5}Mn_{1.5}O_4$/polymer/Li] at 333 K, (○) Polymer [$LiNi_{0.5}Mn_{1.5}O_4$/polymer/Li] at 333 K, and (---) liquid [$LiNi_{0.5}Mn_{1.5}O_4$/Liquid/Li] at 298 K [27].

6.3 Layered Cathode Materials

Transition metal oxides with layered structure have played a crucial role in the development of the lithium-ion batteries. Many layered structures with general formula $LiMO_2$ (M: Co, Mn, Ni) have been proposed. In the past two decades, the replacement of Co with some cheap and environment-friendly transition metal ion remains the focus of research. Many research groups have tried the combination of Ni, Co, and Mn in order to design the material that has better electrochemical performance and that is safe and environment friendly. $LiMnO_2$ is reported to be a promising cathode material for lithium-ion battery. However, it suffers from some serious structural changes to the spinel $LiMn_2O_4$ after some cycles. Very high specific capacity values can be achieved by considering MnO_2 as host structure for the lithium. It has been reported that MnO_2 nanowires can also hold a specific capacity of >300 mAh/g for 25 cycles with a voltage range in between 3.5 and 2.0 V [28]. Therefore, Mn-based layered electrode materials are considered to be helpful for achieving high specific capacity in cathodes for lithium-ion batteries. Success has been achieved by combining the three transition metal ions, i.e., Mn, Ni, and Co in equal ratio [5]. $LiNi_{1/3}Mn_{1/3}Co_{1/3}O_2$ showed discharge capacity of 220 mAh/g in the voltage range of 5.0–2.5 V at a rate of 0.17 mA/cm^2. The charge and discharge curves of a $Li/LiCo_{1/3}Ni_{1/3}Mn_{1/3}O_2$ cell operated at a rate of 0.17 mA cm^{-2} in the voltage range of 2.5 to 4.2 V at 30°C are shown in Fig. 6.7. The oxidation of the electrolyte remains one problem due to charge/discharge at high voltage, but still the capacity retention was shown to be better than $LiCoO_2$ and $LiNi_{1/2}Co_{1/2}O_2$. Therefore, it has been proposed as an alternative (GEN2) to $LiCoO_2$. Since then much focus has been made on the layered compounds with the combination of different transition metal ions. Further studies on $LiNi_{1/3}Mn_{1/3}Co_{1/3}O_2$ with even lesser or no Co content have been carried out by many research groups [29–31]. Over-lithiated compositions have also been tried by many researchers. Researchers have studied the electrochemical performance of the $Li_{1+x}(Co_{1/3}Ni_{1/3}Mn_{1/3})_{1-x}O_2$ with graphite as anode material with different electrolytes. Cell performance was studied at different temperatures. Electrolytes with high salt concentration and high ethylene carbonate content are observed to behave well at room temperature. However, the converse was found to be true at low temperature [32].

The effect of reduced cobalt contents in the form of co-doping of $LiNiO_2$ with Co and Al has been investigated [33]. The discharge capacities for $LiNi_{0.8}Co_{0.2}O_2$ and $LiNi_{0.8}Co_{0.15}Al_{0.05}O_2$ cathodes were measured and found to be 100 and 136 mAh/g, respectively, in the voltage window of 4.2 to 3.2 V. However, in some

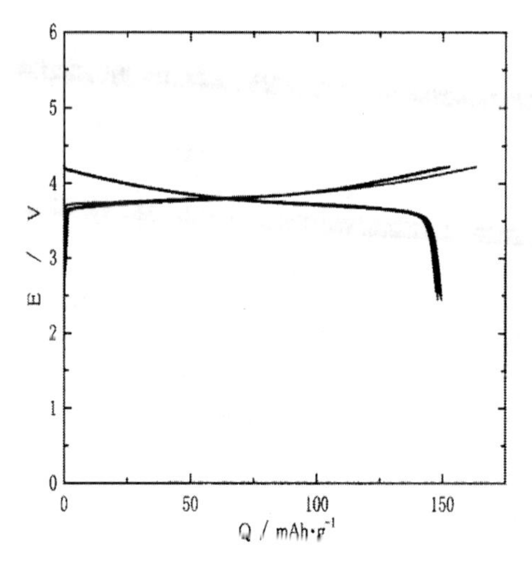

Figure 6.7 Charge and discharge curves of a $Li/LiCo_{1/3}Ni_{1/3}Mn_{1/3}O_2$ cell operated at a rate of 0.17 mA cm^{-2} in voltages between 2.5 and 4.2 V at 30°C. The electrolyte used was 1 M LiPF6 dissolved in EC/DMC (3/7 by volume). Electrode consisted of 88 wt% $LiCo_{1/3}Ni_{1/3}Mn_{1/3}O_2$, 6 wt% acetylene black, and 6 wt% PVDF [5].

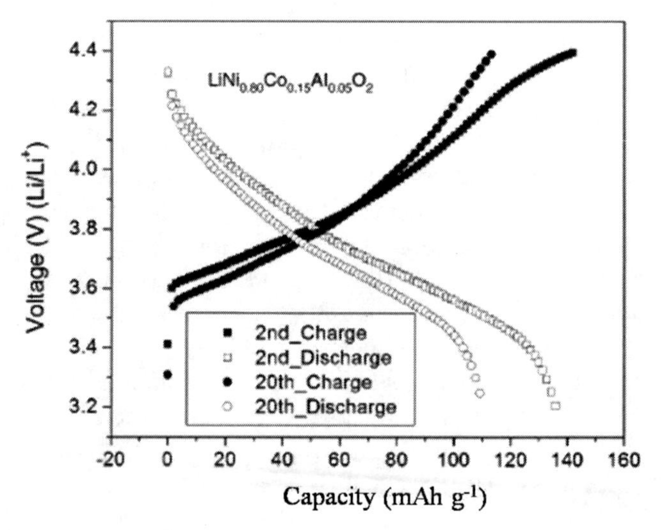

Figure 6.8 Charge discharge profiles of $LiNi_{0.80}Co_{0.15}Al_{0.05}O_2$ composite cathode after 2nd and 20th cycles in the cutoff voltage range of 4.2 to 3.2 V [33].

other reports on the same composition the discharge capacities has been reported to be in between 160–180 mAh/g at C/5 rate depending on the doping content of Al [34]. An improvement in the capacity retention was found when 5 atomic percent of Al was doped along with Co. Figure 6.8 shows the charge–discharge profiles of $LiNi_{0.8}Co_{0.15}Al_{0.05}O_2$ composite cathodes after 2nd and 20th cycles in the cutoff voltage range of 4.2 to 3.2 V. However, the lower capacity was reported because of the presence of the secondary phase, such as Li_2CO_3.

Layered $LiNi_{0.45}Mn_{0.45}Co_{0.1}O_2$ compound with low cobalt contents has been reported and effect of the lithium content in the composition $Li_{1+z}(Ni_{0.45}Mn_{0.45}Co_{0.1})_{1-z}O_2$, $0.8 \leq 1 + z \leq 1.2$ was studied [35]. In order to maintain the Ni and Mn oxidation state at +2 and +4, Co has been doped at Ni and Mn sites in equal quantity. The lithium-deficient $Li_{0.9}[Ni_{0.45}Mn_{0.45}Co_{0.1}]_{1.1}O_2$ compound shows the best electrochemical properties with discharge capacity of 190 mAh/g when cycled between 4.6 and 2.5 V at a current density of 0.5 mA/cm^2. Synthesis temperature of 800°C gives rise to perfect stoichiometry when 5% excess lithium is used. However, if the synthesis temperature is lower some unreacted product is obtained with lithium deficiency. Presence of Ni in the lithium layers is independent of the synthesis temperature; however, lithium content plays an important factor in it. It has been shown that synthesis temperature plays an effective role in the electrochemical performance of the material.

Many researchers have carried out electrochemical studies by optimizing Mn, Ni, and Co contents in the layered structure, $LiMn_yNi_yCo_{1-2y}O_2$ (y = 0.33, 0.4, 0.5) [36]. The best performance was observed in $LiMn_{0.4}Ni_{0.4}Co_{0.2}O_2$. This composition shows the optimum cobalt concentration for stabilization of the structure and electrochemical performance of the material during deep discharge. Synthesis temperature of 800°C is considered to be optimum for the performance. It was shown that above 800°C nickel disorder takes place. Discharge capacity of 150 mAh/g after 70 cycles of charging and discharging at a rate of 0.15 mA/cm^2 between voltage range 4.4 to 2.5 V has been reported. $LiNi_yMn_yCo_{1-2y}O_2$ structure has also been studied by some of other researchers and discharge capacity of 135 mAh/g has been achieved for 1C rate in the voltage range of 4.4 to 2.5 V [37].

The effect of Mn on the layered oxide $LiNi_{0.9-y}Mn_yCo_{0.1}O_2$ ($0.45 \leq y \leq 0.60$) has also been studied [38]. Even though the first discharge capacity of 180 mAh/g has been achieved at a current density of 0.5 mA/cm^2, but in the subsequent cycles capacity fade is found to be severe. Therefore, it has been concluded that manganese-rich compositions are not good for the performance of the material. Researchers have studied the effect of Mn contents on the electrochemical performance of $LiNi_{1-x-y}Co_xMn_yO_2$ by keeping the Co content fixed at 0.25 (x = 0.25, y = 0.1, 0.2, 0.3, and 0.4) [39]. Among various investigated compositions, $LiNi_{0.65}Co_{0.25}Mn_{0.1}O_2$ synthesized at 850°C showed the best characteristics in terms of the initial capacity (198 mAh/g) and the capacity retention (92%) at C-rate of 0.1C in the voltage range of 4.5 to 3.0 V. For $y \geq 0.3$ the material showed high capacity fade revealing that higher Mn content is not good for the structural stability and hence electrochemical performance. Therefore, $LiNi_{0.65}Co_{0.25}Mn_{0.1}O_2$ composition has been proposed to be an alternative to $LiNiO_2$ as cathode material for lithium-ion battery. Figure 6.9 shows the discharge capacity versus cycle number curves of the $LiNi_{0.75-y}Co_{0.25}Mn_yO_2$ for various contents of Co and Mn.

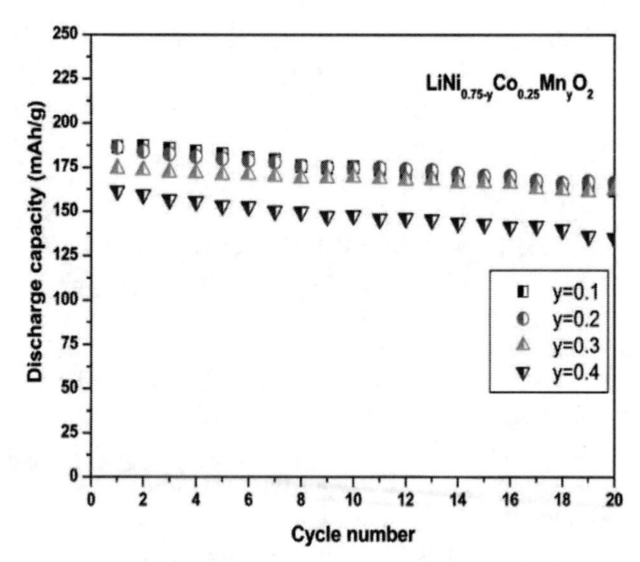

Figure 6.9 Discharge capacity vs. cycle number curves of the $LiNi_{0.75-y}Co_{0.25}Mn_yO_2$ (y = 0.1, 0.2, 0.3, 0.4) materials sintered at 900°C for 12 h [39].

First-principles calculations were also carried out in order to optimize the contents of Ni, Mn and Co in the layered structure, $LiNi_{1-x-y}Co_xMn_yO_2$, for the best electrochemical performance utilizing VASP under LDA approximation scheme [40]. The formation energy at half of the lithium content, i.e., $Li_{0.5}Ni_{0.66}Co_{0.17}Mn_{0.17}O_2$ is most negative among the various compositions under study. Among the fully lithiated compounds under study $LiNi_{0.8}Co_{0.1}Mn_{0.1}O_2$ shows the most negative formation energy. In another report after optimizing the content of Ni, Co and Mn, the synthesis of $LiNi_{0.8}Co_{0.1}Mn_{0.1}O_2$ using the solid-state route has been made [41]. Electrochemical performance of the material was checked in the voltage range of 4.5 to 3.0 V. Discharge capacity of 132 mAh/g was achieved with capacity retention of ~86% after 20 charge–discharge cycles. Some of the research group has focused on the cobalt-free layered compositions. Cr has been used as replacement for Co in layered cathode materials. $Li[Mn_{0.5-x}Cr_{2x}Ni_{0.5-x}]O_2$ ($0 < 2x < 0.2$) as possible cathode material for lithium-ion battery has also been tried [42]. Maximum discharge capacity was observed for $2x = 0.05$ in the voltage range of 4.3 to 3.0 V. Lowest capacity was observed for $2x = 0.2$. $LiMn_{0.5-x}Cr_{2x}Ni_{0.5-x}O_2$ ($0 \leq 2x \leq 0.2$) in the 4.3 to 3.0 V and 4.8 to 3.0 V regions has been studied. The composition with $2x = 0.05$ shows the highest discharge capacity in the 4.3 to 3.0 V range. When the material cycled between 4.8 and 3.0 V, discharge capacity decreases with the increase in chromium content. It has been shown that the "lost" capacity for the cells cycled between 4.3 and 3.0 V could be recovered by discharging the cells to voltages below 1.5 V.

Even though the efforts to reduce the cobalt content in the final composition without compromising on the performance is under study and is focus of various research group, cobalt-free composition, i.e., $LiNi_{0.5}Mn_{0.5}O_2$ has also attained much attention by the scientific community. Synthesis of $LiNi_{0.5}Mn_{0.5}O_2$ by mixed hydroxide method has been reported [43]. Synthesized materials showed capacity of ~160 mAh/g when cycled between the voltage range of 4.3 to 2.5 V. Ab initio methods have been used to identify the mechanisms of the lithium ion hoping in the structure and concluded that the low valence transition metal ions and low strain in the activated state are the crucial parameters in enhancing the rate capability of these materials [44].

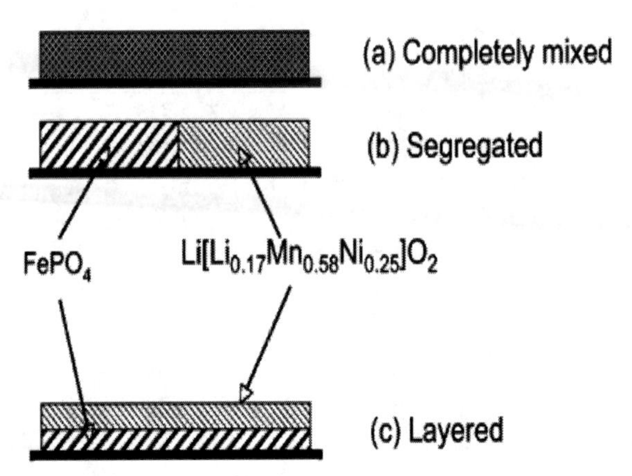

Figure 6.10 Schematic of the three different electrode configurations used in this study: (a) completely intermixed, (b) completely segregated (but on the same current collector foil), and (c) layered, with the high rate $LiFePO_4$ layer in contact with the current collector [45].

Composite cathodes of different layered materials such as $LiCoO_2$ and $Li[Li_{0.17}Mn_{0.58}Ni_{0.25}]O_2$ along with other well-known cathode material such as $LiFePO_4$ has been tried by considering a different design for the final cathode preparation. This approach consisted of three different designs for the cathodes, namely completely mixed, segregated, and layered. The various designs are shown in Fig. 6.10. Performance of the cathodes designed in this way is found to be highly dependent on the rate at which material is charged/discharged. A good electrochemical performance has been achieved when $LiCoO_2$ was used with $LiFePO_4$ [45]. Merging the Li_2MnO_3, which is known to be electrochemically inactive material, in the layered $LiNi_{0.5}Mn_{0.5}O_2$ has explored a new direction to the materials advancement for lithium-ion batteries. It has been shown that Li_2MnO_3 can be made electrochemically active by either treating it with acids such as H_2SO_4 and HNO_3 or charging it above 4.5 V [4]. It has been shown that the removal of Li_2O from the host structure leaves the MnO_2 host electrochemically active, which in turn combines with the other layered structure to give rise to enhanced capacity of the final composite material. In some of the initial reports in this direction, it has been reported that Li_2MnO_3 can be rewritten as $Li[Li_{1/3}Mn_{2/3}]O_2$, and it has been shown that the substitution of some of the lithium and manganese sites with Ni or Cr can enhance the discharge capacity to >200 mAh/g when cycled in the higher voltage range such as 4.8 to 2.5 V [46]. Later it has been reported that such compositions can be rewritten as combination of the Li_2MnO_3 and $LiNi_{0.5}Mn_{0.5}O_2$ [47]. Various compositions can be synthesized using different contents of the two components. The capacity of the finally designed composition is highly dependent on the synthesis route followed. A group of researchers from ANL (Argonne National Lab) have designed the material so that high tap density can be obtained in the finally obtained compositions. The tap density is a crucial parameter in order to get the higher capacity of the layered materials. Hence, it is quite clear that in such kind of composite cathode materials the synthesis parameters are equally crucial to the compositions of the final material.

6.4 Carbon-Coated $LiFePO_4$

Since the report on $LiFePO_4$ by Goodenough and coworkers, this material has been the one of the main stream for research in cathode materials apart from the layered and spinel-based compositions for lithium-ion batteries. Thousands of reports have been published in the scientific journals. This material meets the basic requirement for the EV and HEV. Presence of environment-friendly and cheaper Fe in the structure makes it safer for the environment. First report on this material showed the discharge capacity ~110 mAh/g with flat discharge plateau at ~3.5 V [7]. This material suffered from low electronic conductivity of ~10^{-9} S/cm. On this view point an article highlighted in the *Nature Materials*, in which huge improvement in the electronic conductivity of the material by doping it with the ions such as Mg^{2+}, Al^{3+}, Ti^{4+}, Zr^{4+}, Nb^{5+}, and W^{6+} on lithium sites and combining it with the cation nonstoichiometry has been reported [48]. Room-temperature conductivity was observed to be in excess of 10^{-3} S/cm. High coulombic efficiency along with much improved discharge capacity has been observed at high rates of charging and discharging. Another way to improve the electronic conductivity and electrochemical performance of the material is coating this material with carbon. $LiFePO_4$ has high thermal stability and it makes the material safer to use in the high temperature regions. It also shows superior performance when used at high temperatures.

Much improved performance of iron-based cathode material for lithium-ion battery has been shown by addition of electronically conductive substance [7]. However, the electronically conductive substance is not highlighted in that study. It has been reported that electronically conductive substance enhances the kinetics of redox reactions. Discharge capacity improved to 160–170 mAh/g at 1C rate at 80°C, which is the theoretical capacity of the material. Therefore, it has been shown that it is possible to extract all lithium from the structure when an electronically conductive substance is added. Later it has been shown that composite of $LiFePO_4$ with carbon enhances the electrochemical performance of the material [49]. A 10 wt% addition of carbon enhances the capacity of $LiFePO_4$ to 170 mAh/g with a cycling rate of C/10 at 80°C. The cell showed a specific capacity ~95 mAh/g up to 230 cycles of charging/discharging, when cycled at room temperature at C/2 rate. Synthesis conditions for the $LiFePO_4$ have been optimized [50]. Sintering temperature $500 < T < 600°C$ is considered the ideal synthesis temperature. Material has shown >95% of theoretical capacity for charging/discharging rate of 0.12 mA/cm^2 at room temperature when synthesized under these sintering conditions. It has been reported that volumetric and gravimetric energy densities decrease with the increase in carbon content in $LiFePO_4$/C composite [51]. Therefore, carbon content in the composite plays a crucial role in the overall performance of the material. It has been shown that addition of 3.5 wt% of carbon in the form of sugar at the initial steps of synthesis of $LiFePO_4$/C gives good rate capability of the cathode material. Carbon contents as low as 1 wt% causes a significant decrease in tap density. Coating of $LiFePO_4$ particles with carbon using vapor deposition technique has also been reported [52]. Decomposing propylene gas was considered to be carbon source in the coating. Using this technique, carbon can penetrate in the pores of the particles; 3.4 wt% of carbon content have shown good cyclic performance for 70 cycles at a cycling rate of C/3 in the voltage range of 2.0 to 4.3 V. Discharge capacity of ~150 mAh/g have been reported at 37°C of temperature. In continuation to the same work, the performance of the vapor deposited carbon on $LiFePO_4$ particles against MCMB as counter electrode instead of Li has been reported [53]. It has been shown that carbon-coated particles show sever capacity fade at higher temperature when used with MCMB as counter electrode. Carbon-coated $LiFePO_4$

electrodes were found to release iron ions into the electrolyte when aged at 37 and 55°C. It has been reported that the interfacial films that were produced on the graphite electrodes as a result of possible catalytic effects of the metallic iron particles give rise to the capacity fade. However, when $LiPF_6$ was replaced with lithium *bis*-oxalatoborate $LiB(C_2O_4)_2$ salt MCMB/C-$LiFePO_4$ showed improved performance. The effect of temperature on carbon-coated $LiFePO_4$ has also been studied [54]. Discharge capacities of 160, 158, and 138 mAh/g have been for 60, 25, and –10°C, respectively, for low discharge rates. However, at 2C rate, the discharged capacities were 153.4, 136.7 and 93.5 mAh/g, respectively, for 60, 25, and –10°C. In the recent past, the structural and electrochemical performance of pristine $LiFePO_4$ and C–$LiFePO_4$ composite has been studied in detail [55]. Carbon was coated on $LiFePO_4$ particles and was found to be in the amorphous state as shown by TEM and Raman studies. It has been shown that in case of C–$LiFePO_4$ composite, the Li-ion diffusion coefficient was ~7.13×10^{-14} cm^2s^{-1}, whereas in the case pure $LiFePO_4$, it was merely ~1.28×10^{-15} cm^2s^{-1}. Excellent cycleability (~97% retention after 50 cycles) was attained for C-$LiFePO_4$ compared to pure $LiFePO_4$ (only 69% retention after 25 cycles). It has been shown that rate performance of the material can be improved by lowering the size of the synthesized particles and avoiding the agglomeration.

Breakthrough in $LiFePO_4$ has been achieved in 2009 with a report on ultrafast charging of the material by MIT researchers, [8]. By controlling the surface chemistry and hence changing the stoichiometry of the surface of $LiFePO_4$ particles along with high bulk lithium mobility very high charge and discharge rate of the material can be achieved. Discharge capacity of ~65 mAh/g after 100 cycles was achieved at a rate of 397C. Rate capability of $LiFe_{0.9}P_{0.95}O_{4-\delta}$ synthesized at 600°C under various conditions is shown in Fig. 6.11. This material has shown very promising results for the use in lithium-ion batteries in the upcoming electrical vehicle technology. As mentioned earlier, this study gained much attention from the scientific community and was criticized by Goodenough and coworker on the technical grounds in a letter to the editor of journal of power sources. However none the less to mention that $LiFePO_4$ opened the gateways to look upon the olivine structures as possible hosts for lithium ions for their use as electrode materials in lithium ion batteries.

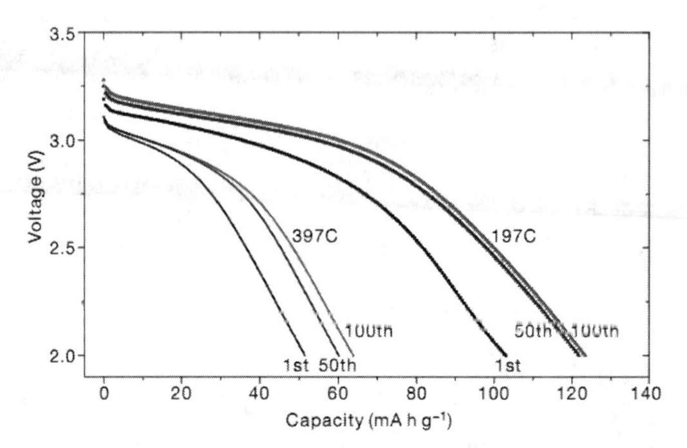

Figure 6.11 Discharge capability at very high rate for $LiFe_{0.9}P_{0.95}O_{4-\delta}$ synthesized at 600°C. Full charge-discharge cycles at constant 197C and 397C current rates without holding the voltage. The loading density of the electrode is 2.96 mg cm^{-2}. The 1st, 50th, and 100th discharges are shown for each rate. The electrode formulation is active material (30 wt%), carbon black (65 wt%), and binder (5 wt%) [8].

6.5 Conclusions

Research efforts are, however, far from concluded in order to improve the lithium-ion battery technology, which holds the promise for the future generation of electric vehicles and other portable devices. Much advancement in the research related to cathode materials for lithium-ion batteries toward their required demand for society has been achieved. However, goals such as implementation of this technology in the vehicular application are still under vigorous research. Three different structural chemistries namely layered, spinel and olivine have been tried successfully. Further advancement in the materials related to materials based on these structures is underway. Among the spinels, $LiNi_{0.5}Mn_{1.5}O_4$-based compositions are one of the promising candidates for the future applications. This is because of the high energy density of the materials based on this composition. Related to the olivine compositions LFP-NCO (LFP: Lithium Iron Phosphate and NCO: Nano Co-crystalline Olivine) are the main candidates for their application Materials based on this chemistry are known to be

Table 6.1 Summary of battery technologies

Name	Description	Electrodes: positive or negative	Companies	Automotive status	Power	Energy	Safety	Life	Cost
LCO	Lithium cobalt oxide	$LiCoO_2$ (Graphite)	Various consumer applications (Not automotive)	Limited Auto applications (due to safety)	Good	Good	Low, Mod	Low	Poor
NCA	Lithium Nickel, cobalt and Aluminum	$Li(Ni_{0.85}Co_{0.1}Al_{0.05})O_2$ (Graphite)	JCI_Saft	Pilot	Good	Good	Mod	Good	Mod
			GAIA						
			Matsuhita						
			Toyota						
LFP	Lithium iron phosphate	$LiFePO_4$ (Graphite)	A123	Pilot	Good	Mod	Mod	Good	Mod, Good
			Valence						
			GAIA						

(*Continued*)

Table 5.1 (*Continued*)

Name	Description	Electrodes: positive or negative	Companies	Automotive status	Power	Energy	Safety	Life	Cost
NCM	Lithium nickel, cobalt and manganese	$Li(Ni_{1/3}Mn_{1/3}Co_{1/3})O_2$ (Graphite)	Litcel (Mitsubishi)	Pilot	Mod	Mod, Good	Mod	Poor	Mod
			Kokam						
			NEC Lamillion						
LMS	Lithium manganese spinel	$LiMnO_2$ or $LiMn_2O_4$ ($Li_4Ti_5O_{12}$)	GS Yuasa	Devel	Mod	Poor	Excel, Good	Excel, Mod	Mod
			Litcel (Mitsubishi)						
			NEC Lamillion						
			EnerDel						
LTO	Lithium titanium	$LiMnO_2(LiTiO_2)$	Altairnano	Devel	Poor, Mod	Poor	Good	Good	Poor
			EnerDel						
MNS	Manganese titanium	$LiMn_{1.5}Ni_{0.5}O_4$ ($Li_4Ti_5O_{12}$)		Research	Good	Mod	Excel	Unknown	Mod
MN	Manganese titanium	$Li_{1.2}Mn_{0.6}Ni_{0.2}O_2$		Research	Excel	Excel	Excel	Unknown	Mod

much safe and to the vehicles based on the lithium-ion battery technology.low cost. A123, Valence, and GAIA are using LFP-NCO–based materials as active cathode materials for lithium-ion batteries [56]. Apart from all the above-mentioned materials, $LiCoO_2$ still remain the state of art material and is still in use related to the portable devices. In order to reduce the cobalt content without compromising the performance, Ni, Mn and Al doped compositions have been the prime focus in this category of materials. $LiNi_{1/3}Mn_{1/3}Ni_{1/3}O_2$ or $LiNi_{0.8}Co_{0.15}Al_{0.05}O_2$ (NCA) is the set examples in this concern, which have shown the promise for future generation of materials for lithium-ion battery. NCA batteries perform quite well in terms of power density, energy density, and longevity. However, this technology faces limitations in safety and cost. Researchers have focused further on the advancement of the materials from specific energy and power point of view. Composite cathode materials based on the layered crystal chemistry are under vigorous research at various research laboratories. Table 6.1 summarizes the various battery technologies under study and already in market. Stability of the electrolyte against these high voltage materials and at high temperatures also remains a challenge. Improvement in the performance of cathode materials and their application in lithium-ion batteries are in the near horizon.

Acknowledgement

The financial support from NASA-EPSCoR (NNX08AB12A) and NASA-URC (NNX08BA48A) grants is gratefully acknowledged.

References

1. Mizushima K., Jones P. C., Wiseman P. J., and Goodenough J. B., Li_xCoO_2 ($0 < x \leq 1$): A new cathode material for batteries of high energy density, *Mater. Res. Bull.*, **15** (1980) 783–789.

2. Thackeray M. M., Johnson P. J., de Picciotto L. A., Bruce P. G., and Goodenough J. B., Electrochemical extraction of lithium from $LiMn_2O_4$, *Mater. Res. Bull.*, **19** (1984) 179–187.

3. Xiaohua Ma, Byoungwoo Kang, and Gerbrand Ceder, High rate micron-sized ordered $LiNi_{0.5}Mn_{1.5}O_4$, *J Electrochem. Soc.*, **157** (2010) A925–A931.

4. Rossouw M. H., and Thackeray M. M., Lithium manganese oxides from Li_2MnO_3 for rechargeable lithium battery applications, *Mater. Res. Bull.*, **26** (1991) 463–473.
5. Ohzuku T., and Makimura Y., Layered lithium insertion material of $LiCo_{1/3}Ni_{1/3}Mn_{1/3}O_2$ for lithium-ion batteries, *Chem. Lett.*, (2001) 642–643.
6. Padhi A. K., Nanjundaswamy K. S., and Goodenough J. B., Phospho-olivines as positive-electrode materials for rechargeable lithium batteries, *J. Electrochem. Soc.*, **144** (1997) 1188–1194.
7. Ravet N., Goodenough J. B., Besner S., Simoneau M., Hovington P., and Armand M., Improved iron based cathode material, 196th Meeting of the electrochemical society, Oct 1999, Honolulu Hawai, Abstract No. 127.
8. Kang B., and Ceder G., Battery materials for ultrafast charging and discharging, *Nature*, **458** (2009) 190–193.
9. Das S. R., Istevao R. Fachini, Majumder S. B., and Katiyar R. S., Structural and electrochemical properties of nanocrystalline $Li_xMn_2O_4$ thin film cathodes (x = 1.0–1.4), *J. Power Sources*, **158** (2006) 518–523.
10. Das S. R., Majumder S. B., and Katiyar R. S., Kinetic analysis of the Li^+ ion intercalation behavior of solution derived nano-crystalline lithium manganate thin films, *J. Power Sources*, **139** (2005) 261–268.
11. Zhong Q. M., Bonakclarpour A., Zhang M. J., Gao Y., and Dahn J. R., Synthesis and electrochemistry of $LiNi_xMn_{2-x}O_4$, *J. Electrochem. Soc.*, **144** (1997) 205–213.
12. Lazarraga M. G., Pascual L., Gadjov H., Kovacheva D., Petrov K., Amarilla J. M., Rojas R. M., Martin-Luengo M. A., and Rojo J. M., Nanosize $LiNi_yMn_{2-y}O_4$ ($0 < y \leq 0.5$) spinels synthesized by a sucrose-aided combustion method. Characterization and electrochemical performance, *J. Mater. Chem.*, **14** (2004) 1640–1647.
13. Myung S.-T., Komaba S., Kumagai N., Yashiro H., Chung H.-T., Cho T.-H., Nano-crystalline $LiNi_{0.5}Mn_{1.5}O_4$ synthesized by emulsion drying method, *Electrochim. Acta*, **47** (2002) 2543–2549.
14. Idemoto Y., Narai H., and Koura N., Crystal structure and cathode performance dependence on oxygen content of $LiMn_{1.5}Ni_{0.5}O_4$ as a cathode material for secondary lithium batteries, *J. Power Sources*, **119–121** (2003) 125–129.
15. Kovacheva D., Markovsky B., Salitra G., Talyosef Y., Levi M. G., Elena, R. M., Kim H.-J., and Aurbach D., Electrochemical behavior of electrodes comprising micro- and nano-sized particles of $LiNi_{0.5}Mn_{1.5}O_4$: a comparative study, *Electrochim. Acta*, **50** (2005) 5553–5560.

16. Fang X., Lu Y., Ding N., Feng X. Y., Liu C., and Chen C. H., Electrochemical properties of nano- and micro-sized $LiNi_{0.5}Mn_{1.5}O_4$ synthesized via thermal decomposition of a ternary eutectic Li–Ni–Mn acetate, *Electrochim. Acta*, **55** (2010) 832–837.

17. Park S. B., Eom W. S., Cho W. I.I., and Jang H., Electrochemical properties of $LiNi_{0.5}Mn_{1.5}O_4$ cathode after Cr doping, *J. Power Sources*, **159** (2006) 679–684.

18. Aklalouch M., Rojas R. M., Rojo J. M., Saadoune I., and Amarilla J. M., The role of particle size on the electrochemical properties at 25 and at 55°C of the $LiCr_{0.2}Ni_{0.4}Mn_{1.4}O_4$ spinel as 5V-cathode materials for lithium-ion batteries, *Electrochim. Acta*, **54** (2009) 7542–7550.

19. Sun Y., Wang Z., Huang X., and Chen L., Synthesis and electrochemical performance of spinel $LiMn_{2-x-y}Ni_xCr_yO_4$ as 5-V cathode materials for lithium ion batteries, *J. Power Sources*, **132** (2004) 161–165.

20. Katiyar R. K., Singhal R., Asmar K., Valentin R., and Ram S. K., High voltage spinel cathode materials for high energy density and high rate capability Li ion rechargeable batteries, *J. Power Sources*, **194** (2009) 526–530.

21. Oh S. W., Myung S.-T., Kang H. B., and Sun Y.-K., Effects of Co doping on $Li[Ni_{0.5}Co_xMn_{1.5-x}]O_4$ spinel materials for 5V lithium secondary batteries via Co-precipitation, *J. Power Sources*, **189** (2009) 752–756.

22. Singhal R., Tomar M. S., Das S. R., Burgos J. G., Singh S. P., Kumar A., and Katiyar R. S., Li-I., Rechargeable battery with $LiMn_{1.5}Ni_{0.46}Rh_{0.04}O_4$ spinel cathode material, *Electrochem. Solid State Lett.*, **10** (2007) A163–A165.

23. Singhal R., Das S. R., Tomar M. S., Ovideo O., Nieto S., Melgarejo R. E., and Katiyar R. S., Synthesis and characterization of Nd doped $LiMn_2O_4$ cathode for Li-ion rechargeable batteries, *J. Power Sources*, **164** (2007) 857–861.

24. Sun Y.-K., Yoon C. S., and Oh I.-H., Surface structural change of ZnO-coated $LiNi_{0.5}Mn_{1.5}O_4$ spinel as 5 V cathode materials at elevated temperatures, *Electrochim. Acta*, **48** (2003) 503–506.

25. Sun Y.-K., Hong K.-J., Prakash Jai, and Amine K., Electrochemical performance of nano-sized ZnO-coated $LiNi_{0.5}Mn_{1.5}O_4$ spinel as 5 V materials at elevated temperatures, *Electrochem. Commun.*, **4** (2002) 344–348.

26. Singhal R., Tomar M. S., Burgos J. G., and Katiyar R. S., Electrochemical performance of ZnO-coated $LiMn_{1.5}Ni_{0.5}O_4$ cathode material, *J. Power Sources*, **183** (2008) 334–338.

27. Kobayashi Y., Miyashiro H., Takei K., Shigemura H., Tabuchi M., Kageyama H., and Iwahori T., 5 V class all-solid-state composite lithium battery with Li_3PO_4 coated $LiNi_{0.5}Mn_{1.5}O_4$, *J. Electrochem. Soc.*, **150** (2003) A1577–A1582.

28. West W. C., Myung N. V., Whitacre J. F., and Ratnakumar B. V., Electrodeposited amorphous manganese oxide nanowire arrays for high energy and power density electrodes, *J. Power Sources*, **126** (2004) 203–206.

29. Singh S. P., Tomar M. S., and Ishikawa Y., Experimental and theoretical studies of $LiNi_{1/3}Mn_{1/3}M_{1/3}O_2$ [M = Mo and Rh] for cathode material, *Microelectron. J.*, **36** (2005) 491–494.

30. Karan K., Saavedra-Arias J. J., Pradhan D. K., Melgarejo R., Kumar A., Thomas R., and Katiyar R. S., Structural and electrochemical characterizations of solution derived $LiMn_{0.5}Ni_{0.5}O_2$ as positive electrode for Li-ion rechargeable batteries, *Electrochem. Solid State Lett.*, **11** (2008) A135–A139.

31. Karan N. K., Abraham D. P., Balasubramanian M., Furczon M. M., Thomas R., and Katiyar R. S., Morphology, structure, and electrochemistry of solution-derived $LiMn_{0.5-x}Cr_{2x}Ni_{0.5-x}O_2$ for lithium-ion cells, *J. Electrochem. Soc.*, **156** (2009) A553–A562.

32. Smart M. C., Whitacre J. F., Ratnakumar B. V., and Amine K., Electrochemical performance and kinetics of $Li_{1+x}(Co_{1/3}Ni_{1/3}Mn_{1/3})_{1-x}O_2$ cathodes and graphite anodes in low-temperature electrolytes, *J. Power Sources*, **168** (2007) 501–508.

33. Majumder S. B., Nieto S., and Katiyar R. S., Synthesis and electrochemical properties of $LiNi_{0.80}(Co_{0.20-x}Al_x)O_2$ (x = 0.0 and 0.05) cathodes for Li ion rechargeable batteries, *J. Power Sources*, **154** (2006) 262–267.

34. Han C. J., Yoon J. H., Cho W. Il, and Jang H., Electrochemical properties of $LiNi_{0.8}Co_{0.2-x}Al_xO_2$ prepared by a sol–gel method, *J. Power Sources*, **136** (2004) 132–138.

35. Xiao J., Chernova N. A., and Whittingham M. S., Layered mixed transition metal oxide cathodes with reduced cobalt content for lithium ion batteries, *Chem. Mater.*, **20** (2008) 7454–7464.

36. Ma M., Chernova N. A., Zavalij P. Y., and Whittingham M. S., Structural and electrochemical properties of $LiMn_{0.4}Ni_{0.4}Co_{0.2}O_2$, *Mater. Res. Soc. Symp. Proc.*, **835** (2005) K11.3.1–K11.3.6.

37. Kamel K. Ben, Amdouni N., Abdel-Ghany A., Zaghib K., Mauger A., Gendron F., and Julien C. M., Local structure and electrochemistry

of $LiNi_yMn_yCo_{1-2y}O_2$ electrode materials for Li-ion batteries, *Ionics*, **14** (2008) 89–97.

38. Jie J., Chernova N. A., and Stanley W. M., Influence of manganese content on the performance of $LiNi_{0.9-y}Mn_yCo_{0.1}O_2$ $(0.45 \leq y \leq 0.60)$ as a cathode material for Li-ion batteries, *Chem. Mater.*, **22** (2010) 1180–1185.

39. Hwang B. J., Tsai Y. W., Chen C. H., and Santhanam R., Influence of Mn content on the morphology and electrochemical performance of $LiNi_{1-x-y}Co_xMn_yO_2$ cathode materials, *J. Mater. Chem.*, **13** (2003) 1962–1968.

40. Saavedra-Arias J. J., Venkateswara Rao C., Shojan J., Manivannan A., Torres L., Ishikawa Y., and Katiyar R. S., A combined first-principles computational/experimental study on $LiNi_{0.66}Co_{0.17}Mn_{0.17}O_2$ as a potential layered cathode material, *J. Power Sources*, **211** (2012) 12–18.

41. Saavedra-Arias J. J., Karan N. K., Pradhan D. K., Kumar A., Nieto S., Thomas R., and Katiyar R. S., Synthesis and electrochemical properties of $Li(Ni_{0.8}Co_{0.1}Mn_{0.1})O_2$ cathode material: Ex situ structural analysis by Raman scattering and X-ray diffraction at various stages of charge–discharge process, *J. Power Sources*, **183** (2008) 761–765.

42. Karan N. K., Balasubramanian M., Abraham D. P., Furczon M. M., Pradhan D. K., Saavedra-Arias J. J., Thomas R., and Katiyar R. S., Structural characteristics and electrochemical performance of layered $Li[Mn_{0.5-x}Cr_{2x}Ni_{0.5-x}]O_2$ cathode materials, *J. Power Sources*, **187** (2009) 586–590.

43. Ohzuku T., and Makimura Y., Layered lithium insertion material of $LiNi_{0.5}Mn_{0.5}O_2$: a possible alternative to $LiCoO_2$ for advanced Lithium Ion Batteries, *Chem. Lett.*, (2001) 744–745.

44. Kang K., Meng Y. S., Breger J., Grey C. P., and Ceder G., Electrodes with high power and high capacity for rechargeable batteries, *Science*, **311** (2006) 977–980.

45. Whitacre J. F., Zaghib K., West W. C., and Ratnakumar B. V., Dual active material composite cathode structures for Li-ion batteries, *J. Power Sources*, **177** (2008) 528–536.

46. Lu Z., Chen Z., and Dahn J. R., Lack of cation clustering in $Li[Ni_xLi_{1/3-2x/3}Mn_{2/3-x/3}]O_2$ $(0 < x \leq 0.5)$ and $Li[Cr_xLi_{(1-x)/3}Mn_{(2-2x)/3}]O_2$ $(0 < x < 1)$, *Chem. Mater.*, **15** (2003) 3214–3220.

47. Johnson C. S., Kim J.-S., C. Lefief, N. Li, Vaughey J. T., and Thackeray M. M., The significance of the Li_2MnO_3 component in "composite"

$_x$$Li_2MnO_3 \cdot (1-x)LiMn_{0.5}Ni_{0.5}O_2$ electrodes, *Electrochem. Commun.*, **6** (2004) 1085–1091.

48. Chung S.-Y., Jason T. B., and Chiang Y.-M., Electronically conductive phospho-olivines as lithium storage electrodes, *Nat. Mater.*, **1** (2002) 123–128.

49. Prosini P. P., Zane D., and Pasquali M., Improved electrochemical performance of a $LiFePO_4$-based composite cathode, *Electrochim. Acta*, **46** (2001) 3517–3523.

50. Yamada A., Chung S. C., and Hinokuma K., Optimized $LiFePO_4$ for lithium battery cathodes, *J. Electrochem. Soc.*, **148** (2001) A224–A229.

51. Chen Z. and Dahn J. R., Reducing carbon in $LiFePO_4$/C composite electrodes to maximize specific energy, volumetric energy, and tap density, *J. Electrochem. Soc.*, **149** (2002) A1184–A1189.

52. Belharouak I., Johnson C., and Amine K., Synthesis and electrochemical analysis of vapor-deposited carbon-coated $LiFePO_4$, *Electrochem. Commun.*, **7** (2005) 983–988.

53. Amine K., Liu J., and Belharouak I., High-temperature storage and cycling of C-$LiFePO_4$/graphite Li-ion cells, *Electrochem. Commun.*, **7** (2005) 669–673.

54. Guerfi A., Ravet N., Charest P., Dontigny M., Petitclerc M., and Zaghib K., Temperature effect on $LiFePO_4$ cathode performance, 210th ECS Meeting, Abstract No. 171.

55. Kumar A., Thomas R., Karan N. K., Saavedra-Arias J. J., Singh M. K., Majumder S. B., Tomar M. S., and Katiyar R. S., Structural and electrochemical characterization of pure $LiFePO_4$ and nanocomposite C-$LiFePO_4$ cathodes for lithium ion rechargeable batteries, *J. Nanotech.*, 2009, Article ID 176517.

56. Axsen, J., Burke, A., Kurani, K., 2008. Batteries for plug-in hybrid electric Vehicles (PHEVs): goals and the state of technology circa 2008. Research Report UCD-ITS-RR-08-14, Institute of Transportation Studies, University of California, Davis.

Chapter 7

Nanomaterials in Regenerative Fuel Cells

Thomas I. Valdez,[a] Ileana González-González,[b] and Carlos R. Cabrera[b]

[a]*NASA Jet Propulsion Laboratory,*
California Institute of Technology, Pasadena, CA 91101, USA
[b]*NASA-URC Center for Advanced Nanoscale Materials, University of Puerto Rico,*
Rio Piedras Campus, San Juan, PR 00931, USA

thomas.i.valdez@jpl.nasa.gov

7.1 Introduction

The National Aeronautics and Space Administration (NASA), as part of its Exploration Technology Development Program (ETDP), has led a recent thrust in the development of regenerative fuel cell systems [1–2]. The goal of this program is to develop both primary and regenerative fuel cell systems to support the energy storage requirements for human and robotic space exploration missions. The primary and regenerative fuel cell systems are based on proton exchange membrane (PEM) technology, replacing the alkaline fuel cell technology presently used for space missions. The research focus for primary fuel cell systems is the development of a 5.5 kW peak power fuel cell to provide power to a Lunar Lander during descent. The research focus for the regenerative fuel cell systems is to improve both efficiency and durability of fuel cells and

Advanced Nanomaterials for Aerospace Applications
Edited by Carlos R. Cabrera and Félix A. Miranda
Copyright © 2014 Pan Stanford Publishing Pte. Ltd.
ISBN 978-981-4463-18-8 (Hardcover), 978-981-4463-19-5 (eBook)
www.panstanford.com

electrolyzers; an added goal for electrolyzer system development is operation at pressures of approximately 2500 PSIG. Although the power level for the landed lunar system has not been determined, the goal based on the current mission architecture is to develop a regenerative fuel cell module that can deliver 2 kW of power.

To meet the goals for the development of an advanced space-rated regenerative fuel cell system, materials must be developed that improve the efficiency and durability of membrane electrode assemblies (MEAs) for both fuel cells and electrolyzers. This chapter discusses the development of nanomaterials used as catalysts for fuel cells and electrolyzers in regenerative fuel cell systems for NASA applications.

7.2 The Regenerative Fuel Cell System

There are four main components in a regenerative fuel cell system: the power generation unit, the reactant generation unit, reactant storage, and fuel storage. The fuel in the case of a hydrogen/oxygen regenerative fuel cell is water. A functional schematic of a regenerative fuel cell system is shown in Fig. 7.1. In this system, fuel cell reactants are produced at the electrolyzer. The electrolyzer requires power to produce the reactants. For a space-based system, the power would be supplied via a solar array or nuclear source. For terrestrial systems, electrolyzer power is typically supplied via solar arrays but can be powered from any energy source such as wind, tidal or a city power grid. As the fuel cell reactants are produced, they are typically stored in pressurized cylinders or, in the case of hydrogen storage; hydrogen can be stored in hydride beds. The reactants can then be fed to a fuel cell, and the fuel cell produces power from the stored reactants. The byproduct from the fuel cell reaction is water; the water is stored and serves to fuel the electrolyzer.

The state when a regenerative fuel cell system is producing reactants can be referred to as the charge cycle. The electrolyzer subsystem in the regenerative fuel cell is active during the charge cycle. In an electrolyzer, water is fed into the anode, cathode, or both electrodes during electrolysis. An electrolyzer system is named by the method of water feed, and can be referred to as anode feed, cathode feed, or flooded, for a system that has water feed to the anode, cathode, or both electrodes, respectively. During the water electrolysis reaction, oxygen is produced at the anode and hydrogen is produced at the cathode of the electrolyzer ("Charge Cycle," Fig. 7.1). One advantage of operating an electrolyzer in anode feed mode is that high current densities are possible (>2 A/cm^2). An advantage of operating an electrolyzer in cathode feed mode is that the water feed can be manipulated such that the product gasses contain lower water content. This can minimize both the electrolyzer system and reactant storage complexity. It has been shown that, at elevated pressures (~2000 PSIG); an electrolyzer operating in a flooded configuration can operate at a higher efficiency [1].

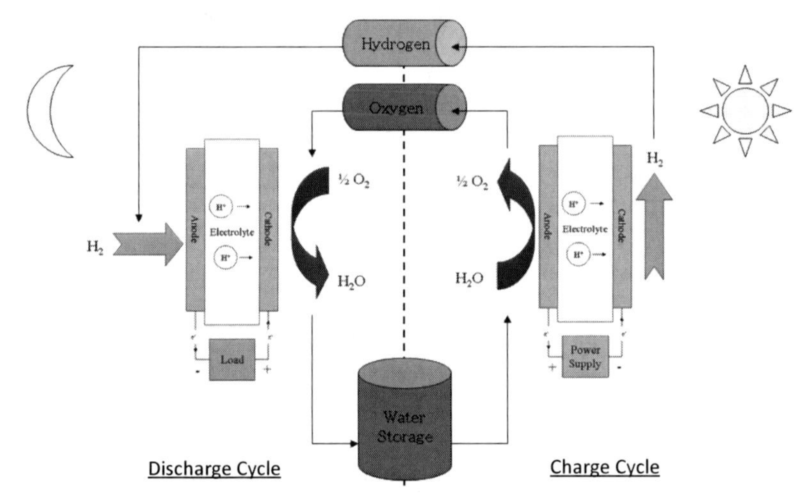

Figure 7.1 Functional schematic of a regenerative fuel cell system.

When the regenerative fuel cell system is producing power, it is operating in the discharge cycle. During this discharge cycle, the fuel cell subsystem is active. In a fuel cell, hydrogen is fed to the anode and oxygen is fed to the cathode, and as power is drawn from the fuel cell, hydrogen is oxidized and oxygen is reduced to water ("Discharge Cycle," Fig. 7.1). The byproduct of a hydrogen/oxygen fuel cell is water, which in this case is stored for later electrolysis and regeneration back to these gases. Hydrogen/oxygen fuel cells can operate in either a flow-through or non-flow-through mode [2]. Non-flow-through fuel cells are also referred to as "dead-ended," which means that the reactants are not allowed to sweep through the fuel cell. In a terrestrial fuel cell system, the hydrogen feed (anode compartment) is typically operated dead-ended. The oxidant feed (cathode compartment), which in a terrestrial system typically operates on air, is allowed to flow through the cell. When the oxidant is allowed to flow through the cathode, it will sweep product water out of the cathode compartment. A flow-through fuel cell system would require components such as gas-liquid separators and condensers for water handling, which can increase system complexity [2]. Non-flow-through fuel cell systems are being developed for space use; these systems use passive components to remove product water from the fuel cell reactions [2]. Some advantages of using non-flow-through fuel cells include: passive balance of plant components, operation at stoichiometric feed rates of reactants, and minimal system complexity [2].

Several options for the development of regenerative fuel cell systems have been considered, including unitized and discrete systems. In a discrete regenerative fuel cell system, the fuel cell and electrolyzer stacks (and their associated MEAs) are separate units, each performing a single function (power generation or reactant production). The fuel cell stack is the power-producing unit and the electrolysis stack is the reactant-producing unit of a regenerative fuel cell system. In a unitized regenerative fuel cell system, the fuel cell and electrolysis stacks are combined into one unit and as such must serve both functions of producing power and regeneration of reactants. Unitized regenerative fuel cells require MEAs that can sustain both the fuel cell and electrolysis reactions. An advantage of a unitized regenerative fuel cell system is that it can be made very compact. An advantage of a discrete regenerative fuel cell system is that the individual fuel cell stack and the electrolyzer stack can each be optimized to maximize system efficiency. Discrete regenerative fuel cell systems are being pursued by NASA [3].

7.3 Space-Rated Regenerative Fuel Cell Systems

The round-trip (charge/discharge cycle) efficiency goal for a regenerative fuel cell system developed by the NASA Energy Storage Program is 54% with a success threshold of 43% [3]. The regenerative fuel cell systems efficiencies include 5% and 10% allocations for parasitic inefficiencies for the program goal and threshold values, respectively. To meet the program threshold values, the fuel cell and electrolyzer MEAs would need to operate at 0.9 and 1.46 V, respectively, at an operating current of 200 mA/cm^2.

To meet the program goal values, the MEAs would need to operate at 0.92 and 1.44 V at an operating current of 200 mA/cm^2 for the fuel cell and electrolyzer, respectively. For the fuel cell MEA, this represents an improvement of 40–60 mV in cell voltage or a 4–7% improvement in voltage efficiency over state-of-the-art, respectively. In commercial low-cost fuel cell systems, hydrogen/oxygen fuel cell MEAs have been shown to operate at 0.79 V at 200 mA/cm^2 [4]. Similarly, for the electrolyzer MEA, this represents a 90 to 110 mV improvement in cell voltage or a 6–8% improvement in voltage efficiency over state-of-the-art, respectively. Finally, a space-rated regenerative fuel cell system is required to operate maintenance-free for duration of up to 10,000 h [3].

7.4 Fuel Cell and Electrolyzer Reactions

In order to improve the operating efficiencies of fuel cells and electrolyzers, one must understand the reactions occurring on the catalyst surfaces. The following sections are a brief overview of the hydrogen oxidation reaction (HOR), oxygen reduction reaction (ORR), the oxygen evolution reaction (OER), and the hydrogen evolution reaction (HER).

7.5 Fuel Cell, Hydrogen Oxidation Reaction

When power is being drawn from a fuel cell, hydrogen molecules introduced into the fuel cell anode compartment dissociate into protons and electrons at the catalysts surface. The electrochemical reaction for the HOR is given as Eq. 7.1.

$$H_2 \rightarrow 2H^+ + 2e^- \quad E^0_{\text{anodic}} = 0 \text{ V vs. SHE,} \tag{7.1}$$

where E^0_{anodic} is the reversible electrochemical potential for the anodic fuel cell reaction and SHE is the standard hydrogen electrode potential. Pathways to enhance the HOR are similar for both terrestrial and space-rated fuel cell systems, with the exception that in space-rated fuel cell systems, durability would supersede performance enhancements. Catalysts supports such as carbon and porphyrins can increase the electrochemical surface area of platinum and thus enhance the HOR. To address durability concerns, platinum black is the anode catalyst of choice for space-rated fuel cell systems. Because the HOR is fast compared to the ORR, the focus to improving fuel cell efficiency is to enhance the ORR at the fuel cell cathode.

7.6 Fuel Cell, Oxygen Reduction Reaction

When oxygen enters the cathode compartment of an operating fuel cell, the oxygen molecules dissociate and bond to the catalyst surface. If the catalyst is platinum, then a platinum–oxygen bond is formed, enabling the oxygen reduction to proceed. Each oxygen atom then leaves the platinum catalyst site, combining with two electrons (which have travelled through the external circuit) and two protons (which have travelled through the membrane) to form one molecule of water ("Discharge Cycle," Fig. 7.1). The cathodic fuel cell reaction is given as Eq. 7.2.

$$\frac{1}{2}O_2 + 2H^+ + 2e^- \rightarrow H_2O \quad E^0_{\text{cathodic}} = 1.23\,\text{V vs. SHE}, \tag{7.2}$$

where E^0_{anodic} is the reversible electrochemical potential for the cathodic fuel cell reaction. The ORR pathway is schematically shown in Fig. 7.2.

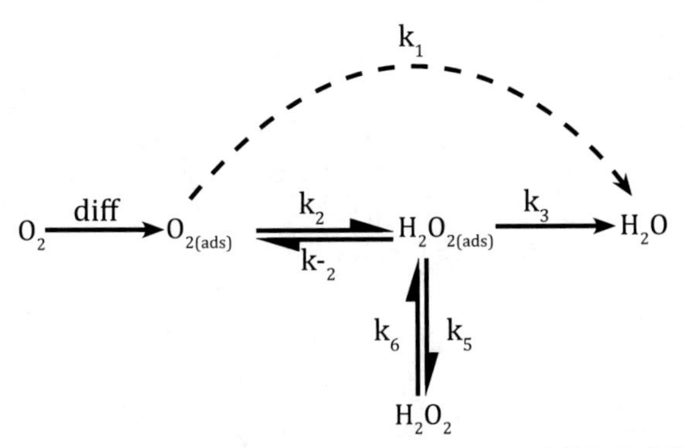

Figure 7.2 The oxygen reduction reaction pathway on a metal surface [5].

The ORR on metallic surfaces is a multiple electron reaction that includes many possible intermediate steps [5]. Oxygen can be electrochemically reduced to water by the reaction with the rate constant k_1, involving a four-electron reduction, or through a series of two electron reductions. Once the oxygen molecule is reduced to peroxide (k_2), it can be desorbed from the metal surface (k_5) or it can be further reduced to water (k_3). An incomplete reduction of oxygen to H_2O_2 in a fuel cell results in inefficient energy conversion and reactive intermediates that can convert to free radicals. These reactive species can then damage the membrane of the cell [6]. The optimization of the most promising catalyst must address both non-catalytic and catalytic factors: maximization of the catalyst surface area while exposing the most active microstructures and modification of the intrinsic activity of the catalyst. The enhancement of catalytic properties by a second component may occur through a change in local bonding geometry (structure effects), the distribution of active sites (ensemble effects) or directly by modification of the reactivity of the catalyst surface atoms (electronic effects) [8–10].

Platinum remains the catalyst of choice, specifically for acid-based PEM fuel cells. The key to finding a promising electrocatalyst for the ORR is to understand the reaction at the platinum-electrolyte interface, which has been fundamentally studied [7]. Grgur and coworkers have presented polarization curves that show that the ORR activity in H_2SO_4 increases in the sequence Pt(111) < Pt(100) < Pt(110) [8], but deactivation issues occur due to bisulfate adsorption to Pt(111) [9]. Density functional calculations have been used to develop a detailed description of the free-energy landscape of the electrochemical oxygen reduction reaction over Pt(111) as a function of applied bias. This allowed the identification of the origin of the overpotential found for this reaction. Adsorbed oxygen and hydroxyl have been found to be very stable intermediates at potentials close to equilibrium, and the calculated rate constants for the activated proton/electron transfer to adsorbed oxygen or hydroxyl can account quantitatively for the observed kinetics [10].

7.7 Electrolysis, Oxygen Evolution Reaction

When water is being feed to an electrolyzer, and the electrolyzer is being powered, the water will be electrochemically oxidized. The anodic reaction for an acid-water electrolysis cell is given as Eq. 7.3.

$$H_2O \rightarrow \frac{1}{2}O_2 + 2H^+ + 2e^- \quad E^0_{\text{anodic}} = -1.23 \text{ V vs. SHE}, \tag{7.3}$$

where E^0_{anodic} is the reversible electrochemical potential for the anodic electrolysis reaction. Similar to the ORR, the OER can occur via various pathways; the following pathways have been proposed:

(i) The electrochemical path [15,16]

$$S + H_2O \leftrightarrow S\text{-}OH_{ads} + H^+ + e^- \tag{ia}$$

$$S\text{-}OH_{ads} \leftrightarrow S\text{-}O + H^+ + e^- \tag{ib}$$

$$2S\text{-}O \rightarrow 2S + O_2(g) \tag{ic}$$

(ii) The Oxide Path [15,16]

$$S + H_2O \leftrightarrow S\text{-}OH_{ads} + H^+ + e^- \tag{iia}$$

$$2S\text{-}OH_{ads} \leftrightarrow S\text{-}O + S + H_2O \tag{iib}$$

$$2S\text{-}O \rightarrow 2S + O_2(g) \tag{iic}$$

(iii) The Krasil'shchikov path [16–18]

$$S + H_2O \leftrightarrow S\text{-}OH_{ads} + H^+ + e^- \tag{iiia}$$

$$S\text{-}OH_{ads} \leftrightarrow S\text{-}O^- + H^+ \tag{iiib}$$

$$S\text{-}O^- \leftrightarrow S\text{-}O + e^- \tag{iiic}$$

$$2S\text{-}O \rightarrow 2S + O_2(g), \tag{iiid}$$

where S, in the above pathways, denotes a catalytically active site. In the first step, in each proposed oxygen evolution pathway, water dissociates into a proton and hydroxide group on a catalytically active site (S-OHads). The second oxygen evolution step occurs by either the oxidation of the hydroxide species ("Electrochemical Path"), the recombination of hydroxide species ("Oxide Path"), or the acid/base dissociation of the hydroxide species ("Krasil'shchikov Path") to form an oxide with the catalytically active site. In the Krasil'shchikov Path, oxide formation on the catalytically active site occurs in two steps, the final step being the oxidation of a charged oxide (S-O-) on the catalytically active site. The final step in each proposed oxygen evolution pathway occurs with the recombination of adsorbed oxygen on the catalytically active site.

7.8 Electrolysis, Hydrogen Evolution Reaction

In electrolysis, the Hydrogen Evolution Reaction (HER) occurs when two electrons (which have travelled through the external circuit) and two protons (which have travelled through the membrane) meet to form a hydrogen molecule at the cathode ("Charge Cycle," Fig. 7.1). The cathodic reaction of an acid-based electrolyzer is given as Eq. 7.4.

$$2H^+ + 2e^- \rightarrow H_2 \quad E^0_{\text{cathodic}} = 0 \text{ V vs. SHE,} \tag{7.4}$$

where E^0_{cathodic} is the reversible electrochemical potential for the cathodic electrolysis reaction. The HER is a fast reaction as compared to the OER, thus the focus to improve the efficiency of a water electrolyzer is to enhance the OER at the electrolyzer anode.

7.9 Fabrication of Nanomaterials for Regenerative Fuel Cells

Two important factors in the fabrication of catalysts for regenerative fuel cells are particle size and alloying effects [11]. The catalyst particle size impacts surface area, and thus electrochemically activity. Alloying can influence a catalysts electronic structure; this can impact reactant and byproduct binding on the catalysts surface and also improve electrochemical activity [12]. Catalysts are typically produced via the reduction of platinum precursor salts. The reduction of precursor salts can include via sol-gel routes, thermal processing, reactive ball milling, and catalyst precipitation. Advanced catalysts can also be fabricated via sputter deposition; these materials typically need to be deposited on a high-surface-area substrate to maximize activity. An outline for the fabrication of catalysts for regenerative fuel cell systems is shown in Fig. 7.3.

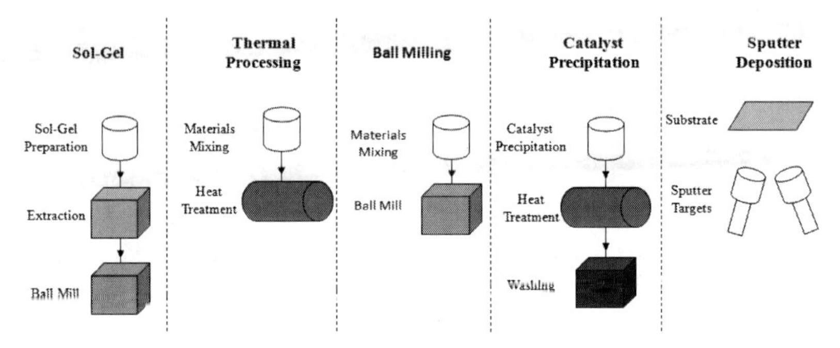

Figure 7.3 Typical fabrication technique for nanomaterials in regenerative fuel cell systems.

For fuel cell systems, catalysts must be tailored for use as either the anode or cathode of the fuel cell. At the anode, an advanced catalyst must be able to enhance the HOR. At the cathode, an advanced catalyst must enhance the oxygen ORR. In a fuel cell, the anode is typically platinum-black and the cathode is typically a platinum alloy. Various studies have shown that alloying platinum with a non-noble metal such as cobalt, nickel, or iron can increase the oxygen reduction reaction at the cathode [21–26].

Similar to fuel cell catalysts, advanced catalysts must be tailored for use as either the anode or the cathode in electrolyzer systems. In an electrolyzer, the oxygen evolution reaction (OER) is slower than the hydrogen evolution reaction (HER). Thus, in an electrolyzer, the OER catalysts are typically alloys, whereas the HER catalysts is typically platinum black. Some OER catalyst alloys include platinum–iridium, iridium–ruthenium, and ternary systems such as iridium–platinum–ruthenium.

7.10 Materials Development for Proton Exchange Membrane-Based Hydrogen/Oxygen Fuel Cells

Regenerative fuel cell systems require fuel cell catalysts that can operate in a pure oxygen environment where supports such as carbon are difficult to implement in this type of environment. To demonstrate the ORR enhancements of alloying platinum, ball-milled platinum–cobalt was selected as the platinum alloy system to study. Three platinum–cobalt alloys were fabricated via high-energy ball milling. An example of the platinum–cobalt alloy is shown in Fig. 7.4. The goal was to determine the effect of starting materials on ORR activity. A sample of commercial platinum-black, platinum powder used for milling and a milled platinum powder were included to compare the various catalysts activities. The platinum and cobalt powders used for milling are 325 mesh (~40 microns). A plot of specific current at 0.9 V vs. NHE as a function of catalysts is shown in Fig. 7.5. As shown in Fig. 7.5, the commercial platinum black could sustain the highest currents of all catalysts tested. The current sustained with the commercial platinum black catalysts is 1.24×10^{-3} A/mg-Pt. The commercial platinum black powder has a particle size of approximately five nanometers and thus a much larger surface area than the other materials tested. The effect of cobalt on platinum activity can best be seen when the ball-milled alloy catalysts is compared to the 325 mesh platinum powder. The current sustained by the sample with the ball-milled platinum powder is 2.1×10^{-4} A/mg-Pt as compared to 5.7×10^{-4} A/mg-Pt for the platinum–cobalt alloy fabricated with the 325 mesh powders. The performance of the

Figure 7.4 Scanning electron micrographs of a platinum–cobalt alloy. Reproduced with permission from *Meeting Abstr. Electrochem. Soc.*, **1002**, 275 (2010). Copyright 2010, The Electrochemical Society [13].

platinum–cobalt alloy can be further improved if starting materials with smaller particle sizes are used in ball milling. The catalyst fabricated with ball-milled platinum-black and nano-cobalt can sustain a current of 8.2 × 10^{-4} A/mg-Pt, a performance that is approaching that of the commercially available platinum black.

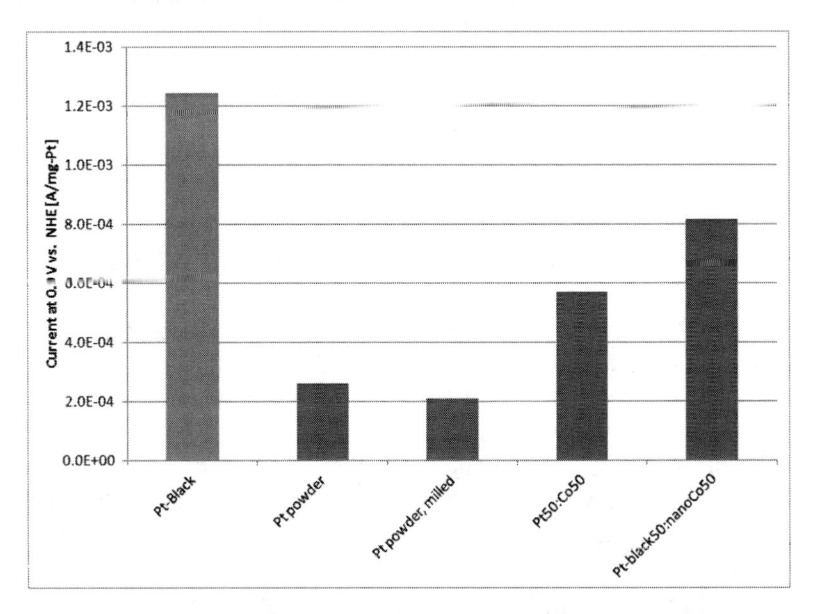

Figure 7.5 Comparison of various platinum–cobalt alloys fabricated via high energy ball-milling. Reproduced with permission from *Meeting Abstr. Electrochem. Soc.*, **1002**, 275 (2010). Copyright 2010, The Electrochemical Society [13].

7.11 Oxygen Evolution Catalysts

The production of hydrogen via electrolysis is carried out predominately via alkaline electrolysis technology. The advantages of PEM electrolyzers over alkaline technologies include increased system longevity, the use of non-caustic electrolytes, reduced complexity, and high-pressure capabilities, as have been discussed in the literature [1]. Some issues with PEM electrolyzers include a lower operating efficiency relative to alkaline systems, lower durability and higher costs, and these are in part related to the catalysts. Catalysts for the efficient electrolysis of water in PEM-based electrolyzers have been studied by various organizations [6,16,28–32]. The current focus in electrolyzer research is to develop systems for NASA space applications. Terrestrial applications such as energy storage from renewable power sources are also being pursued by the U. S. Department of Energy.

A desirable oxygen evolution catalyst should have a low overpotential for oxygen evolution, high surface area, and long-term chemical and mechanical stability [14]. These attributes of a good oxygen evolution catalyst translate to both an efficient and long-life electrolyzer. Ruthenium-based materials are known to exhibit some of the lowest overpotentials for oxygen evolution [33–37] but the use of ruthenium is limited by its stability in acidic media at oxygen evolution potentials [38–40]. An approach to stabilize ruthenium oxide is to dope it with a second metal component that could have the general stoichiometry shown in Eq. 7.5:

$$M + RuO_2 \rightarrow MO_xRuO_{(2-x)}, \tag{7.5}$$

where M, is the dopant metal [15]. A dopant in the range of 5% has been shown to produce a stable and active oxygen evolution catalyst [15].

In the thermal processing technique, the starting materials were mixed and then heat treated in a furnace under flowing argon. The argon in the thermal processing technique was used to remove any oxygen from the furnace. The processing temperature in the thermal processing technique was in the range of 200–300°C, and the total time for processing the oxygen evolution catalysts was at least 12 h [32,41]. Iridium-doped and lead-doped ruthenium oxide catalysts were fabricated by the thermal processing technique. Scanning electron micrographs of the thermally processed iridium-doped ruthenium oxide are shown in Fig. 7.6. The iridium-doped ruthenium oxide catalyst particle size ranges from 80 nm to a few microns, with no evidence for particle sintering based on the heat treatment used. In the high-energy ball mill technique, the base materials were mixed and ball-milled for approximately 4 h. Only the lead oxide/ruthenium oxide oxygen evolution catalyst was fabricated by this technique.

Figure 7.6 Scanning electron micrograph for the thermally processed iridium-doped ruthenium oxide catalyst. Reproduced with permission from *Meeting Abstr. Electrochem. Soc.*, **902**, 1044 (2009). Copyright 2003, The Electrochemical Society [16].

The catalysts tested in the three-electrode cell included the standard ruthenium oxide, the thermally processed iridium-doped ruthenium oxide, a ball milled lead–ruthenium oxide and a lead-doped ruthenium oxide. The lead-doped ruthenium oxide catalyst was fabricated by a similar thermal processing technique as the iridium-doped ruthenium oxide catalyst. The results of the potentiodynamic characterization of the four anode catalysts are shown in Fig. 7.7. The catalyst referred to as RuO_2 is commercially

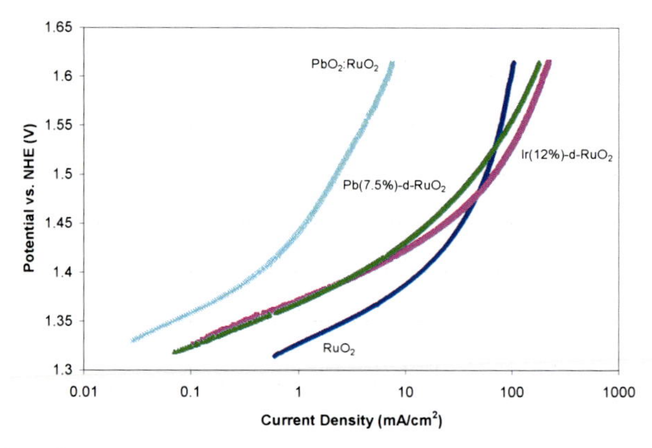

Figure 7.7 Potentiodynamic polarization curves for various ruthenium-based catalysts tested in a three-electrode electrochemical cell at a scan rate of 0.2 mV/s in 25°C 1N sulfuric acid. Reproduced with permission from *Electrochem. Soc. Trans.*, **25**, 1371 (2009). Copyright 2003, The Electrochemical Society [15].

available from Johnson Matthey. The sample labeled PbO_2:RuO_2 represents a catalyst that was prepared in-house via the ball milling technique. The samples labeled Pb(7.5%)-d-RuO_2 and Ir(12%)-d-RuO_2 are catalysts prepared in-house by the thermal processing technique.

The commercial ruthenium oxide catalyst outperformed all the catalysts at current densities below 50 mA/cm^2. At higher current densities, the lead- and iridium-doped catalyst performed the best in the three-electrode cell. The electrode potentials at 100 mA/cm^2 for the lead- and iridium-doped catalyst were 1.53 and 1.56 V versus NHE, respectively. The lead–ruthenium catalysts did not perform better than iridium–ruthenium catalysts, but suggest an opportunity for noble metal reduction in commercial electrolysis catalyst systems.

7.12 MEA Development for NASA Non-Flow-Through Hydrogen/Oxygen Fuel Cell Stack

PEM MEA development was initiated with a survey of commercially available cells. Membrane electrode assemblies were purchased from various vendors and tested for hydrogen-oxygen performance. It was determined that MEA development for hydrogen oxygen fuel cells should focus on membrane and cathode improvements. A fuel cell cathode (catalyst/gas diffusion layer) was developed and optimized using Nafion 115 as the polymer electrolyte. Membrane electrode assemblies based on the advanced fuel cell cathode were fabricated using Nafion 117, 115, 105, 112, and 212. The findings from these MEA studies have been reported in the literature [17]. The current–voltage polarization performance of a “Commercial,” “Standard,” and “Advanced” MEAs are shown in Fig. 7.8. Each MEA was fabricated with a 2 mil Nafion membrane; the “standard” and “advanced” MEAs were fabricated with Nafion 212 and the “commercial” MEA was fabricated with Nafion 112. Both the “standard” and the “advanced” MEAs used a commercially available platinum black catalyst; the catalyst used to fabricate the “commercial” MEA is proprietary and is not reported. The current–voltage polarization performance of the commercial MEA is reported to be 0.85 V (69% voltage efficiency) at 200 mA/cm^2 and 30 PSIG balanced hydrogen/oxygen, 70°C. This performance was typical for the commercial MEAs tested. The current–voltage polarization performance of the standard MEA is reported to be 0.9 V at 200 mA/cm^2 and 30 PSIG balanced hydrogen/oxygen at 70°C. The maximum performance obtained for a fuel cell MEA fabricated with Nafion 212 is 0.91 V (74% voltage efficiency) at 200 mA/cm^2 and 30 PSIG balanced hydrogen/oxygen at 70°C. A fuel cell performance of 0.92 V at 200 mA/cm^2 can be reached under the same conditions but at 80°C. Membrane and cathode improvements in fuel cell MEAs have resulted in a 60 mV improvement in cell voltage over commercially available MEAs for cells operation at 200 mA/cm^2 and 30 PSIG balanced hydrogen and oxygen pressure.

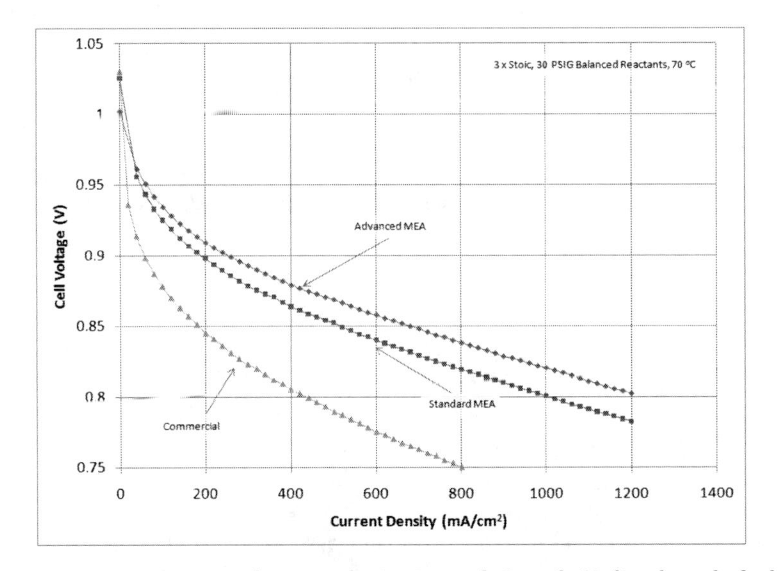

Figure 7.8 Current–voltage polarization of 2 mil Nafion-based fuel cell MEAs operating on 30 PSIG balanced oxygen and hydrogen pressure, 70°C. Reproduced with permission from *J. Electrochem. Soc.*, **156**, B152 (2009). Copyright 2003, The Electrochemical Society [43].

7.13 MEA Development for NASA High-Pressure Electrolyzers

Similarly, electrolyzer MEA development also initiated with a survey of commercially available electrolysis cells. The focus of electrolyzer MEA development was on advancing the performance and stability of oxygen evolution (anode) catalysts. The iridium–ruthenium oxide catalyst system was identified as the most promising oxygen evolution catalyst to investigate. Studies of oxygen evolution catalysts and advanced electrolysis MEAs have been reported in the literature [32,37,44]. Electrolyzer MEA development has focused on Nafion 115 as the polymer electrolyte, as electrolyzer MEAs are designed to operate at high pressures and thus cannot take advantage of thinner commercially available polymer electrolytes. A commercially available electrolyzer MEA fabricated with an iridium–ruthenium oxide anode was used as the standard. The current–voltage polarization performance of both commercially available and "in-house" fabricated electrolyzer MEAs with iridium–ruthenium oxide anode catalysts is shown in Fig. 7.9. The performance of the commercially available electrolysis cell is reported to be 1.48 V (83% voltage efficiency) at 200 mA/cm^2, 70°C, and ambient pressure operation. The performance of the electrolysis cell fabricated with the "in-house" catalyst is reported to be 1.42 V (87% voltage efficiency) for a cell operating at 200 mA/cm^2, 70°C, and ambient pressure. Improvement in the activity of the oxygen

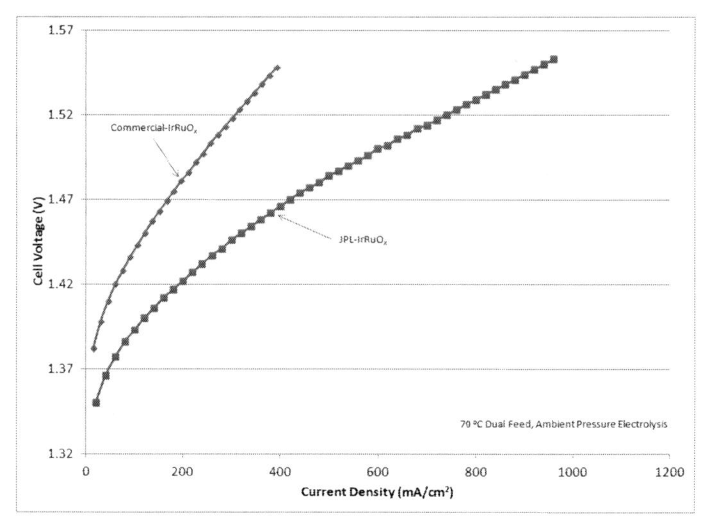

Figure 7.9 Current–voltage polarization performance of an electrolysis cell fabricated with an advanced iridium–ruthenium oxide catalyst as compared to a commercially available electrolysis cell fabricated with a proprietary iridium–ruthenium oxide catalyst. Reproduced with permission from *Electrochem. Soc. Trans.*, **25**, 1371 (2009). Copyright 2003, The Electrochemical Society [32].

evolution catalysts have resulted in a 60 mV reduction in the oxygen evolution overpotential, and thus a performance improvement for electrolyzer MEAs operating at 200 mA/cm^2, 70°C and ambient pressure. The "in-house" fabricated iridium–ruthenium oxide catalyst has been supplied to vendors developing electrolysis MEAs on various NASA SBIR programs. Electrolysis MEAs were fabricated using a novel Nafion-based polymer electrolyte. The current–voltage polarization performance of the "in-house" iridium–ruthenium oxide catalyst has been verified and has been shown to be stable in the range of 100 h.

7.14 Conclusion

To meet the efficiency requirements for a space-rated regenerative fuel cell system, improvements in catalysts for fuel cells and electrolyzers are required. The NASA program threshold values of 0.9 V for fuel cells and 1.46 V for electrolyzers have been met or exceeded. The performance of a fuel cell, based on a 212 Nafion membrane, has demonstrated an operating voltage of 0.91 V at an applied load of 200 mA/cm^2, operating at 70°C, 30 PSIG with balanced pressure reactants. The fuel cell voltage efficiency at these conditions is calculated to be 74%. The performance of an electrolyzer, based on a 115 Nafion membrane, demonstrated 1.42 V at 200 mA/cm^2, operating at 70°C and ambient pressure. The electrolyzer voltage efficiency at these conditions is calculated to be 87%. It is estimated that a regenerative fuel cell can operate with an efficiency of 58% assuming a 10% allocation for parasitic inefficiencies. A regenerative fuel cell system that can operate at an efficiency of 58% would exceed the NASA program goals. The focus of future work will be to address the durability of the fuel cell and electrolyzer catalysts. To date, durability has been demonstrated in the 100 h time frame. The ultimate objective of this work is to meet the NASA program goal of 10,000 h of operation.

Acknowledgment

The work presented here was carried out at the Jet Propulsion Laboratory, California Institute of Technology for the National Aeronautics and Space Administration.

References

1. M. A. Hoberecht, NASA's planned fuel cell development activities for 2009, NASA/TM-2010-216106, 2009.
2. C. R. Mercer, A. L. Jankovsky, C. M. Reid, T. B. Miller, and M. A. Hoberecht, Energy Storage Project, final report, NASA/TM-2011-216963, 2011.
3. S. R. Narayanan, A. Kindler, A. Kisor, T. Valdez, R. J. Roy, C. Eldridge, B. Murach, M. Hoberecht, and J. Graf, *J. Electrochem. Soc.*, **158**, B1348, 2011.
4. M. A. Hoberecht, A comparison of flow-through versus non-flow-through proton exchange membrane fuel cell systems for NASA's exploration missions, NASA/TM—2010-216107, 2010.
5. K. C. Neyerlin, W. Gu, J. Jorne, A. Clark, Jr., and H. A. Gasteiger, *J. Electrochem. Soc.*, **154**, B279, 2007.
6. H. S. Wroblowa, C. P. Yen, and G. Razumney, *J. Electroanal. Chem. Interfacial Electrochem.*, **69**(2), 195, 1976.
7. B. Wang, *J. Power Sources*, **152**, 1, 2005.
8. T. Toda, H. Igarashi, H. Uchida, and M. Watanabe, *J. Electrochem. Soc.*, **146**(10), 3750, 1999.
9. E. J. Lamas and P. B. Balbuena, *J. Chem. Theory Comput.*, **2**(5), 1388, 2006.
10. J. L. Fernández, D. A. Walsh, and A. J. Bard, *J. Am. Chem. Soc.*, **127**(1), 357, 2004.
11. N. M. Marković, T. J. Schmidt, and V. Stamenković, P. N. Ross, *Fuel Cells*, **1**(2), 105, 2001.
12. B. N. Grgur, N. M. Marković, and P. N. Ross, *Langmuir*, **13**(24), 6370, 1997.
13. N. M. Marković, N. S. Marinković, and R. R. Adžić, *J. Electroanal. Chem. Interfacial Electrochem.*, **241**(1–2), 309, 1988.
14. J. K. Nørskov, J. Rossmeisl, A. Logadottir, L. Lindqvist, J. R. Kitchin, T. Bligaard, and H. Jónsson, *J. Phys. Chem. B*, **108** (46), 17886, 2004.
15. J. O' M. Bockris, *J. Chem. Phys.*, **24**(4), 817, 1956.
16. E. Rasten, Electrocatalysis in water electrolysis with solid polymer electrolyte, Thesis, Norwegian University of Science and Technology, 2001.
17. C. P. De Pauli and S. Trasatti, *J. Electroanal. Chem.*, 538–539, **145**, 2002.
18. R. Berenguer, C. Quijada, and E. Morallon, *Electrochim. Acta*, **54**, 5230, 2009.

19. M.-K. Min, J. Cho, K. Cho, and H. Kim, *Electrochim. Acta*, **45**, 4211, 2000.

20. W. E. Kaden, T. Wu, W. A. Kunkel, and S. L. Anderson, *Science*, **326** (5954), 826, 2009.

21. M. Neergat, A. K. Shukla, and K. S. Gandhi, *J. Appl. Electrochem.*, **31**(4), 373, 2001.

22. U. A. Paulus, A. Wokaun, G. G. Scherer, T. J. Schmidt, V. Stamenkovic, V. Radmilovic, N. M. Markovic, and P. N. Ross, *J. Phys. Chem. B*, 2002, **106**(16), 4181, 2002.

23. H. Yang, W. Vogel, C. Lamy, and N. Alonso-Vante, *J. Phys. Chem. B*, **108**(30), 11024, 2004.

24. S. Koh, C. Yu, P. Mani, R. Srivastava, and P. Strasser, *J. Power Sources*, **172**(1), 50, 2007.

25. C. C. Hays, J. Kulleck, B. Haines, A. Kisor, and S.R. Narayan, *Meeting Abstr. Electrochem. Soc.*, **1002**, 274, 2010.

26. C. C. Hays, P. Bahrami, M. Errico, and J. G. Kulleck, *Meeting Abstr. Electrochem. Soc.*, **1201**, 1432, 2012.

27. J. Ma, S. Firdosy, G. Grüner, J.-P. Fleurial, and S. R. Narayanan, *Meeting Abstr. Electrochem. Soc.*, **1002**, 275, 2010.

28. J. M. Sedlak, R. J. Lawrance, and J. F. Enos, *Int. J. Hydrogen Energy*, **6**, 159, 1981.

29. P. Millet, M. Pineri, and R. Durand, *J. Appl. Electrochem.*, **19**(2), 162, 1989.

30. P. Millet, F. Andolfatto, and R. Durand, *Int. J. Hydrogen Energy*, **2**(2), 87, 1996.

31. E. Rasten, G. Hagen, and R. Tunold, *Electrochim. Acta*, **48**(25–26), 3945, 2003.

32. T. I. Valdez, K. J. Billings, J. Sakamoto, F. Mansfeld, and S. R. Narayanan, *Electrochem. Soc. Trans.*, **25**(1), 1371, 2009.

33. S. Trasatti, *Electrochim. Acta*, **29**(11), 1503, 1984.

34. A. Marshall, M. Tsypkin, B. Børresen, G. Hagen, and R. Tunold, *J. New Mater. Electrochem. Syst.*, **7**, 197, 2004.

35. J. Rossmeisl, A. Logadottir, and J. K. Nørskov, *Chem. Phys.*, **319**(1–3), 178, 2005.

36. J. Rossmeisl, Z.-W. Qu, H. Zhu, G.-J. Kroes, and J. K. Nørskov, *J. Electroanal. Chem.*, **607**, 83, 2007.

37. T. I. Valdez, J. Sakamoto, K. J. Billings, S. A. Firdosy, F. Mansfeld, and S. R. Narayanan, *Meeting Abstr. Electrochem. Soc.*, **801**(9) 306, 2008.

38. C. Iwakura, K. Hirao, and H. Tamura, *Electrochim. Acta*, **22**(4), 329, 1977.

39. R. S. Yeo, J. Orehotsky, W. Visscher, and S. Srinivasan, *J. Electrochem. Soc. Elect. Sci. Tech.*, **9**, 1900, 1981.

40. P. Piela, C. Eickes, E. Brosha, F. Garzon, and P. Zelenay, *J. Electrochem. Soc.*, **151**(12), A2053, 2004.

41. T. I. Valdez and S. R. Narayanan, Ir-doped ruthenium oxide catalyst for oxygen evolution, US Patent No: 8,183,174 B2, 2012.

42. T. I. Valdez, K. J. Billings, F. Mansfeld, and S. R. Narayanan, *Meeting Abstr. Electrochem. Soc.*, **902**(10), 1044, 2009.

43. S. R. Narayanan, T. I. Valdez, and S. Firdosy, *J. Electrochem. Soc.*, **156**, B152, 2009.

44. T. I. Valdez, K. J. Billings, S. A. Firdosy, F. Mansfeld, and S. R. Narayanan, *Meeting Abstr. Electrochem. Soc.*, **802**(4), 415, 2008.

Chapter 8

Nanotechnology for Nanoelectronic Devices

Félix A. Miranda[a] and Harish M. Manohara[b]

[a]*NASA John H. Glenn Research Center at Lewis Field,*
21000 Brookpark Rd., Cleveland, OH 44135, USA
[b]*NASA Jet Propulsion Laboratory,*
California Institute of Technology, Pasadena, CA 91101, USA

felix.a.miranda@nasa.gov

8.1 Introduction

The growth of the nanotechnology field has been fast, and its potential impact has permeated many diverse areas. This has prompted multiple professional technical societies to quickly organize working groups and technical committees to explore means to accelerate advancements in this field toward practical applications [1]. The areas encompassed by the field of nanotechnology are certainly very broad. For example, the unique properties of carbon nanotubes (CNTs) are attractive and desirable for sensor applications, energy storage, instrumentation, and electronics, just to name a few. While all these areas are of extreme relevance for technological development as evidenced by the other chapters comprising this book, in this chapter we focus on the impact of nanotechnology

Advanced Nanomaterials for Aerospace Applications
Edited by Carlos R. Cabrera and Félix A. Miranda
Copyright © 2014 Pan Stanford Publishing Pte. Ltd.
ISBN 978-981-4463-18-8 (Hardcover), 978-981-4463-19-5 (eBook)
www.panstanford.com

in the area of nanoelectronic devices. Furthermore, we will discuss how advances in nanoelectronics could be tailored for NASA applications.

8.2 What Makes Nanotechnology So Attractive for Nanoelectronics Applications?

In general, the advances of the last several decades on semiconductor-based electronics circuits have been the result of many factors such as better materials processing and synthesis methods, better design and modeling tools, and maintaining pace and compliance with the Moore's Law [2]. According to Moore's Law, the number of transistors on a chip roughly doubles every two years. As a result the device scale progressively gets smaller and maintaining this pace through conventional approaches continues to become harder. Therefore, a new dimensional realm is desired to continue advancements in the electronic device technology. Nanotechnology offers such a realm. There are several building blocks upon which the new nanoelectronic components and devices could be developed, the one most studied being CNTs.

In addition to excellent structural properties (e.g., strength 10–60 GPa as compared to 4.1 GPa for high-speed (HS) steel, and 3.6 to 4.1 for Kevlar), CNTs have excellent electrical and thermal conductivity properties. That is, CNTs have almost an order of magnitude higher thermal conductivity and hundreds of times higher electrical conductivity than copper (i.e., CNTs thermal conductivity >3000 W/m·K versus copper's 400 W/m·K; CNTs' electrical conductivity $>200 \times 10^7$ S/m versus copper's 6×10^7 S/m) [3]. Therefore, with diameters ranging from <1 nm to 50 nm, CNTs are really ideal to develop nanoelectronic devices such as field-effect transistors (FETs), Schottky diodes, switches, nanosensors, vacuum microelectronics, and other essential as well as promising building blocks for communications and computational applications. In the next section, we will discuss some examples of the aforementioned components.

8.3 Examples of Nanoelectronic Devices

In the field of nanoelectronics, the recent advances have been remarkable. While a variety of devices have been demonstrated, it is fair to say that the bulk of the development of nanoelectronics proof-of-concepts is rooted in a considerable number of research efforts dedicated to the demonstration of CNT FETs [4,5]. In this particular device, single-wall or multiwall CNTs form the channel between the drain and the source of the nanoFET, while a silicon substrate is used as the back gate. A typical illustration of this concept is shown in Fig. 8.1.

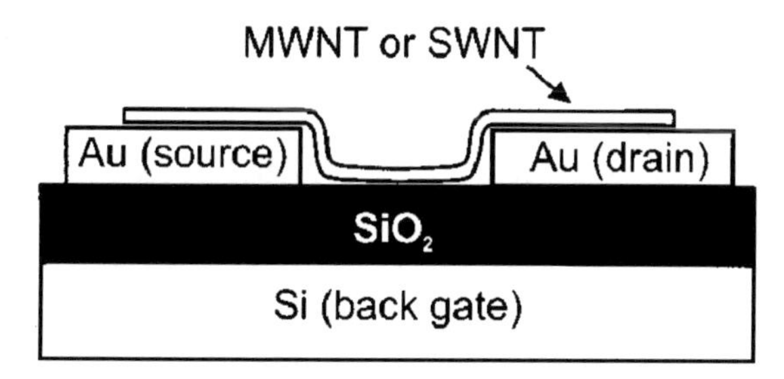

Figure 8.1 Schematic cross section of the nanoFET devices. A single CNT of either multiwall (MW) or single-wall (SW)-type bridges the gap between two gold (Au) electrodes separated from the silicon (Si) substrate by a thin silicon dioxide (SiO_2) layer. The Si substrate is used as back gate. Reprinted with permission from R. Martel, T. Schmidt, H. R. Shea, T. Hertel, and P. Avouris, Single- and multi-wall carbon nanotube field-effect transistors, *Appl. Phys. Lett.*, **73**(17), 2447–2449, 26 October 1998. Copyright (1998), American Institute of Physics.

These CNT FETs or nanoFETs can then in turn be customized for many specific applications. For example, some recent advances and demonstrations of nanoelectronic devices and circuits include NanoFET bioprobes capable of detecting intracellular potentials with high spatial and temporal resolution without irreversibly altering or damaging the cells [6]. This work uses FETs made from kinks or bends in silicon nanowires (SiNW). Schematics of this concept as well as a micrograph of the fabricated device are shown in Fig. 8.2.

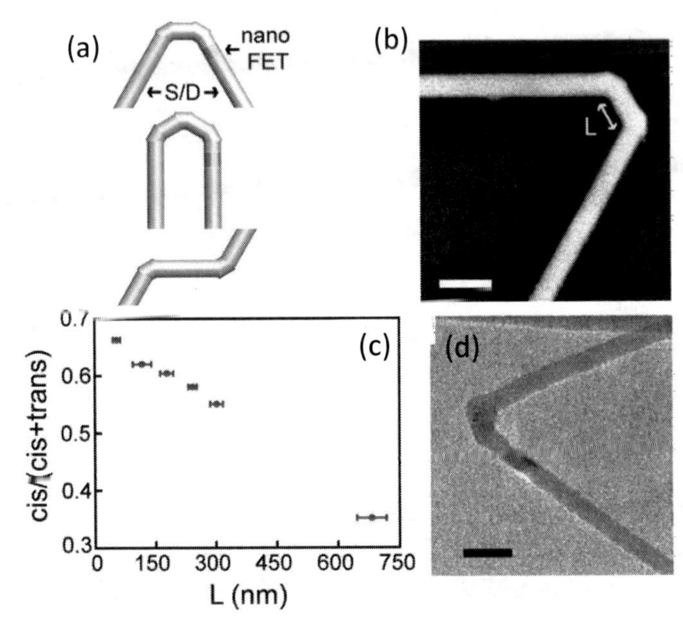

Figure 8.2 Synthesis of kinked SiNW probes. (a) Schematics of 60° (top) and 0° (middle) multiply kinked nanowires and *cis* (top) and *trans* (bottom) configurations in nanowire structures. The blue and pink regions designate the source/drain (S/D) and nanoscale FET channel, respectively. (b) SEM image of a doubly kinked nanowire with a *cis* configuration. L is the length of segment between two adjacent kinks. (c) *cis*/(*cis* + *trans*) versus L plot. Error bars indicate ± 1 SD from the mean. (d) Transmission electron microscopy image of an ultrathin 60° kinked nanowire. Scale bars, 200 nm (b); 50 nm (d). Reprinted with permission from B. Tian, T. Cohen-Karni, Q. Qing, X. Duan, P. Xie, and C. M. Lieber, Three-dimensional flexible nanoscale field-effect transistors as localized bio-probes, *Science*, **329**, 830–834, August 2010. Copyright (2010), AAAS.

Another example of the advantages of CNT-based electronic devices can be also seen in Schottky diodes [7]. It is stated that because of the narrow dimensions of the CNTs (diameters in the range of 1 to 3 nm), Schottky diodes made with single-walled CNT (SWNT) have very small junction areas (i.e., a few square nanometers) and consequently exhibit junction capacitances in the order of the atto-farads (i.e., 10^{-18} Farad). Accordingly, the aforementioned capacitance values translate into cutoff frequencies larger than

5 THz as compared to the current upper limit of 1.5 THz at which the state of practice Schottky diodes exhibit extremely low efficiencies. A schematic and a picture of the SWNT-based Schottky diodes as discussed in Manohara's work are shown in Fig. 8.3.

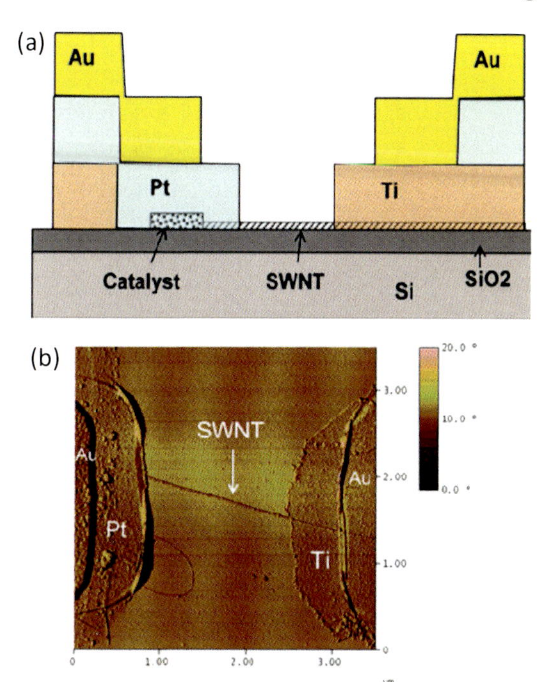

Figure 8.3 (a) Schematic representation of the SWNT-Schottky diode showing the Ti-Schottky and the Pt-Ohmic metal layers deposited through angled evaporation. (b) Atomic force microscopy (AFM) phase plot image of a typical SWNT-Schottky diode. The semiconducting SWNT diameter varied from 1 to 3 nm while its length varied from 1.7 to 2.5 μm. Reprinted with permission from H. M. Manohara, E. W. Wong, E. Schlecht, B. D. Hunt, P. H. Siegel, Carbon Nanotube Schottky diodes using Ti-Schottky and Pt-Ohmic contacts for high frequency applications, *Nano Lett.*, **5**(7), 1469–1474 (2005). Copyright (2005) American Chemical Society.

Similar device structures have been shown to function as directionally dependent FETs [8]. In addition, Schottky contacts on organic single crystals have been demonstrated, offering yet another stepping-stone for the attainment of practical organic-based electronic devices [9].

The impact of CNTs has also been observed in their integration with elastic polymers creating elastic materials whose conductivity exceeds that of any other type of elastometer (e.g., graphite-based elastometers) by two orders of magnitude (i.e., 0.1 S/cm versus 57 S/cm, respectively) [10]. Such materials could form the basis for transistor arrays that enable flexible nanoelectronics circuits stretchable up to 134% without suffering adverse mechanical damages. Applications to artificial skins with embedded sensors with diverse sensing capabilities (e.g., pressure, temperature, etc.) have also been shown to be feasible [11]. Research efforts, such as those by Jonathan Viventi et al. have also explored the possibility of using these materials as embedded biosensors to detect tiny potentials in the brain associated with the onset of epileptic seizures or to trigger electrical pulses to suppress such seizures [12]

Carbon nanotubes, silicon (Si), and silicon carbide (SiC) nanowires have also been used to demonstrate fast electromechanical switches [13–14]. For example, electrostatically actuated microelectromechanical systems (MEMS) switches have been extensively investigated [15]. While great progress has been made in the optimization of these switches, their performance still requires application of high activation voltages (i.e., of the order of 50–70 V dc) and their switching speed or response time is still in the order of microseconds. In contrast, Si, SiC nanowires, and CNT-based switches exhibit response times of the order of nanoseconds and switching potentials of a few volts DC, making them ideal for applications requiring highly miniaturized and high-speed electronic switches under low bias condition (e.g., battery powered) [16–17]. In particular, SiC switches have also shown great potential for operation in extreme environments as in high temperatures (e.g., Venus) or in high radiation scenarios (e.g., Jovian environment), both of which represent NASA mission relevance.

The motivation to enhance the performance of the aforementioned concepts and devices has sparked off several methods for synthesizing CNT beyond those traditionally employed to grow these materials. Aside from chemical vapor deposition (CVD), which remains the most popular method to synthesize CNT, other approaches such as solution deposition methods, which can be conducted at room temperature (contrary to the 900°C required for CVD), have been tried in an attempt to enhance the optimization and manufacturability of CNT FETs [18]. Clear understanding of the correlation between different methods and materials for producing CNT-based electronics circuits is critical to maximize the performance of such circuits for diverse electronic applications. Studies have been performed where the correlation between electronic device architecture and performance has been investigated. Some of the parameters studied in CNT FETs, both as a function of circuit geometry and/or architecture, as well as in terms of fabrication processes (e.g., annealing) are noise, hysteresis, and threshold voltage, among others [19]. The outcome of these research efforts has contributed to determine which properties will be advantageous for a given practical application. For example, suppression of hysteresis through manipulation of device geometry is highly relevant for electronic applications, while the opposite, i.e., greater hysteresis resulting from different level of environmental exposure could be harnessed for sensor applications [19]. Similarly, it is possible to develop other electronic devices such as differential amplifiers, which are of great interest for optimizing small signals from a variety of sensors in order to increase their signal-to-noise ratio. Research efforts along these lines are currently underway by Chin et al. [20].

8.4 Characterization Techniques for Nanoelectronics Applications

The realization of the aforementioned devices, particularly for performance at radio frequencies (RF), entails the implementation and constant refinement of techniques to evaluate the properties of CNTs and other type of nanowires or nanofibers. Demonstration of nanoelectronic circuits first requires that the relevant properties of their fundamental units (i.e., CNTs or nanowires) be characterized in an accurate and reliable fashion. This is necessary not only to determine their electrical and radiofrequency (RF) properties but also to determine ways of optimizing those properties. Such comprehensive device characterization also helps optimize fabrication processes that increase the production yield. Development of characterization approaches and techniques requires clever implementation of metrology. Several groups have concentrated their efforts in developing a framework of characterization approaches of individual nanowires at RF frequency ranges. For example,

Fig. 8.4 shows a schematic and a scanning electron micrograph of an approach to characterize platinum (Pt) nanowires for broadband microwave and millimeter wave applications [21]. This technique uses the Pt nanowire to bridge the gap between the center conductors of a coplanar wave guide (CPW) transmission line structure. The center conductor of the CPW is typically made of gold or copper. Since the transmission properties of the center conductor of the CPW transmission lines are well known, by measuring the reflection and transmission scattering parameters (i.e., S_{11} and S_{21}, respectively) with the nanowire in place, the transmission losses and therefore the electrical properties of the nanowire can be measured. Note that this approach is generic and could be used for the characterization of other types of nanostructures such as metallic nanowires, CNTs, and polymer nanowires, among others.

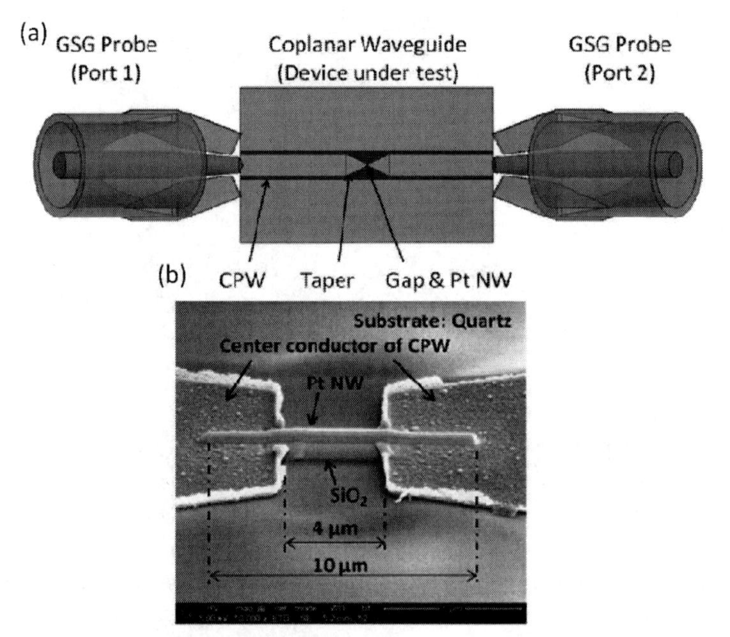

Figure 8.4 (a) Drawing of a coplanar waveguide (CPW)-based measurement setup; (b) SEM of 250 nm diameter Pt NW deposited on a signal trace of the host CPW line. Reprinted with permission from K. Kim, T. M. Wallis, P. Rice, C.-J. Chiang, A. Imtiaz, P. Kabos, and D. Filipovic, A framework for broadband characterization of individual nanowires, *IEEE Microwave Wireless Comp. Lett.*, **20**(3), 178–180, March 2010. Copyright (2005) IEEE.

Another example of a characterization approach is shown in Fig. 8.5. This approach is based on a waveguide metrology technique developed by Challa et al. (University of Mississippi) to measure the dielectric constant of multiwalled carbon nanotubes (MWCNTs) [22].

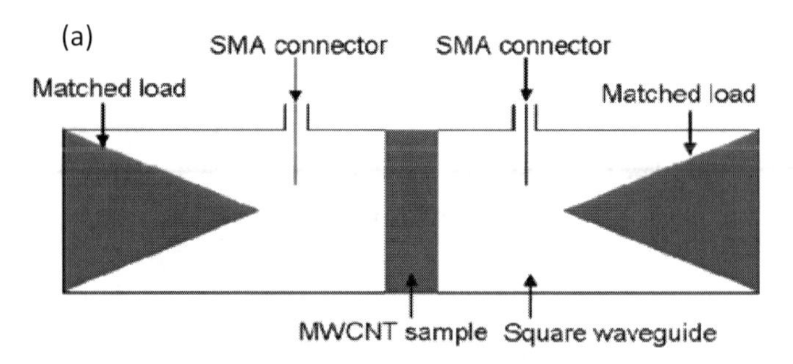

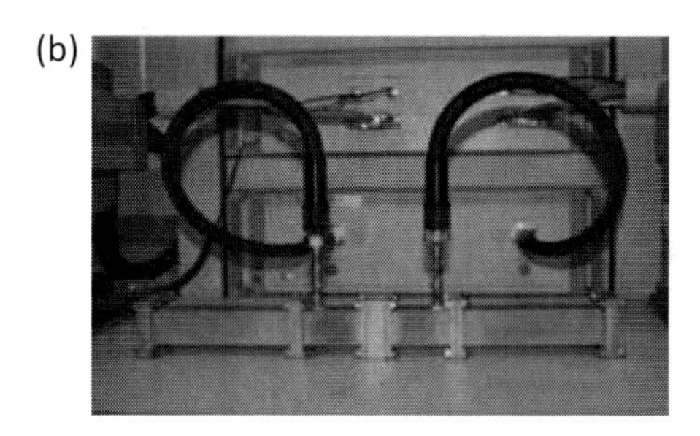

Figure 8.5 (a) Waveguide fixture and matched loads with a multiwalled carbon nanotube (MWCNT) composite sample of thickness t, placed at the center of sample holder. (b) Implementation of the experimental setup. Reprinted with permission from R. K. Challa, D. Kajfez, V. Demir, J. R. Gladden, and A. Z. Elsherbeni, Characterization of multiwalled carbon nanotube (MWCNT) composites in a waveguide of square cross section, *IEEE Microwave Wireless Comp. Lett.*, **18**(3), 161–163, March 2008. Copyright (2008) IEEE.

It is also important to note that part of the optimization of CNT-based electronics requires customizing these nanotubes to increase their robustness. For example, in a reliability study conducted by Petkov [23], it was demonstrated that the functionality of CNTs in a CNT FET could in principle be disabled by exposure to electron beams of the order of 1 keV. The outcome of this study is shown in Fig. 8.6.

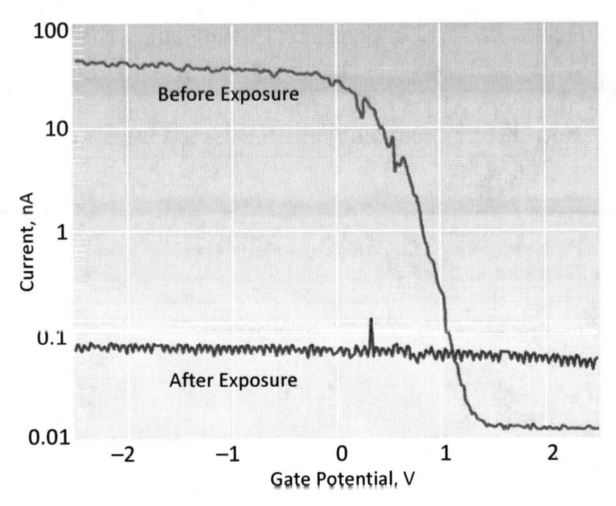

Figure 8.6 Carbon nanotube field-effect transistor's functionality destroyed by exposure to a 1 keV electron beam. Reprinted with permission from Mihail Petkov; *Source*: *NASA Tech Briefs*, **32**(3), March 2008, 34.

Consequently, such sensitivity to low energy electron beams could jeopardize the performance of CNT-based electronic devices (e.g., CNT-based nanoFETS or nanonatennas) the consequences of which will vary depending on the theater of operation of these devices.

The most significant challenge facing nanotube electronics to make them commercially viable or even attractive to specific applications requiring low-level production is the ability to synthesize CNTs in a controlled manner. For example, the growth of CNTs at precise locations, with predetermined orientations and dimensions (length and diameter), and desired chirality is still a difficult problem that is under investigation [24,25]. Additionally, good quality interconnects and electrical contacts, which are especially important for RF applications, are still under development [26]. Other issues or challenges such as developing approaches to control the growth of the individual nanotubes in a self-aligned fashion and not in randomly oriented bundles are also being undertaken [27].

It is important to remark that recent developments in the field have brought to focus novel materials such as graphene (shown in Fig. 8.7). Graphene is being investigated aggressively to capitalize on its excellent electrical and physical properties such as the ability to sustain current densities a million times higher than that of copper, record strength and thermal conductivity, 97% transparency at optical frequencies, and elastic stretching capabilities of up to 20% [28–29]. Professors Andre Geim and Kontanstin Novoselov of the University of Manchester, UK, were responsible for isolating this material in 2004 and subsequently were awarded the Nobel Prize in Physics in 2010. Yet there are still challenges to address such as understanding the influence of graphene's defects on electron-transport properties in order to optimize graphene for practical applications.

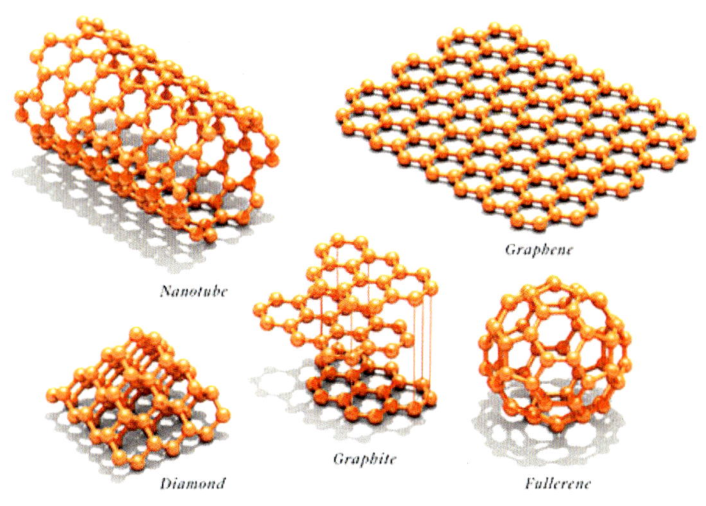

Figure 8.7 The carbon family has a new member: graphene. It comes on top of plain old graphite, now understood as a layer cake of graphene sheets; diamond, a cubic crystal; nanotubes, which consist of hexagonal cells shaped like a straw; and fullerenes, in which five cells are pentagons and the rest are hexagons, an arrangement that causes the structure to close up into a soccer-ball shape. Reprinted with permission from A. Sinitskii, J. M. Tour, Graphene electronics, unzipped, *IEEE Spectrum*, 30–33, November 2010. Copyright (2008) IEEE.

A very detailed discussion on the potential for CNTs for electronic and photonic devices is provided by Avouris [30]. Likewise, other materials such as boron nitride nanotubes (BNNTs) are being studied because of their excellent structural and electrical properties. For example, BNNTs exhibit a field-tunable bandgap that could be exploited for optical and electronic devices. They have the additional attribute of being piezoelectric- an ideal feature for serving as a basis of energy harvesting nanoelectronic devices [31]. For NASA applications involving space exploration to Mars, planetary systems, and beyond, robust electronics that can be utilized as a building block of energy harvesting systems operating in the "power starving mode approach" are highly coveted. This and other applications of interest to NASA will be discussed in the next section of this chapter.

8.5 Nanotechnology/Nanoelectronics in Future NASA Communications Systems

As mentioned previously, the suitability of nanotechnology for electronic and communication systems is being assessed closely because it offers the potential to enhance or replace silicon-based technology for high-speed, high-computing-power devices that operate at low voltages. However, to attain the aforementioned benefits some key factors must be addressed. One of them pertains to mitigating the following shortcomings of the conventional complimentary metal–oxide semiconductor (CMOS) device technology: (i) power dissipation, (ii) difficulty in maintaining control over the device geometry as circuit dimensions decrease (e.g., channel length, gate thickness, among others), and (iii) global interconnect delays resulting from the limited capability to reduce device dimensions. Accordingly, nanotechnology-based devices offer potential solutions to each of these problems. Below we provide some examples.

8.5.1 Logic Devices

Logic devices that utilize single-electron tunneling across charged island, or molecular tunneling between source and drain contacts, offer a means of accomplishing logic functions using small dimensions, low-voltage devices. For high-speed operation, the high carrier mobility of CNTs, which is greater than 100,000 $cm^2/V\text{-}s$ at room temperature, may be significant. For logic devices, the dynamic power dissipated (P_{dis}) per logic device is given by

$$P_{\text{dis}} = fCV^2, \tag{8.1}$$

where f is the operating frequency, C is the device capacitance, and V is the supply voltage. The relationship illustrated in Eq. (8.1) highlights several advantages of nanotechnology relevant to nanoelectronics. First, minimizing device capacitance and supply voltage is critical to reducing power dissipation and increasing computing power. Second, smaller active device dimensions also minimizes interdevice delays, thereby overcoming another key barrier facing conventional CMOS technology. The device scaling enabled by nanotechnology achieves these objectives and is one of the key drivers pushing current research in nanotechnology, particularly in nanoelectronics.

NASA's needs for nano-based computing technologies are in many aspects similar to those of the wider commercial and military market, but there are significant requirements that are specific to NASA. Nano-based communication devices, unlike the mono-functional device characteristic of today's CMOS-based devices, will have evolvable sensor capabilities for autonomous, self-reconfigurable performance as required, for example, in future space-platforms based on software defined radio (SDR) [32]. Also desirable to NASA are high-performance parallel processors featuring: (i) low power consumption, (ii) capacity to adapt and reconfigure, (iii) high fault tolerance (e.g., radiation resistant), (iv) neural network processing capability, and (v) distributed network sensing and decision making capabilities. In order to make nanoelectronic devices the natural choice to further these space applications, they must be realized by maintaining tradeoff advantages in size, weight, and power (SWAP) while simultaneously increasing communication capacity and computational capability. Below we provide some examples of nanoelectronic devices that are of interest to NASA, and under development in different laboratories both within and outside NASA.

8.5.2 Nano Field-Effect Transistors

At NASA we have investigated the use of nano-based logic devices for low-power applications. Our goal is to understand the mechanisms that enable logic behavior in nanodevices, with the goal of reducing device size, power consumption, and increasing circuit density and complexity. While this chapter revolves fundamentally around CNTs, it is worthwhile to discuss a joint research program between NASA Glenn Research Center and the University of Puerto Rico that has resulted in the successful demonstration of nanoFETs devices using electrospun nanofibers (see Fig. 8.8) comprising polyaniline (PANi) and polyethylene oxide (PEO) (see Fig. 8.9). In this program, nanofiber transistors with 10 micron-long channels displayed saturation drain currents using drain voltages of less than 1.0 V [33]. By decreasing the channel length, we anticipate that the drain voltage and device capacitance can be substantially reduced, thus providing a cornerstone for the types of logic devices that NASA needs. Note that attainment of such applications is not without challenges. For example, these polymeric materials are temperature-limited and could start to decompose primarily through oxidation at temperatures above 300°C. Furthermore, they could be sensitive to the radiation permeating the space environment. Consequently, further work is needed, either via packaging or material treatment, to protect them from ionizing radiation and extreme temperatures.

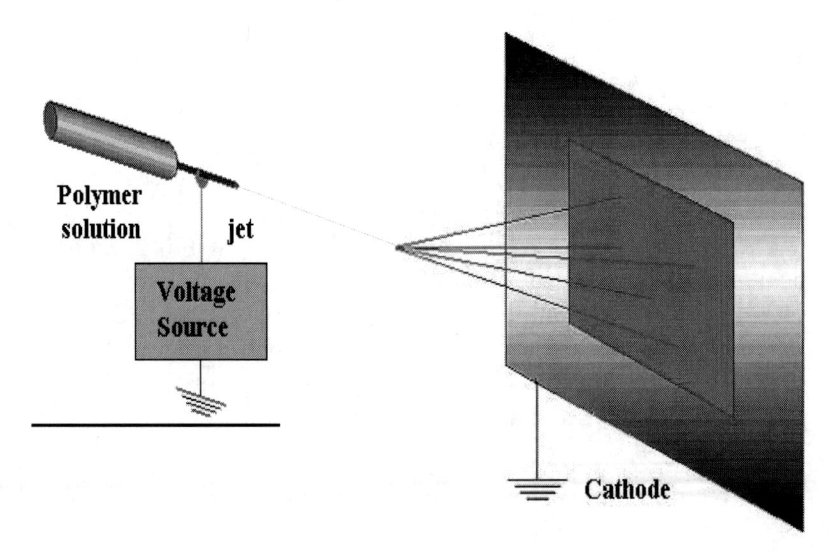

Figure 8.8 Basic electrospinning apparatus. Figure courtesy of Prof. N. Pinto, University of Puerto Rico-Humacao.

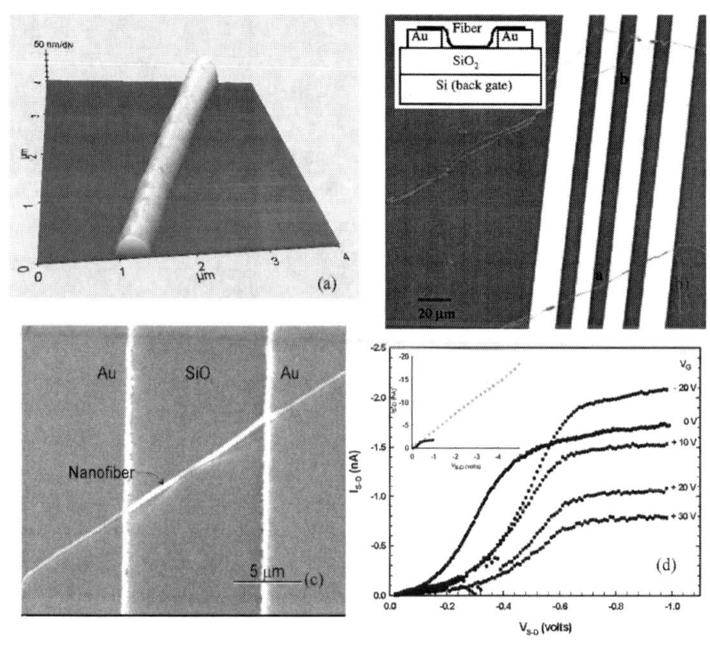

Figure 8.9 (a) Atomic force microscopy (AFM) image of polyaniline/polyethylene nanofiber. (b) Scanning electron microscope (SEM) image of electrospun PANi/PEO fiber over a pre-patterned SiO_2/Si substrate. Only fibers a and b contacting the two inner electrodes make up the device. Inset shows the schematic cross section of the device showing a fiber bridging the gap between the source and the drain gold (Au) electrodes. (c) SEM image of a nanofiber deposited on metalized SiO_2/Si substrate, (d) current–voltage characteristics of a nanofiber FTE. Figures 8.9b,d are reprinted with permission from N. J. Pinto, A. T. Johnson, A. G. MacDiarmid, C. H. Mueller, N. Theofylaktos, D. C. Robinson, and F. A. Miranda, *Appl. Phys. Lett.*, **83**, 4244–4246 (2003). Copyright 2003, American Institute of Physics.

The PANi/PEO nanofibers explored in this program are in the same class of conjugated polymer materials that have been proposed for extremely power-efficient, small-size single-electron tunneling devices. One of the challenges of this effort is to establish whether the nanofibers under study can be scaled to dimensions less than 10 nanometers. Hence, additional experiments to study the feasibility of preparing nanofibers of established high mobility material such as pentacene, and observing quantum-controlled conduction mechanisms and possible transistor behavior are necessary. In the interim, pentacene thin films are being developed for ultrahigh-density and low-power electronic logic switches and sensors. As shown in Fig. 8.10, logic "AND" gate-like function in pentacene-based thin films has already been demonstrated in an attempt to research device suitability for space communication applications [34,35].

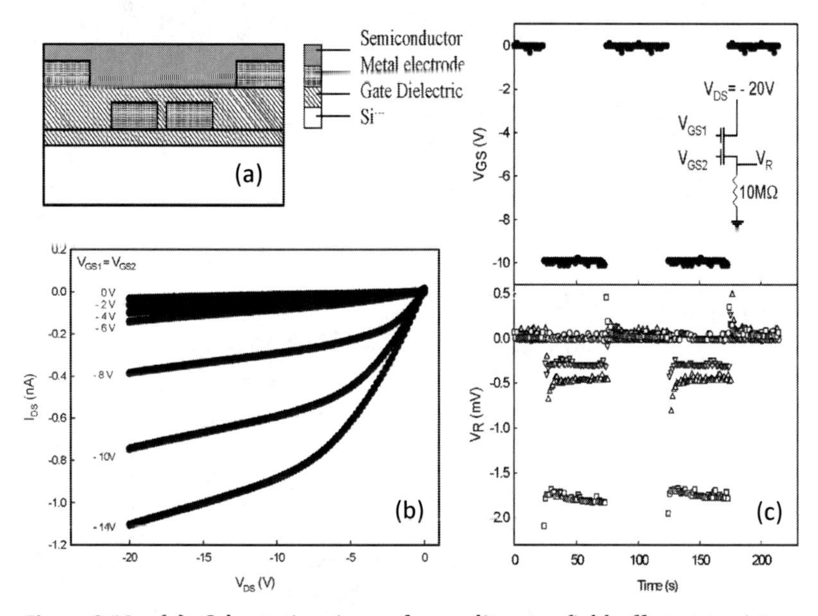

Figure 8.10 (a) Schematic view of a split-gate field-effect transistor. (b) Drain source current versus drain-source voltage (I_{DS}–V_{DS}). (c) Electrical performance of the split gate field-effect transistor as a logic AND circuit. Reprinted with permission from N. J. Pinto, R. Pérez, C. H. Mueller, N. Theofylaktos, and F. A. Miranda, *J. Appl. Phys.*, **99**, 084504-1–5 (2006). Copyright 2006, American Institute of Physics.

8.5.3 Graphene-Based Nanocircuits

As discussed in Section 8.4, graphene is a carbon sheet that is a single atom thick. Accordingly, graphene has potential for nanoelectronics that could be faster and consume less power than their silicon counterparts. The challenge ahead relies in synthesis methods leading to the reproducibility of graphene-based nanoelectronics structures. Yet some demonstrations of the promise this material holds have already been made in the areas of nanoswitches and nanosensors. For example, Fig. 8.11 shows a schematic of a graphene-based nanoswitch/nanosensing diode developed by researchers at the University of Puerto Rico-Humacao and NASA Glenn Research Center (patent-pending) [36]. This proof-of-concept device consists of a thin film of graphene deposited on an electrodized, highly doped ($\rho \sim 10\ \Omega$-cm) silicon wafer. The graphene film acts as a conductive path between a gold electrode deposited on the top of a silicon dioxide layer and the silicon wafer (on the exposed side surface or on the backside), so as to form a Schottky diode. By virtue of the two-dimensional nature of graphene, this device has extremely high sensitivity to different gaseous species, thereby serving as a building block for a volatile species sensor with the attribute of

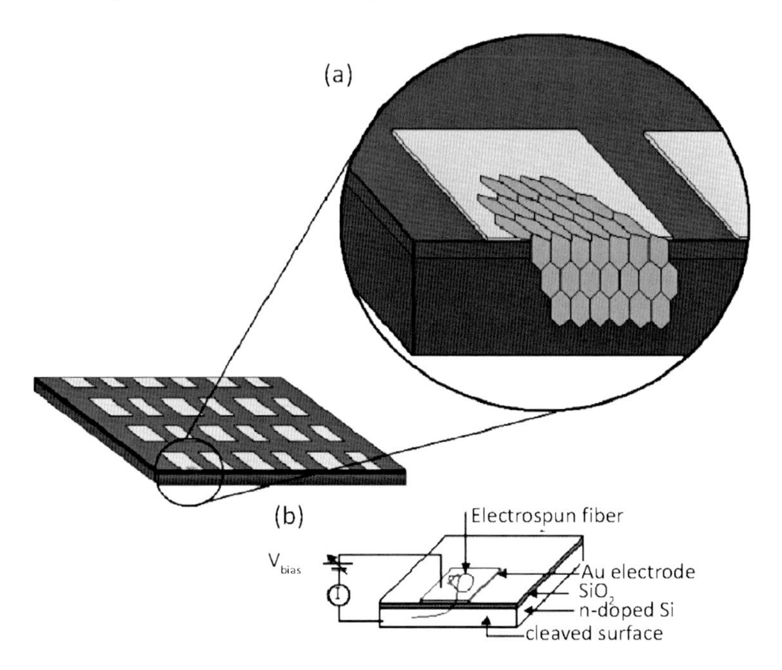

Figure 8.11 Schematic of graphene-based nanoswitch/nanosensing diode. Figure 8.11a courtesy of Prof. N. Pinto, University of Puerto Rico-Humacao Campus. Figure 8.11b reprinted with permission from N. J. Pinto, R. González, A. T. Johnson, Jr., and A. G. MacDiarmid, *Appl. Phys. Lett.*, **89**, 033505-1–3 (2006). Copyright 2006, American Institute of Physics.

having reversibility properties. That is, as shown in Fig. 8.12, when activated by applying DC voltage between the top metal electrode and the silicon substrate the sensor cycles between active and passive sensing states in response to the presence or absence of the gaseous species. Upon exposing the device to a volatile species environment (e.g., NH_3), the diode response is unequivocally different from that manifested under normal ambient conditions. More relevant yet, the behavior is reversible with the performance of the diode returning to its normal operational mode as the volatile species is removed. The device can be used as a switch where its operational stages (i.e., open/closed, on/off) could be controlled by a given gaseous species. This feature forms the basis for the functional operation of the device resulting in a reliable, long mean time between failure (MTBF) nanoswitch/sensor, ideal for applications where frequent replacement of the device is not a viable option.

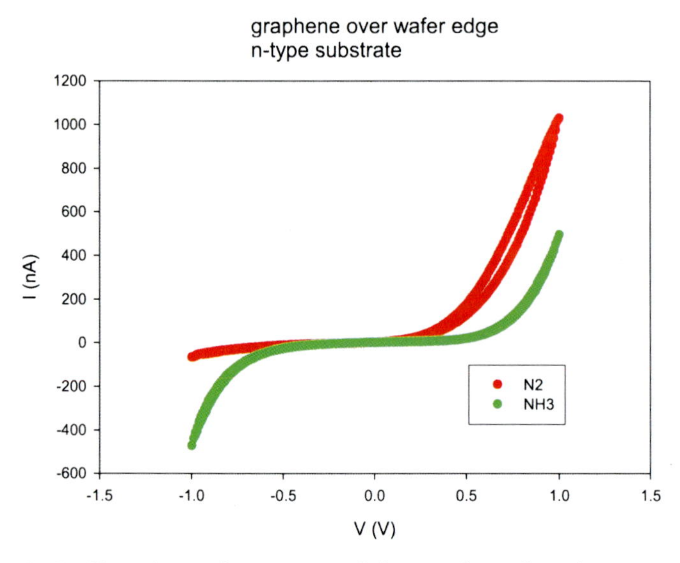

Figure 8.12 Experimental response of the graphene-based nanoswitch/nanosensor n-type doped silicon when atmosphere is changed from nitrogen (N_2) to ammonia (NH_3).

That is, this proof-of-concept has great potential as a building block for implementation of a switch/sensor device for harsh, embedded, or enclosed environments (e.g., space-based habitats, airplanes, subways, etc.) where the longevity and reusability of the circuit are critical for reliable operation. The sensing performance of this device has been experimentally tested in ambient atmosphere, as well as under ammonia gas (NH_3). The experimental data demonstrate the dual switching/sensing nature of the nano Schottky diode.

8.5.4 Nanoantennas

An area that could benefit from the emergence of nanotubes and nanowires is that of nanoantennas. Nanoantennas are required for the development of nano-wireless circuits comprised of nanosensors, biosensors, and nanopumps, among others. However, in order to optimize nanoantennas for the aforementioned applications, reliable nanostructures and nanomaterials are required. Besides the ubiquitous CNTs, silicon nanowires are already being optimized for many nanoelectronic applications such as FETs and sensors, as well as for other areas such as photovoltaics, batteries, and catalysts [37]. Note that because of the biocompatibility of silicon, silicon nanowires are very good candidates for biosensors and embedded drug delivery systems (e.g., smart nanopumps). Approaches and processes for manipulating these nanowires must be optimized to produce nanoscale antennas with the same accuracy and reliability as that of other more established techniques. For example, inkjet printing has been used to produce a variety of planar antennas for ultrahigh-frequency (UHF) applications, such as UHF dipole antennas, as well as CNT-based phased arrays for potential use in planetary surface-to-surface and surface-to-orbit communications [38–40].

What is the benefit? The integration of nanoantennas with other nanoelectronics circuitry on the same carrier (i.e., monolithically) should impact favorably the coveted tradeoff (particularly for aerospace electronics) of size, weight, and power (SWaP). This is because its realization should help in making front ends of communication systems smaller (i.e., more compact), and less lossy due to proximity and cableless nature of the antenna and receiver front end manifold. A good discussion on the several embodiments of nanoantennas and their advantages in communication systems is provided by Russer et al. in [41].

8.5.5 Nanoionics

Another area of interest at least from the standpoint of nanoelectronic devices is nanoionics. As stated by Nessel et al. [42], this technology concerns itself with materials and devices that rely on ion transport and chemical change at the nanoscale. This technology leverages on the properties of specific materials such as chalcogenides, which through low energy-induced chemical changes/reactions can enable the creation of conductive, "metallic-like" paths either on the surface or in the interior of the films. Capitalizing on these effects, researchers have demonstrated a variety of applications in the area of memory cells, tunable devices, nanovalves, and others. In particular, Nessel et al. [42] have demonstrated a novel nanoionics-based switch for microwave applications. Figure 8.13 illustrates the operation of such a coplanar series switch and Fig. 8.14 shows a microphotograph of the actual nanoionic switch.

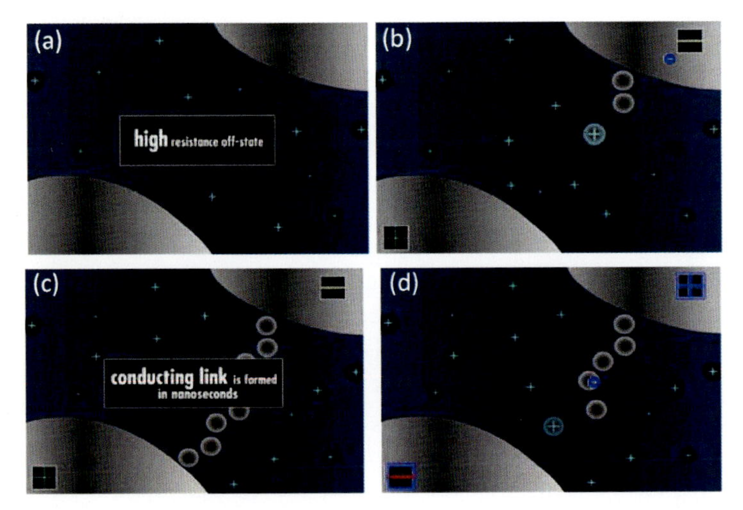

Figure 8.13 Device operation for a coplanar series nanoionics-based switch. (a) The switch is off. (b) Small bias forces electrons in cathode to reduce silver ions in substrate. (c) The process halts when path is formed. (d) The process is reversed with application of reverse bias. Reprinted with permission from J. A. Nessel, R. Q. Lee, C. H. Mueller, M. N. Kozicki, M. Ren, and J. Morse, A novel nanoionics-based switch for microwave applications, *2008 IEEE MTT-S International Microwave Symposium Digest*, June 2008, pp. 1051–1054. Copyright (2008) IEEE.

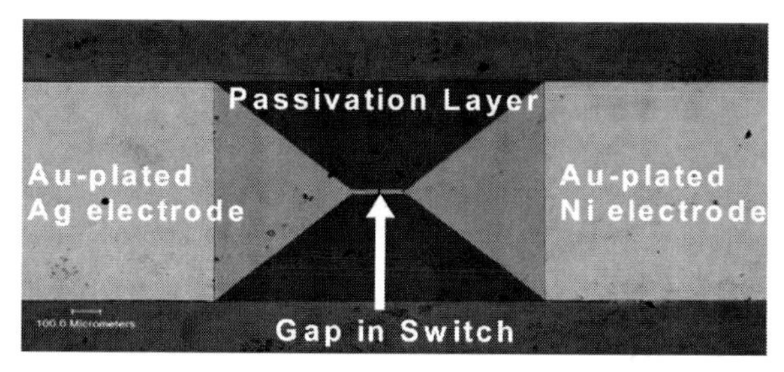

Figure 8.14 Microphotograph of coplanar nanoionic switch design. Reprinted with permission from J. A. Nessel, R. Q. Lee, C. H. Mueller, M. N. Kozicki, M. Ren, and J. Morse, A novel nanoionics-based switch for microwave applications, *2008 IEEE MTT-S International Microwave Symposium Digest*, June 2008, pp. 1051–1054. Copyright (2008) IEEE.

The performance of the switch as a function of frequency is shown in Fig. 8.15, demonstrating excellent RF properties in the frequency range from 1 to 6 GHz. The switch RF performance and its no-moving parts nature, make it suitable for space communication systems where reliability and long mean time before failure (MTBF) are highly desired.

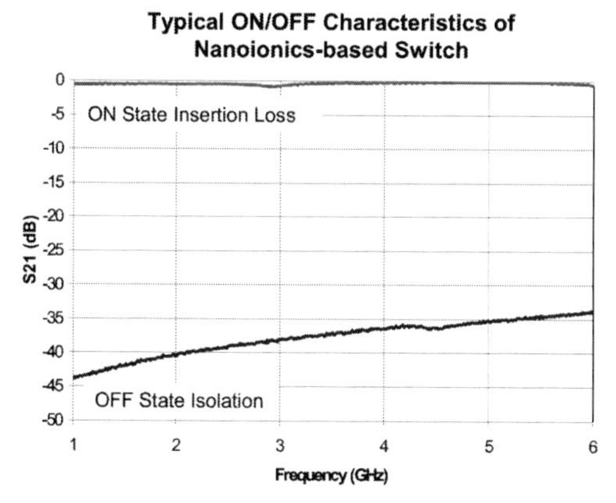

Figure 8.15 The insertion loss (S_{21}) of a switch in the ON state is better than 0.5 dB over the 1 to 6 GHz frequency range (ISM band), while the isolation in the OFF state is better than –35 dB. Reprinted with permission from J. A. Nessel, R. Q. Lee, C. H. Mueller, M. N. Kozicki, M. Ren, and J. Morse, A novel nanoionics-based switch for microwave applications, *2008 IEEE MTT-S International Microwave Symposium Digest*, June 2008, pp. 1051–1054. Copyright (2008) IEEE.

8.5.6 CNT Vacuum Microelectronics

Advances in nanotechnology and silicon micromachining techniques have resulted in the resurgence of vacuum tube technology for RF frequencies. In order to realize high-power sources at high frequencies—from W-Band (94 GHz) to THz—several groups are researching different vacuum tube designs that take advantage of silicon and metallic micromachined structures with CNT field emission electron sources [43,44]. Significant challenges remain in developing highly reliable, high-current density, robust cold electron sources that can be seamlessly integrated with three-dimensional vacuum-packaged microstructures. Several field emission CNT cathode architectures have been reported in the literature to produce high-current densities [45,46], but long lifetime operation with minimal performance degradation is still a challenge.

Other than for RF sources, vacuum microelectronics is being developed for digital applications as well. Recently, a vacuum electronic-based programmable logic gate operation was demonstrated at high temperatures (>700°C) [47]. These "digital" vacuum electronic devices have been shown to be potentially comparable in device density and switching speeds to those of their solid-state counterparts (CMOS devices). Such devices are desirable to NASA's extreme environment applications involving high temperature and pressures (e.g., Venus), and high radiation (e.g., Jovian neighborhood).

8.6 Conclusions

In this chapter, we have presented a brief overview and some examples of the opportunities offered by nanotechnology in the development of nanoelectronic devices for NASA as well as non-NASA applications. The progress made so far through prototype demonstrations based on CNTs, graphene and polymer nanowires, among others, render credibility of the potential of this field in advancing nanoelectronics circuits for sensing, communications, and control applications, among others. Yet, to fully harvest the benefits of this technology, there are still challenges that must be addressed and surpassed. For example, identifying and optimizing techniques leading to mass manufacturability of dense, aligned, and purified semiconducting nanotube networks are essential toward the growth of nanoelectronics. Also, the attainment of fully integrated nanosystems calls for further optimization of individual nano-based electronic components such as nanoFETs, sensors, antennas, amplifiers, signal processing devices, mixers, etc. as well as for the development of design and fabrication protocols for their full integration and compatibility with state of the art integrated circuitry. We believe that the future of these endeavors is a bright and promising one.

References

1. L. Pierantoni, RF Nanotechnology-Concept, Birth, Mission and Perspectives, *IEEE Microw. Mag.*, 130–137, June 2010.
2. http://download.intel.com/museum/Moores_Law/Printed_Materials/Moores_Law_2pg.pdf.
3. http://www.nanocyl.com/CNT-Expertise-Centre/Carbon-Nanotubes.
4. R. Martel, T. Schmidt, H. R. Shea, T. Hertel, and P. Avouris, Single- and Multi-Wall Carbon Nanotube Field-Effect Transistors, *Appl. Phys. Lett.*, **73**(17), 2447–2449, 26 October 1998.
5. A. Javey, J. Guo, D. B. Farmer, Q. Wang, D. Wang, R. G. Gordon, M. Lundstrom, and H. Dai, Carbon Nanotube Field-Effect Transistors with Integrated Ohmic Contacts and High-K Gate Dielectrics, *Nanoletters*, **4**(3), 447–450, 2004.
6. B. Tian, T. Cohen-Karni, Q. Qing, X. Duan, P. Xie, and C. M. Lieber, Three-Dimensional Flexible Nanoscale Field-Effect Transistors as Localized Bio-probes, *Science*, **329**, 830–834, August 2010.
7. H. M. Manohara, E. W. Wong, E. Schlecht, B. D. Hunt, P. H. Siegel, Carbon Nanotube Schottky Diodes Using Ti-Schottky and Pt-Ohmic Contacts for High Frequency Applications, *Nano Lett.*, **5**(**7**), 1469–1474 (2005).
8. M. H. Yang, K. B. K. Teo, W. I. Milne, and D. G. Hasko, Carbon Nanotube Schottky Diode and Directionally Dependent Field-Effect Transistor Using Asymmetrical Contacts, *Appl. Phys. Lett.*, **87**, 253116 (2005).
9. T. Kaji, T. Takenobu, A. F. Morpurgo, and Y. Iwasa, Organic Single-Crystal Schottky Gate Transistors, *Adv. Mater.*, **21**, 1–5, 2009.
10. J. A. Rogers, T. Someya, and Y. Huang, Materials and Mechanics for Strechable Electronics, *Science*, **327**, 1603–1607, March 26, 2010.

11. M. Wilson, Conductive Elastic Polymers form the Basis for Organic Skinlike Electronics, *Phys. Today*, 18–19, October 2008.
12. J. Viventi, D.-H. Kim, L. Vigeland, E. S. Frechette, J. A. Blanco, Y.-S. Kim, A. E. Avrin, V. R. Tiruvadi, S.-W. Hwang, A. C. Vanleer, D. F. Wulsin, K. Davis, C. E. Gelber, L. Palmer, J. V. der Spiegel, J. Wu, J. Xiao, Y. Huang, D. Contreras, J. A. Rogers, and B. Litt, Flexible, Foldable, Actively Multiplexed, High-Density Electrode Array for Mapping Brain Activity *in vivo, Nat. Neurosci.*, **14**(12), 1599–1605, *November* 13, 2011.
13. T.-H. Lee1, S. Bhunia, and M. Mehregany, Electromechanical Computing at 500°C with Silicon Carbide, *Science*, **329**(5997), 1316–1318 (2010).
14. X. L. Feng, M. H. Matheny, C. A. Zorman, M. Mehregany, and M. L. Roukes, Low Voltage Nanoelectromechanical Switches Based on Silicon Carbide Nanowires, *Nano Lett.*, 2010, **10**, 2891–2896.
15. G. Rebeiz, RF MEMS Switches: Status of the Technology, In *Proceedings of the 12th International Conference on Solid State Sensors, Actuators and Microsystems*, pp. 1726–1729, Boston, June 8–12, 2003.
16. A. B. Kaul, E. W. Wong, L. Epp, B. D. Hunt, Electromechanical Carbon Nanotube Switches for High-Frequency Applications, *Nano. Lett.*, **6**, 942 (2006).
17. A. Kaul, E. Wong, and L. Epp, Fast Electromechanical Switches Based on Carbon Nanotubes, *NASA Tech. Briefs*, p. 32, May 2008.
18. N. Salaets and M. Ervin, Solution Deposition Methods for Carbon Nanotube Field-Effect Transistors, *Defense Tech. Briefs*, 24–26, June 2010.
19. A. M. Dorsey and M. H. Erwin, Effects of Differing Carbon Nanotube Field-Effect Transistors Architectures, *Defense Tech. Briefs*, 26–27, February 2010.
20. M. Chin and S. Kilpatrick, Differential Amplifier Circuits Based on Carbon Nanotube Field Effect Transistors (CNTFETs), *Defense Tech. Briefs*, 37–38, October 2010.
21. K. Kim, T. M. Wallis, P. Rice, C.-J. Chiang, A. Imtiaz, P. Kabos, and D. Filipovic, A Framework for Broadband Characterization of Individual Nanowires, *IEEE Microwave Wireless Comp. Lett.*, **20**(3), 178–180, March 2010.
22. R. K. Challa, D. Kajfez, V. Demir, J. R. Gladden, and A. Z. Elsherbeni, Characterization of Multiwalled Carbon Nanotube (MWCNT) Composites in a Waveguide of Square Cross Section, *IEEE Microwave Wireless Comp. Lett.*, **18**(3), 161–163, March 2008.

23. M. Petkov, Disabling CNT Electronic Devices by use of Electron Beams, *NASA Tech. Briefs*, **34**, March 2008.

24. S. J. Koh, Strategies for Controlled Placement of Nanoscale Building Blocks, *Nanoscale Res. Lett.*, **2**, 519–545 (2007).

25. Y. Maeda, S. Kimura, M. Kanda, Y. Hirashima, T. Hasegawa, T. Wakahara, Y. Lian, T. Nakahodo, T. Tsuchiya, T. Akasaka, J. Lu, X. Zhang, Z. Gao, Y. Yu, S. Nagase, S. Kazaoui, N. Minami, T. Shimizu, H. Tokumoto, R. Saito, Large-Scale Separation of Metallic and Semiconducting Single-Walled Carbon Nanotubes, *J. Am. Chem. Soc.*, 2005, **127**, 10287–10290.

26. Z. Chen, W. Hu, and J. Guo, K. Saito, Fabrication of Nanoelectrodes Based on Controlled Placement of Carbon Nanotubes Using Alternating-Current Electric Field, *J. Vac. Sci. Technol. B*, **22**(2), 776–780 (2004).

27. G. Warwick and R. Wall, Black Magic: Nano-Enhanced Composite Materials Could Displace Metals and Drive Cost Downs, *Aviation Week Space Technol.*, 89–92, November 1/8, 2010.

28. A. Sinitskii and J. M. Tour, Graphene Electronics, Unzipped, *IEEE Spectrum.*, 30–33, November 2010.

29. Mark Wilson, Graphene Production Goes Industrial, *Phys. Today*, 15–16, August 2010.

30. P. Avouris, Carbon Nanotube Electronics and Photonics, *Phys. Today*, 34–40, January 2009.

31. M. L. Cohen and A. Zettl, The physics of boron nitride nanotubes, *Phys. Today*, 34–38, November 30, 2010.

32. NASA's Space Operations Project Office (http://spaceflightsystems.grc.nasa.gov/SpaceOps/CoNNeCT/).

33. N. J. Pinto, A. T. Johnson, A. G. MacDiarmid, C. H. Mueller, N. Theofylaktos, D. C. Robinson, and F. A. Miranda, Electrospun Polyaniline/Polyethylene Oxide Nanofiber Field-Effect Transistor, *Appl. Phys. Lett.*, **83**, 4244–4246 (2003).

34. N. J. Pinto, R. Pérez, C. H. Mueller, N. Theofylaktos, and F. A. Miranda, Dual Input AND Gates Fabricated from a Single Channel Poly(3-Hexylthiophene Thin Film Field Effect Transistor, *J. Appl. Phys.*, **99**, 084504, 1–5 (2006).

35. C. H. Mueller, N. Theofylaktos, F. A. Miranda, A. T. Johnson, and N. J. Pinto, Demonstration of AND Logic Circuit Using Split-Gate Pentacene Transistors, *Thin Solid Films*, 496/2, 494–499 (2006).

36. F. A. Miranda, M. A. Meador, O. Theofylaktos, N. J. Pinto, C. H. Mueller and J. S Peìrez, Graphene-based Reversible Nano-switch/sensor Schottky Diode, *NASA Tech. Brief*, **34**(10), 24–25, October 2010.

37. Volker Schmidt, Joerg V. Wittemann, Stephan Senz, and Ulrich Gösele, "Silicon Nanowires: A Review on Aspects of their Growth and their Electrical Properties," *Adv. Mater.* 2009, **21**, 2681–2702; "Silicon Nanowires, in Aldrich Material Science "*Material Matters: Methods for Nanopatterning and Lithography*," **6**(1), 11.

38. A. Sridhar, T. Blaudeck, R. R. Baumann, Inkjet Printing as a Key Enabling Technology for Printed Electronics, *Aldrich Material Science Material Matters: Methods for Nanopatterning and Lithography*, **6**(1), 12–15.

39. M. Y. Chen, X. Lu, H. Subbaraman, and R. T. Chen, Fully Printed Phased-Array Antenna for Space Communications, *Proc. SPIE*, 7318, 731814-1–6 (2009).

40. Yihong Chen and Xuejun Lu, Fully Printed High-Frequency Phased Array Antenna on Flexible Susbtrate, *NASA Tech. Brief*, **34**(6), 40, June 2010.

41. P. Russer, N. Fitchner, P. Lugli, W. Porod, J. Russer, and H. Yordanov, Nanoelectronics-based Integrated Antennas, *IEEE Microwave Magazine*, 58–71, December 2010.

42. J. A. Nessel, R. Q. Lee, C. H. Mueller, M. N. Kozicki, M. Ren, J. Morse, A Novel Nanoionics-Based Switch for Microwave Applications, 2008 IEEE MTT-S International Microwave Symposium Digest, June 2008, pp. 1051–1054.

43. H.M. Manohara, R. Toda, R.H. Lin, A. Liao, M.J. Bronikowski, P.H. Siegel, Carbon Nanotube Bundle Array Cold Cathodes for THz Vacuum Tube Sources, J Infrared Milli Terahz Waves, 30, pp. 1338–1350 (2009).

44. P. H. Siegel, THz Technology, *IEEE Transactions on Microwave Theory and Techniques*, **50**(3), MARCH 2002.

45. H. M. Manohara, M. J. Bronikowski, M. Hoenk, B. D. Hunt, P. H. Siegel, High-Current-Density Field Emitters Based on Arrays of Carbon Nanotube Bundles, *J. Vacuum Sci. Technol. B*, **23**(1), 157–161 (2005).

46. D. S. Y. Hsu, J. Shaw, Integrally Gated Carbon Nanotube-on-Post Field Emitter Arrays, *Appl. Phys. Lett.*, **80**(1), 7 January 2002.

47. H. Manohara, R. Toda, R. H. Lin, A. Liao, M. Mojarradi, Carbon Nanotube-Based Digital Vacuum Electronics and Miniature Instrumentation for Space exploration, MOEMS and Miniaturized Systems IX, Photonics West 2010, *Proc. SPIE*, 7594, 75940Q (2010).

Chapter 9

Brief Introduction to Nanocomposites for Electromagnetic Shielding

Stefano Bellucci and Federico Micciulla

Laboratori Nazionali di Frascati, Instituto Nazionale di Fisica Nucleare, Via Enrico Fermi, 40 00044 Frascati, Rome, Italy

bellucci@lnf.infn.it

9.1 Introduction

Through the centuries, the history of human civilization has been related to the development of scientific discoveries into new technologies and materials and vice versa. Civilization evolved from the Stone Age to the Bronze Age, the Iron Age, the Steel Age, and to, nowadays, the Nano Age. Each age is marked by the advent of certain materials. For example, the Iron Age brought tools and utensils; the Steel Age gave rails and the Industrial Revolution [1]. The Nano Age is a promising "New Age" where different kinds of scientific fields and technologies will contribute, together, to realize high-performance materials. It is used to call it as Nanotechnology Age. This term was neologized, for the first time, by Eric Drexler, in 1986: "...a technology at molecular level that can let us set every atom where we want it to be. We called this ability 'Nanotechnology,'

Advanced Nanomaterials for Aerospace Applications
Edited by Carlos R. Cabrera and Félix A. Miranda
Copyright © 2014 Pan Stanford Publishing Pte. Ltd.
ISBN 978-981-4463-18-8 (Hardcover), 978-981-4463-19-5 (eBook)
www.panstanford.com

because it works on nanometric scale, one billionth of meter." Nanotechnology is a pervasive and enabling technology and promises breakthroughs in different areas; due to this, it is spreading to several fields: medicine, aerospace, electronic, mechanical, manufacturing, biology, materials, information technology, national security, and go on. Nanotechnology is the control of matter at the nanoscale and the exploitation of the novel properties and phenomena developed at that scale in the creation of practical materials, devices, and systems. It will have a profound impact on worldwide economy and society in the 21st century, comparable to that of semiconductor technology, information technology, or cellular and molecular biology. It is easy to believe that nanotechnology will be the next Industrial Revolution. The discovery of novel materials, processes, and phenomena at the nanoscale and the development of new experimental and theoretical techniques for research provide new opportunities for the development of innovative nanosystems and nanostructured materials. The real hit of nanotechnology can be found in the study of the matter and wanted modification of properties of materials at the nanoscale, they can be very different from those at a larger scale. Nanomaterials especially biologically based ones, have always been around us, but it has only been in recent years that we have had the tools to identify these materials, to study and characterize them, and to manipulate in a controlled process. Development of new systems of investigation, as electronic microscopes, has helpful on discovering and characterization of nanomaterials.

When the dimension of a material is reduced from a large size, the properties remain the same at first, and then small changes occur, until finally when the size drops below 100 nm, dramatic changes in properties can occur. Materials can be nanostructured for new properties and advanced performances. This field is opening new avenues in science and technology [2].

Here we have focused our analysis on a specific class of nanomaterials: nanocomposites, for Electromagnetic shielding application.

9.2 Nanocomposites

In the past 70 years, it is now when the field of macro- and the nanocomposites is growing very much. Different kinds of composites have been designed and employed in wide fields of applications in these years: aerospace and aeronautics industry, automotive industry, civil engineering, electronics, medical equipment, and sport tools. The demand for a major employment of composites increases in the coming years, gaining insight on major market drivers, supply and demand, price movements, market opportunities and competitive forces proves critically important. Composite materials are interesting because they are a multifunctional, and joining of different phases of them creates unique and high-performance materials. As a simple definition, it is possible to relate to a composite when two or more different materials are combined together to create a superior and unique material. Composite materials are multiphase materials, with high properties. Usually the single phases that make up a composite material do not possess properties similar to that of the composite materials. It is possible to design a composite material possessing desired properties and with precise and specific combination of phases. Usually it is easy to identify the different phases, making up the composite: one as matrix, it includes inside other phase, and, the other or others one as filler, the material embeddeds into the matrix. It is useful to classify composite materials as follows:

- composites with metal matrix
- composites with ceramic matrix
- composites with polymeric matrix

Each kind of composite has its own properties and field of application. This chapter focuses on polymer composites. Polymer composites are important commercial materials and find applications such as filled elastomers for damping, electrical insulators, thermal conductors, and high-performance composites for use in aircraft. It is possible to achieve high-stiffness but lightweight composite materials, for example, with addition of carbon fiber into the polymer matrix. Polymeric composite materials' high performance is a compromise of right equilibrium between the matrix and the filler. It is important to achieve a homogenous dispersion of the filler materials inside the polymeric matrix. Phases have to be well homogenized to create composite materials with high and uniform properties. It is necessary to take care of this during the designing procedure of the material. On the other hand, non-uniform dispersion or exceeding the amount of filler into the matrix can decrease the performance of composite materials.

In the past years, the efforts to obtain high-performance composite materials have given rise to a new class of composites: nanocomposites.

Nanocomposites are a class of composites in which one or more separate phases have one dimension in the nanoscale. The field of nanocomposites involves the study of multiphase materials in which at least one of the constituent phases' dimension is less than 100 nm [3], Fig. 9.1.

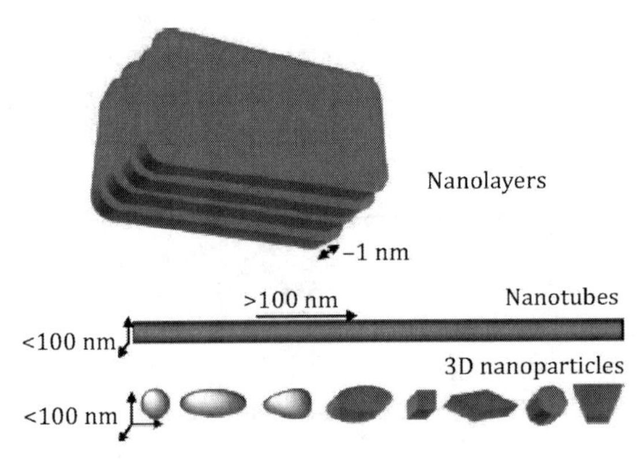

Figure 9.1 Various types of nanostructured materials [8].

Nanostructure materials can be classified as three main categories [4,5]:

- **Nanoparticles:** When the three dimensions of particulates are in the order of nanometers, they are related to equi-axed (isodimensional) nanoparticles or nanogranules or nanocrystals; e.g., silica.
- **Nanotubes:** When two dimensions are in the nanometer scale and the third is larger, forming a lengthened structure, they are normally identified to "nanotubes" or nanofibers/whiskers/nanorods; e.g., carbon nanotubes (CNTs), cellulose whiskers.
- **Nanolayers:** The particulates that are characterized by only one dimension in the nanometer scale are nanolayers/nanosheets. They are present in the form of sheets of one to a few nanometer thick to hundreds to thousands nanometers

long e.g., clay (layered silicates), layered double hydroxides (LDHs), graphene.

Nanocomposites improve the good properties of the matrix with the enhanced performance of the nanostructures. The reason for applying nanoscale fillers into polymer matrices is the attainment of potentially unique properties, as a result of nanometric dimensions. In particular, it is possible to achieve high results using a low volume or weight fraction of nanomaterials, comparing to microfillers. For example, the inclusion of equi-axed nanoparticles in thermoplastics and particularly in semicrystalline thermoplastic, increases the yield stress, the tensile strength, and Young's modulus compared to pure polymer. A volume fraction of only 0.04 wt% mica-type silicates in epoxy increases the modulus below the glass transition temperature by 58% and modulus in the rubbery region by 450% [6,7]. The extreme aspect ratios of the nanofiller, compared to the conventional micro-fillers, make the nanocomposite successful. It is possible to distinguish the nano-effects in six interrelated characteristics:

(1) low-percolation threshold (~0.1–2 vol%)
(2) particle–particle correlation (orientation and position) arising at low-volume fractions ($Ø_C < 0.001$)
(3) large number density of particles per particle volume (10^6–10^8 particles/μm^3)
(4) extended interfacial area per volume of particles (10^3–10^4 m^2/mL)
(5) short distances between particles (10–50 nm at $Ø \sim 1$–8 vol%)
(6) comparable size scales among the inclusion of rigid nanoparticles, the distance between particles, and the relaxation volume of polymer chains [8]

It is good to notice how after downscaling the size to the nanoscale, the materials have an interfacial area per volume many times higher than the same at the microscale. In addition, with this aspect, the use of the nanofiller allows one to reduce drastically the amount to add to the polymer: Even at low contents of nanofiller (below 5% in weight of the nanofiller against greater than percentages of 15% by weight of classic filler), much better performance can be obtained. Therefore, it is an important objective to achieve high performance and low weight, especially, for aerospace or aeronautic applications. This happens substantially because of the "nano-effect" [9], surface effects become important at nanoscale. The interfacial region drives the communication between the nanofiller and polymeric matrix and thus it controls the properties of composite materials. Because of its proximity to the surface of the filler, it shows, conventionally, properties different from the bulk matrix. Locally, it is possible to have different chemistries, curing degrees, polymer chains' mobilities, and crystal order's degrees. This is explained in terms of the radius of gyration of the matrix (R_g), which is key spatial parameter to which the majority of the polymers' static and dynamic properties can be ultimately related and has a value in a few tens of nanometers [8]. So, there exists a strong dependence of interfacial area per volume fraction (μm^{-1} = m^2/mL) of the filler on the aspect ratio (α) of nanoscale particles with varying shapes. The aspect ratio ($\alpha = H/R$) is based on approximating particles as cylinders (area/volume = $1/H + 1/R$). Afterward, the relative volume fraction of interfacial material to bulk is drastically increased as the size becomes smaller (Fig. 9.2).

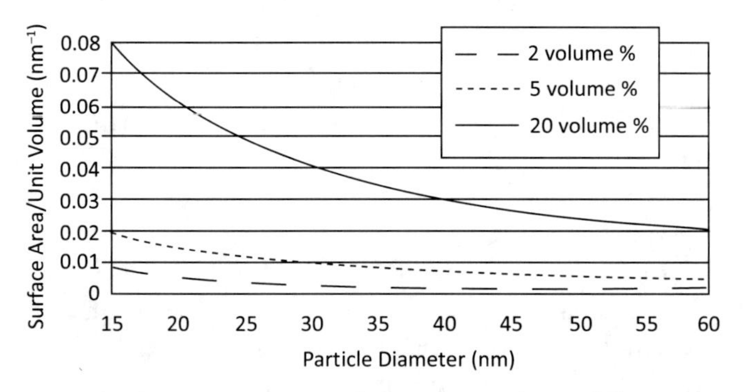

Figure 9.2 Surface area per unit volume vs. particle size [3].

Aspect ratios greater than 1 correspond to rods (length/diameter) and less than 1 correspond to plates (height/diameter) [10]. For example, fully exfoliated and dispersed high-aspect-ratio plates or rods, such as montmorillonite and single-wall carbon nanotubes (SWCNTs), generate internal interfacial area comparable to that of macromolecular structures, such as dendrimers and proteins, and two to three orders of magnitude more than classic mineral fillers.

It is useful to underline how the interaction of the polymer's macromolecules with the nanoparticles gives significant opportunity for changing the polymer mobility and relaxation dynamics. For example, polystyrene chains intercalated between the layers of a smectic clay have more mobility locally than in the bulk polymer [11]. This may create an ordering between layers, and then it leads to the formation of low- and high-density regions with different degrees of mobility of polymeric chains: In the low-density regions, there are more opportunity for mobility. It is clear how the rheological/glass transition temperature of a polymer can be controlled by changing the polymer mobility with nano-composites interfaces. The molecular mobility highly depends on the available free volume, which plays an important role in deciding the mechanical properties, diffusion of small molecules through the polymer, and so on. The glass transition temperature (T_g) is the most important thermal transition that a polymer undergoes. At the T_g, the polymer softens due to the onset of long-range coordinated molecular motion accompanied by a change in the free volume properties in the polymer [12]. In other words, It is a temperature range beyond which the polymer becomes soft and leathery. Below T_g, the polymer becomes hard and cannot be used under dynamic loading conditions [13]. For example, T_g can be eliminated for clay nanocomposites with intercalated polymer chains. This indicates a limited ability for cooperative chain motion when the polymers are confined between layers, and the T_g does not change significantly. If the polymer is tightly bound to the nanofiller, T_g can be increased without help of confinement [14].

Recently, Díez-Pascual et al. showed how it is possible to achieve a remarkable increase in the storage and Young's modulus, T_g and degradation temperatures of the poly(ether ether ketone) (PEEK) matrix with the addition of small amounts of SWCNTs wrapped in compatibilizing agents, such as polysulfone, an amorphous thermoplastic miscible with and structurally similar to PEEK [15,16]. Polysulfone possesses dual affinity with PEEK and the SWCNTs; its phenyl moieties interact with the matrix through p–p stacking as well as with the sp^2-bonded hexagonal networks of the nanofillers. Furthermore, the chemical interaction of polar segments of the compatibilizer (sulfone, ether, etc.) to surface groups located on the SWCNT structure and defect sites (side-walls or open ends) boosts the compatibilization effect, leading to composites with enhanced nanofiller dispersion and CNT–polymer load transfer; as a result, the mechanical properties of these composites are superior to those containing pristine CNTs.

The significant T_g enhancement in the compatibilized systems is attributed to the high T_g of the amorphous polysulfone combined with the strong restrictions on the polymer chain motion induced by the improved adhesion between the matrix and nanofiller phases. In contrast, the non-compatibilized samples show a smaller increase, probably due to partial agglomeration of the SWCNTs at the fiber interface. The agglomerates could alter the flow behavior of the matrix during impregnation, resulting in nanometer-scale porosities [17]; this leads to an increase in the free volume between the polymer molecules, allowing additional mobility, which may explain their lower T_g [18]. Karapappas et al. analyzed the effect of the addition of four different commercially available nanofillers into a state of the art epoxy-based gelcoat. They showed how the inclusion of the nanofillers had no practical impact on T_g, but they found a different improvement in the composite's properties related to nanoinclusions; in fact, the addition of CNTs and the exfoliated nanographite leads to an enhancement of all properties, while the addition of nanoclay and nano-titanium dioxide was beneficial only for the tensile and thermal conductivity properties. The presence of the nanofillers brought a lot of improvements in the properties of the gelcoats. The aforementioned improvements include the tensile and fracture properties as well as the electrical and thermal ones. The presence of the nanofillers also decreased the coefficient of linear thermal expansion and made the nanodoped epoxy gelcoats more resistant to UV degradation. One of the drawbacks of the introduction of the nanofillers into the resin was the increase of viscosity, which can be an issue regarding the workable ability of those materials when composite manufacturing is considered. The most common gelcoats are based on epoxy or unsaturated polyester resin chemistry. Gelcoats are modified resins that are applied to moulds in the liquid state. They are cured to form cross-linked polymers and are subsequently backed up with composite polymer matrices, often mixtures of polyester resin and fiberglass or epoxy resin with glass, kevlar and/or carbon fibers. The manufactured component, when sufficiently cured and removed from the mould, presents the gelcoated surface. This is usually pigmented to provide a colored, glossy surface, which improves the aesthetic appearance of the component. Gelcoats are designed to be durable, providing resistance to ultraviolet degradation and hydrolysis. Specialized gelcoats can be used to manufacture the moulds, which in turn are used to manufacture components. The nanofillers were introduced into the gelcoat in the form of a master batch that was prepared by using a high shear mixing device, i.e., a three-roll mill [19].

Therefore, it is easy to understand how through an appropriate choice of the type of nanoinclusions and of a just combination of them into the polymer matrix and a precise method of dispersion, you can control and transfer the high properties of the nanostructure in the resulting nanocomposite.

The key concept of nanocomposites is focused not only on the shape of the particle but also on the engineering and tailoring of the nanoparticles' exact morphology with specific characteristics to achieve a desired property suite from the resultant composite materials.

The incorporation of nanoscale particles into polymer can generally be done in four ways as follows:

(a) Solution method: It involves the dissolution of polymers in adequate solvent with nanoscale particles and evaporation of solvent or precipitation.
(b) Melt mixing: In this method, the polymer is directly melt-mixed with nanoparticles.
(c) In situ polymerization: In this method, the nanoparticles are first dispersed in liquid monomer or monomer solution. Polymerization is performed in the presence of nanoscale particles.
(d) Template synthesis: This method is completely different from other methods. In this method, using polymers as template, the nanoscale particles are synthesized from precursor solution.

Nanocomposites—as these types of materials are often referred to—have received much attention over the past decade as scientists search for ways to enhance the properties of engineering polymers while retaining their processing ease. Unlike traditional filled-polymer systems, nanocomposites require relatively low-dispersant loadings (~2 wt%) to achieve significant property enhancements. Some of these enhancements include increased Young's modulus, increased mechanical behavior, increased electrical conductivity, increased gas barrier, increased thermal performance, increased atomic oxygen resistance, resistance to small molecule permeation, and improved ablative performance. As a result of these enhancements, nanocomposites have the potential to play a significant role in future space systems, for example, and in many other fields like biology, medicine, electronics, automotive industry, and so on. Launch vehicles would greatly benefit from appropriately designed nanocomposites that could provide improved barrier properties and gradient morphologies enabling linerless composite cryogenic fuel tanks. Self-rigidizing, self-passivating nanocomposite materials could be used to construct space vehicle components that are both highly resistant to space-borne particles and resistant to degradation from electromagnetic radiation, while reducing the overall weight of the spacecraft. Current estimates for launching payloads into space orbit stand at $10,000/lb ($22,000/kg) [20]. Significant weight, and hence cost, reductions can be realized with the use of organic materials, but such materials tend to perform very poorly in the harsh space environment. Organic polymers with uniformly dispersed nanoscale inorganic precursors may enable these materials to withstand the harsh space environment and be used as critical weight reduction materials on current and future space systems [8].

Table 9.1 Estimation of worldwide volume and value for polymer-nanocomposites (2003–2008) [21]

	2003		2008		
Polymer	Volume ($\times 10^6$ kg)	Volume ($\times 10^6$ \$)	Volume ($\times 10^6$ kg)	Volume ($\times 10^6$ \$)	AAGR (%) (2003–2008)
Thermoplastics	5.7	70.7	27.7	178.9	20.4
Thermosets	5.4	20.1	8.2	32.2	9.9
Total	11.1	90.8	35.9	211.1	18.4

This work studies nanostructured epoxy materials that were prepared and used as coatings for electrical equipment in order to protect them from electromagnetic interference (EMI) that can occur when the electromagnetic waves of a device interfere with another device, causing an unwanted response. Carbon Nanotubes have been employed as fillers.

9.3 Dielectric and Magnetic Materials

The world "dielectric" is used as an adjective and covers a wide range of materials. It refers to an electrical insulator that can be polarized by an applied electric field. The electromagnetic (EM) properties of materials at macroscopic level are due to the modes of interactions involved between the materials (atomic and/or molecular level) and electric and/or magnetic fields. From the achieved response of involved materials with interaction, it is possible to make a classification of their electromagnetic properties. A vast wealth of experimental evidence accumulated over the past two centuries shows that large-scale electromagnetic phenomena are governed by Maxwell's equations. Therefore, a roundup of Maxwell's equations is important:

$$\begin{cases} \nabla \times \mathbf{E}(r,t) = -\partial \mathbf{B}(r,t)/\partial t \quad \text{(Faraday Law)} \\ \nabla \times \mathbf{H}(r,t) = \partial \mathbf{D}(r,t)/\partial t + (r,t) \quad \text{(Ampére Law)} \\ \nabla \cdot \mathbf{D}(r,t) = \rho(r,t) \quad \text{(Gauss Law)} \\ \nabla \cdot \mathbf{B}(r,t) = 0 \quad \text{(Gauss Law)} \\ \nabla \cdot \mathbf{J} + (\partial \rho(r,t)/\partial t) = 0 \quad \text{(Conservation of change)} \end{cases}$$

According to the traditional usage, **E** and **H** are known as the intensities, respectively, of the electric and magnetic field, **D** is called the electric displacement, **B** is the magnetic induction, $\boldsymbol{\rho}$ the charge density, and **J** the current density. No other assumptions have been made thus far than that an EM field may be characterized by four vectors **E**, **B**, **D**, and **H**, which at ordinary points satisfy Maxwell's equations, and that the distribution of current, which gives rise to this field, is such as to ensure the conservation of charge. Between the five vectors **E**, **B**, **D**, **H**, and **J**, there are but two independent relations—the Eqs. (9.1) and (9.2). To determinate these set of equations further conditions need to be imposed. If it is assumed we will be in free space, could be written in following the traditional usage:

$$\mathbf{D} = \varepsilon_0 \mathbf{E} \tag{9.1}$$

$$\mathbf{B} = \mu_0 \mathbf{H} \tag{9.2}$$

$$\mathbf{J}_\mathrm{f} = \sigma \mathbf{E}, \tag{9.3}$$

where the values and the dimensions of the constants ε_0 (permittivity in vacuum) and μ_0 (permeability in vacuum) are fixed and they depend upon the system of units adopted, and where $\mathbf{J}_\mathrm{f}$ is free current density.

If the physical properties of a body in the neighborhood of some interior point are the same in all directions, the body is said to be isotropic. At every point in an isotropic medium, **D** is parallel to **E** and **H** is parallel to **B**; so the relationship becomes as follows:

$$\mathbf{D} = \varepsilon \mathbf{E} \tag{9.4}$$

$$\mathbf{B} = \mu \mathbf{H} \tag{9.5}$$

$$\mathbf{J} = \sigma \mathbf{E} \tag{9.6}$$

$$\mathbf{J} = \mathbf{J}_\mathrm{f} + \mathbf{J}_\mathrm{b} \tag{9.7}$$

The medium can be characterized electromagnetically by the three constants ε (permittivity), μ (permeability), **J** (total current density), $\mathbf{J}_\mathrm{b}$ (bound current density) and σ (conductivity). In (9.4), **D** describes the density of flux of electric force line into the materials.

To describe the electromagnetic state of a sample of matter, it will be convenient to introduce two additional vectors. We shall define the electric and magnetic polarization vectors by the equations in the medium. If we assume that the conductivity is other than zero, any initial free charge distribution in the medium must vanish spontaneously.

$$\mathbf{P} = \mathbf{D} - \varepsilon_0 \mathbf{E} \tag{9.8}$$

$$\mathbf{M} = \mathbf{B}/\mu_0 - \mathbf{H} \tag{9.9}$$

The polarization vectors (**P**, **M**) are thus definitely associated with matter and vanish in free space. We do not go into details, but it is important to underline that in isotropic media, the polarization vectors are found experimentally to be proportional to the corresponding field vectors, ferromagnetic materials are excluded. The electric and magnetic susceptibilities χ_e, and χ_m, are defined by the following relations:

$$\mathbf{P} = \chi_e \varepsilon_0 \mathbf{E} \tag{9.10}$$

$$\mathbf{M} = \chi_m \mathbf{H} \tag{9.11}$$

It is possible to identify two new parameters of material: relative permittivity ε_r, and relative permeability, μ_r, of a material.

When ε_r is greater than 1, we refer to materials as dielectric materials, or dielectrics, such as glass, paper, mica, various ceramics, polyethylene, and certain metal oxides. A high permittivity tends to reduce any electric field present. For instance, the capacitance of a capacitor can be raised by increasing the permittivity of the dielectric material. Dielectrics are generally used in capacitors and transmission lines in filtering EMI by alternating currents (AC), audio frequencies (AF), and radio frequencies (RF).

Magnetic fields are given by charges in motion, into the materials. All magnetic effects are able to attribute to the time-varying system of electric charges, which leads the macroscopic aspects of magnetism.

Relative permeability, μ_r, is the ratio of the permeability of a specific medium to the permeability of free space given by the magnetic constant. The permeability is a parameter that gives the magnetic property of medium. When value of μ_r is greater than 1, the materials are called ferromagnetic (as iron); they show a permanent magnetism and it is easy to influence them through the action of the external magnetic fields. When μ_r is equal to 1, we have diamagnetic materials; they are not susceptible to magnetization (such as plastic, polymer, etc.). There are some materials that are weakly magnetizable; they are defined as paramagnetic.

$$\varepsilon_r = (1 + \chi_e) \tag{9.12}$$

$$\mu_r = (1 + \chi_m) \tag{9.13}$$

The relative permittivity, ε_r, of a material for a frequency of zero is known as its static relative permittivity or as its dielectric constant; under given conditions, ε_r reflects the extent to which it concentrates electrostatic lines of flux. Technically, it is the ratio of the amount of electrical energy stored in a material by an applied voltage, relative to that stored in a vacuum. Similarly, it is also the ratio of the capacitance of a capacitor using that material as a dielectric, compared to a similar capacitor that has a vacuum as its dielectric [22].

The electric polarization of a dielectric material may be due to four major types of polarization. When an electric field is applied to dielectric material between two plates of capacitor, it generates a displacement of charge within the material, which is created through a progressive orientation of permanent or induced dipoles. The interaction between permanent or induced electric dipoles with an applied electric field is called polarization, which is the induced dipole moment per unit volume. The four contributions are as follows:

Electronic polarization [$\mathbf{P}_e$] arises from the valence electron cloud after 10^{-14} s of application of a sharp step function of field

$$\mathbf{P}_e = N\alpha_e\mathbf{E}_{loc}, \quad (9.14)$$

where N is the number of atoms or molecules, α_e is the electronic polarizability, and $\mathbf{E}_{loc}$ is the local electric field that an atom or molecule experiences. The local electric field is greater than the average field, **E**, because of the polarization of other surrounding atoms.

Ionic (atomic) polarization [$\mathbf{P}_i$] is due to the elastic deformation inside polar molecules, more generally inside ionic materials, and it occurs within about 10^{-12} s of the being applied. For an ionic solid, the atomic polarization is given by

$$\mathbf{P}_i = N\alpha_i\mathbf{E}_{loc}, \quad (9.15)$$

where N is the number of formula units per unit volume, and α_i is the ionic polarizability. Both ionic and electronic polarization contribute to the formation of induced dipoles.

Orientation polarization [$\mathbf{P}_{or}$] is directly related to materials composed of molecules that have permanent electric dipole moments. The permanent dipoles tend to become aligned with the applied electric field, but entropy and thermal effects tend to counter this alignment. Thus, orientation polarization is highly temperature dependent, unlike the forms of induced polarization, which are nearly temperature independent. In electric fields of moderate intensity, the orientation polarization is proportional to the local electric field, as for the other forms of polarization

$$\mathbf{P}_o = N\alpha_o\mathbf{E}_{loc} \tag{9.16}$$

but in strong electric fields, this proportionality is not maintained since saturation must occur when all the permanent dipoles are aligned. Orientation polarization occurs mostly in gases and liquids where molecules are free to rotate. In some solids, the rotation of polar molecules may be restricted by lattice forces, which can reduce orientation polarization. The orientation polarizability, α_o, can be expressed by

$$\alpha_o = N\mathbf{p}_e^2/3k_BT, \tag{9.17}$$

where $\mathbf{p}_e$ is the permanent electric dipole moment per molecule, which is a measurable quantity, k_B is Boltzmann's constant, and T is absolute temperature.

Space-charge (interfacial) polarization [$\mathbf{P}_{sc}$] results from the accumulation of charge at structural interfaces in heterogeneous materials. The total polarization is possible to express as a sum of these:

$$\mathbf{P} = \mathbf{P}_e + \mathbf{P}_i + \mathbf{P}_o + \mathbf{P}_{sc} \tag{9.18}$$

Also it is possible to define a total polarizability α as

$$\alpha = \alpha_e + \alpha_i + \alpha_o + \alpha_{sc} \tag{9.19}$$

Related to Eqs. (9.4), (9.8), (9.18), and (9.19), it is possible to write

$$\mathbf{P} = \mathbf{D} - \varepsilon_0\mathbf{E} = (\varepsilon - \varepsilon_0)\mathbf{E} = N\alpha\mathbf{E}, \tag{9.20}$$

where ε is the absolute permittivity, and it is related to ε_r according to the following relationship:

$$\varepsilon = \varepsilon_r\varepsilon_0 \tag{9.21}$$

and $\mathbf{E}$ is the external field under the hypothesis that it is possible to think $\mathbf{E} = \mathbf{E}_{loc}$. Generally, the measurement of the local electric field is not easy, because it depends on the external electric field

E and polarization **P**. Normally, the local field, ($\mathbf{E}_{loc}$) is regarded as the field existing in a cavity inside the material and it depends on the shape of this one, too. It is possible to write

$$\mathbf{E}_{loc} = \mathbf{E} + \gamma\mathbf{P}, \tag{9.22}$$

where γ is called the polarization factor.

Lorenz was the first to calculate the polarization factor. Following the Lorenz formulation it is possible to treat a given molecule in a condensed phase in an applied field, as sitting alone in the centre of an empty spherical cavity, on the boundary of which a charge is distributed so that the field lines are not affected by the presence of cavity. Using the boundary condition on electrical displacement at the interface of this artificial cavity, it is possible to find the uniform field inside as

$$\mathbf{E}_{loc} = \mathbf{E} + \mathbf{P}/3\varepsilon_0. \tag{9.23}$$

This formulation of local electric field is named Lorentz field. Under the hypothesis of nonpolar material or dilute solution of polar molecules in vacuum or in polar solvent and comparing expression (9.23) with (9.20), it is possible to write

$$\mathbf{E}_{loc} = \mathbf{E} + (\varepsilon - \varepsilon_0)\mathbf{E}/3\varepsilon_0 = \mathbf{E}(\varepsilon + 2\varepsilon_0)/3\varepsilon_0 \tag{9.24}$$

It is easy to combine the previous formulas and to achieve

$$(\varepsilon - \varepsilon_0)/(\varepsilon + 2\varepsilon_0) = N\alpha/3\varepsilon_0 \tag{9.25}$$

This formula is well known as the Clausius–Mossotti relation. It is a good approximation of local electric field into high density gases and nonpolar liquids. The experiment data showed that this formula describes very well the dielectric behavior of a crystalline solid that is composed of only one kind of atoms and where only the term α_e, the electronic polarizability, is present. In case of the crystalline solid composed of two or more types of atoms, it is not allowed to neglect the ionic polarizability, α_I and formula (9.25) cannot be used [23]. It is important to remark that α_e and α depend on frequency; hence, they show resonance phenomena: α in the infrared range and α_e in the ultraviolet range.

For dense polar material, it is not possible to use the approximation of the Lorentz theory on the local electric field

(Lorentz field). Using the Lorentz field, the results give a higher value of $\mathbf{E}_{loc}$ than the experimental data show. In this case, formula (9.25) becomes

$$(\varepsilon - \varepsilon_0)/(\varepsilon + 2\varepsilon_0) = N(\alpha_e + \alpha_i + N\mathbf{pe}^2/3k_BT + \alpha_{sc})/3\varepsilon_0 \qquad (9.26)$$

The Lorentz theory considers that all the electric dipoles are parallel orientated to $\mathbf{E}_{loc}$. Onsager suggested that outside the influence of $\mathbf{E}_{loc}$, the electric dipoles are not parallel orientated to $\mathbf{E}_{loc}$, but they are orientated in all directions. This explains the high value of $\mathbf{E}_{loc}$ in the Lorentz theory. Osanger equation is given by

$$\mathbf{E}_{loc} = \mathbf{E} + (\varepsilon - \varepsilon_0)\mathbf{E}/(2\varepsilon + \varepsilon_0) = \mathbf{E}(3\varepsilon)/(2\varepsilon + \varepsilon_0) \qquad (9.27)$$

When $\varepsilon_r \approx 1$, this formula becomes formula (9.25) and it is possible to have the Lorentz field. Onsager gave a more refined treatment of $\mathbf{E}_{loc}$ for the presence of permanent dipoles. In both the Lorentz and Onsager models, the given molecule is regarded as at the center of a spherical cavity. In 1954, Buckley and Maryott said in their work: "The Onsager relation thus represents a rather satisfactory average. However, the agreement between observed and calculated values is far from perfect and discrepancies of the order of 10 to 20 percent are relatively common [24]." So, it gives an approximated value of $\mathbf{E}_{loc}$, and then of ε_r, than achieved by Lorentz model.

Moreover, it is necessary to underline as all showed the theories of the local field are semi-empirical—a hybrid of microscopic and macroscopic considerations. The question naturally arises whether there cannot be a more rigorous and fundamental treatment by statistical mechanics of the effect of the dipolar interaction which is responsible for the difference between $\mathbf{E}$ and $\mathbf{E}_{loc}$. Here, the analysis on electric dipoles was made without taking in account their strong dependence on frequency, which has a great influence on the electric behavior of dielectric materials.

As it is done for permittivity, now, it is possible to introduce an absolute permeability

$$\mu = \mu_r \mu_0. \qquad (9.28)$$

In materials, magnetic field ($\mathbf{H}$) can be inducted by the movement of electrons in the orbit or due to spin refers to time-varying charges. When the net vector motion may be neglected (in total it is zero), current loop does not exist or magnetic field induced is zero, these types of materials are called diamagnetic. In so-called ferromagnetic materials, the atomic or molecular arrangement permits a finite set of current loops or dipoles. These permanent dipoles magnets lead the "external permanent magnetism", property of these kinds of materials. If they have a random arrangement, they cancel with each other producing null magnetism. It is important to notice that ferromagnetic materials have a permanent electrical polarization, too, in absence of applied electric field, similar to their permanent magnetization. This is true below the Curie temperature.

Table 9.2 Subclassification of dielectrics, conductors, and magnetic materials [31]

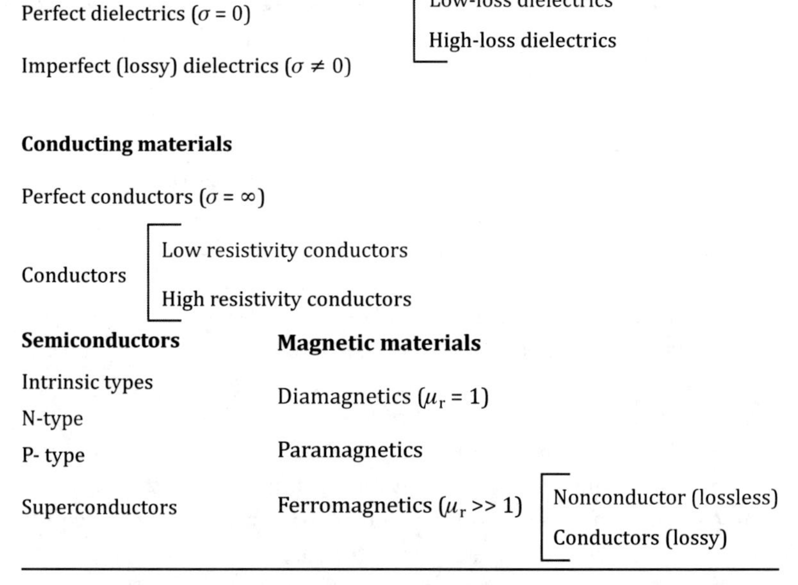

Category	Subclass	Subdivision
Dielectric materials		
	Perfect dielectrics ($\sigma = 0$)	
	Imperfect (lossy) dielectrics ($\sigma \neq 0$)	Low-loss dielectrics
		High-loss dielectrics
Conducting materials		
	Perfect conductors ($\sigma = \infty$)	
	Conductors	Low resistivity conductors
		High resistivity conductors
Semiconductors	Intrinsic types	
	N-type	
	P- type	
	Superconductors	
Magnetic materials	Diamagnetics ($\mu_r = 1$)	
	Paramagnetics	
	Ferromagnetics ($\mu_r >> 1$)	Nonconductor (lossless)
		Conductors (lossy)

Therefore, it is possible to classify electromagnetic materials according to their macroscopic parameters, ε, μ, σ, and q (charge), if any. In the microscopic scale, the electrical properties of a material are mainly determined by the electron energy bands of the material. According to the energy gap between the valence band and the conduction band, materials can be classified into insulators, semiconductors, and conductors. According to conductivity (σ), materials can be classified as insulators, semiconductors, and conductors. Materials can also be classified according to their permeability μ_r values, as shown in Table 9.2.

Other terms used for the zero frequency ε_r include relative dielectric constant and static dielectric constant. Relative permittivity is typically denoted as $\varepsilon_r(\omega)$ (sometimes κ or K) and is defined as

$$\varepsilon_r = \varepsilon(\omega)/\varepsilon_0 \quad \text{or} \quad \varepsilon(\omega) = \varepsilon_r \varepsilon_0, \tag{9.29}$$

where ε_r is a dimensionless number that is generally complex; so it is necessary to introduce $\varepsilon^*(\omega)$ as the complex frequency-dependent absolute permittivity of the material.

The imaginary portion (ε'') of the permittivity corresponds to a phase shift of the polarization **P** relative to **E** and leads to the attenuation of electromagnetic waves passing through the medium. In other words, the complex permittivity identifies the global effect of polarization of material. Absolute permittivity and relative permittivity can be decomposed into real and imaginary parts as follows:

$$\varepsilon^* = \varepsilon'(\omega) - j\varepsilon''(\omega) \tag{9.30}$$

$$\varepsilon_r^* = \varepsilon_r'(\omega) - j\varepsilon_r''(\omega), \tag{9.31}$$

where ε' and ε_r' are the real part of the complex absolute and relative permittivity, which is related to the stored energy within the medium and ε'' and ε_r'' are the imaginary part of the complex absolute and relative permittivity, which is related to the dissipation (or loss) of energy within the medium.

Likewise it is possible to identify a complex absolute and relative permeability:

$$\mu^* = \mu'(\omega) - j\mu''(\omega) \tag{9.32}$$

$$\mu_r^* = \mu_r'(\omega) - j\mu_r''(\omega), \tag{9.33}$$

where μ' and μ_r' are the real part of the complex absolute and relative permeability and μ'' are μ_r'' are the imaginary part of the complex absolute and relative permeability. Equations (9.30–9.33) are used to describe the electric and magnetic behaviors, respectively, of loss conductivity materials [25].

Coming back to system of Maxwell's equations and considering that the physical properties of a body in the neighborhood of some interior point are the same in all directions and in the following, ρ will be put equal to 0 in dielectrics as well as conductors. Also, J will be put equal to 0. Using Eqs. (9.1–9.4), it is possible to achieve this formulation of Maxwell's equations:

$$\begin{cases} \nabla \cdot \mathbf{E} = 0 \\ \nabla \times \mathbf{E} = -(\partial \mathbf{B}/\partial t) \\ \nabla \cdot \mathbf{H} = 0 \\ \nabla \cdot \mathbf{B} = \mu\varepsilon(\partial \mathbf{E}(r, t)/\partial t) \end{cases}$$

It is possible to write this system of equations in this other way:

$$\nabla^2\mathbf{E}(r, t) = \mu\varepsilon(\partial^2\mathbf{E}(r, t)/\partial^2 t)$$

$$\nabla^2\mathbf{B}(r, t) = \mu\varepsilon(\partial^2\mathbf{B}(r, t)/\partial^2 t)$$

Now, it is simpler to see how solutions of this system depends on the time and distance measured along a single axis in space. So we find the wave equation of electromagnetic field. Better, these equations lead directly to **E** and **B** satisfying thewave equation for which the solutions are linear combinations of plane waves traveling at thespeed of light, c. It is defined as [25]

$$c = 1/\sqrt{\varepsilon_0\mu_0} \tag{9.34}$$

When we have a propagation of electromagnetic wave in low-conductivity material, such as polymer, both the surface and inner parts of materials respond to electromagnetic wave. There are two parameters that can describe electromagnetic properties of low-conductivity materials: constitutive parameters and propagation parameters.

Constitutive parameters are driven by the following equations:

$$\mathbf{D} = \varepsilon^*\mathbf{E} = (\varepsilon' - j\varepsilon'')\mathbf{E}$$

$$J = \sigma\mathbf{E}$$

Often it is used to consider an equivalent circuit describing the currents in dielectric material. It is inserted into the capacitor as shown in Fig. 9.3a.

The total current (I) consist of two parts: the charging current (I_c) and loss current (I_l):

$$I = I_c + I_l = jC\omega\, U + GU = (jC\omega + G)U, \tag{9.35}$$

where C is the capacitance of the capacitor loaded with the dielectric material, G is the conductance of the dielectric material and $U = U_0\exp(j\omega t)$ is an ac voltage source. I_l is in phase with the source voltage U. The phase angle between I_c and I is often called loss angle δ.

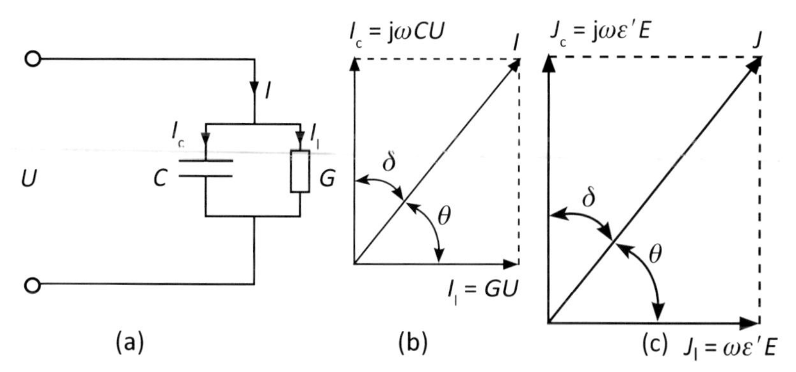

Figure 9.3 (a) Equivalent circuit for analysis on dielectric materials. (b) Complex plane showing charging current and loss current. (c) Complex plane showing the charging current density and loss current density [24].

Writing Eq. (9.35) in terms of ε^* gives

$$C = (\varepsilon^*/\varepsilon_0)\, C_0 = (\varepsilon' - j\varepsilon'')C_0/\varepsilon_0 \tag{9.36}$$

$$I = j\omega(\varepsilon' - j\varepsilon'')\,(C_0/\varepsilon_0)U = (j\omega\varepsilon' + \omega\varepsilon'')\,(C_0/\varepsilon_0)U \tag{9.37}$$

Therefore, the current density J transverse to the capacitor under the applied field strength **E** becomes

$$J = (j\omega\varepsilon' + \omega\varepsilon'')\mathbf{E} = \varepsilon(\partial E/\partial t) \tag{9.38}$$

The product of angular frequency and loss factor is equivalent to a dielectric conductivity:

$$\sigma = \omega\varepsilon'' \tag{9.39}$$

This dielectric conductivity sums over all the dissipative effects of the material. It may represent an actual conductivity caused by migrating charge carriers and it may also refer to an energy loss associated with the dispersion of ε', for example, the friction accompanying the orientation of dipoles [24]. In Fig. 9.3c, two parameters describe the energy dissipation of a dielectric material:

The dielectric loss tangent

$$\tan\delta_e = \varepsilon'/\varepsilon'' \quad (9.40)$$

The dielectric power factor

$$\cos\theta_c = \varepsilon''/\sqrt{(\varepsilon'')^2 + (\varepsilon')^2} \quad (9.41)$$

Equations (9.40) and (9.41) show that for a small loss angle δ_e, cos $\theta_e \approx \tan\delta_e$.

According to Faraday's inductance law, the magnetic case is given as

$$U = L\,(dI/dt) \quad (9.42)$$

Introducing a magnetization current I_m,

$$I_m = -jU/(\omega L_0), \quad (9.43)$$

where U is the magnetization voltage, L_0 is the inductance of an empty inductor, and ω is the angular frequency. Introducing an ideal, lossless magnetic material with relative permeability μ_r', the magnetization field becomes

$$I_m = -jU/(\omega L_0)\mu_r' \quad (9.44)$$

In the complex plane, shown in Fig. 9.4a, the magnetization current I_m lags the voltage U by 90° for no loss of magnetic materials.

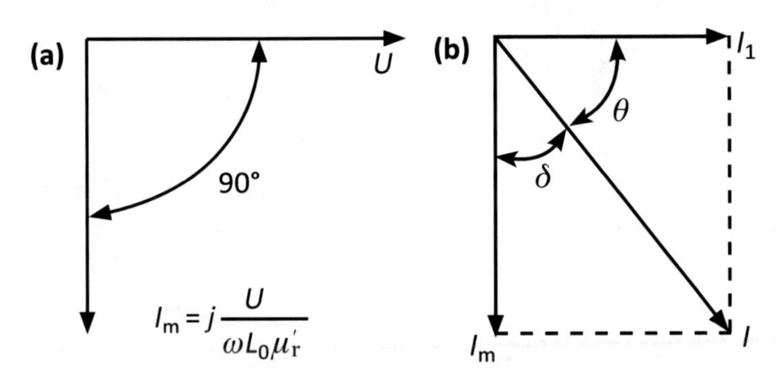

Figure 9.4 (a) The magnetization current in a complex plane. (a) Relationship between magnetization current and voltage and (b) relationship between magnetization current and loss current [24].

As shown in Fig. 9.4b, an actual magnetic material has magnetic loss, and the magnetic loss current I_l caused by energy dissipation during the magnetization cycle is in phase with U. Using Eqs. (9.32) and (9.33), it is possible to achieve the total magnetization current I, in complete analogy to the dielectric case:

$$I = I_m + I_L = U/(j\omega L_0\mu_r^*) = -jU(\mu_r' + j\mu_r'')/\omega(L_0/\mu_0)(\mu_r'^2 + \mu_r''^2) \tag{9.45}$$

Carrying on with the analogy to dielectric casa it is possible introduce two parameters: the magnetic loss tangent given by

$$\tan\delta_m = \mu_r''/\mu_r' \tag{9.46}$$

and the power factor given by

$$\cos\theta_m = \mu_r'/\sqrt{(\mu_r')^2 + (\mu_r'')^2} \tag{9.47}$$

For dielectric materials, the value of conductivity (σ) is small; then, the study is concentrated on permittivity and permeability of material. Generally, both permittivity and permeability are complex numbers, and the imaginary part of permittivity is related to the conductivity of material. To summarize, it is possible to describe the interaction between electro and magnetic field with the material in two ways: energy storage and energy dissipation. Energy storage describes the energy lossless portion of the exchange of energy between the field and material; energy dissipation occurs when the electromagnetic energy is absorbed by material. Both complex permittivity and permeability describe the storage (real part) and the dissipation (imaginary part) effects of each.
Propagation parameters are led by

Characteristic wave impedance of the medium

$$Z_0 = E/H = \sqrt{\mu/\varepsilon} = \left(\sqrt{\mu_r/\varepsilon_r}\right)\left(\sqrt{\mu_0/\varepsilon_0}\right) \tag{9.48}$$

Wave velocity in the medium

$$v = 1/\sqrt{\mu/\varepsilon} = \left(\sqrt{\mu_r/\varepsilon_r}\right)\left(\sqrt{\mu_0/\varepsilon_0}\right) = c/\left(\sqrt{\mu_r/\varepsilon_r}\right) \tag{9.49}$$

It is useful to note that expressing the permittivity and permeability as complex number leads to a complex quantity for the wave velocity v, where the imaginary portion is a mathematic convenience for expressing loss. Sometimes, it is more useful to use the complex propagation coefficient γ [24]. It is possible to write it in two ways:

$$\gamma = j\omega(\sqrt{\mu\varepsilon}) = j(\omega/c)(\sqrt{\mu_r\varepsilon_r}) = j(\omega/c)Z_0 \quad (9.50a)$$

$$\gamma = j(\omega/cn^*) \quad (9.50b)$$

In Eq. (9.50a), the wave impedance (Z_0) appears. Equation (9.50b) shows the complex refractive index (n^*) of the loss material; it depends on the complex relative permittivity (ε_r^*) of the material.

9.4 Electromagnetic Compatibility

With the rapid increase in the use of telecommunication and digital systems, fast processors, more generally, electronic devices and the introduction of new design practices, electromagnetic compatibility (EMC) has become more important and has been brought to the forefront of advanced design. In fact, the proliferation of electronic equipment and various devices in both industrial and domestic environments has given rise to a large number of sources and receptors of EMI, which increases the potential for interference. According to the growth of the number of electronic devices, the miniaturization of components in electronic systems increases their susceptibility to interference. Nowadays, a large number of the units are smaller and more portable equipment, such as cell phones and laptop computers, which can be used anywhere instead of only in a controlled environment such as an office. It is good to note how developments in interconnection technology have lowered the threshold for EMI. It is useful to underline how high-speed digital circuitry generally generates more interference than the traditional analog circuitry. This leads to EMI problems. Reducing or eliminating coupling paths by proper layout; shielding, filtering, and grounding practices—due to these reasons, it is to understand how it becomes important to design hardware with an inherent immunity to EMI and to adopt defensive programming practices to develop software that has a high level of immunity to EMI [26].

Reliability and EMC compliance have become important marketing features of electronic equipment in an increasingly competitive industry; electronic discharge has been a major concern due to the destruction of microchips during handling; Electromagnetic discharge (ESD) also poses problems for aircraft and automobiles.

The necessity to ensure the security of data has been one of the main driving forces in the growth of the shielding market; people are becoming aware of the potential health risks associated with all types of EMI radiation; military systems involve protection against electromagnetic pulse (EMP) and electronic countermeasures including microwave weapons and stealth technology. Considerations of an array of communications and radars in battle conditions are also important.

It is useful to explain what EMC, EMI and susceptibility are.

Electromagnetic compatibility (EMC) is the capability of electrical and electronic systems, equipment, and devices to operate in their intended electromagnetic environment within a defined margin of safety, and at design levels or performance, without suffering or causing unacceptable degradation as a result of EMI. Electromagnetic compatibility can generally be achieved by suppressing EMI and immunizing susceptibility of the systems and devices. Electromagnetic compatibility is a field of science and engineering concerned with the design and operation of electrical and electronic systems/devices in a manner that makes them immune to certain amounts of EMI, while at the same time keeping their EMI emissions within specific limits.

Electromagnetic interference (EMI) is the process by which disruptive electromagnetic energy is transmitted from one electronic device to another via radiated or conducted paths or both. Suppression is the process of reducing or eliminating EMI energy. It may include shielding and filtering.

Susceptibility is a relative measure of a device's or a system's propensity to be disrupted or damaged by EMI exposure to an incident field or signal. It indicates the lack of immunity. Immunity is a relative measure of a device's or system's ability to withstand EMI exposure while maintaining a predefined performance level. Radiated immunity is a product's relative ability to withstand electromagnetic energy that arrives via free space propagation. Conducted immunity is a product's relative ability to withstand electromagnetic energy that penetrates it through external cables, power cords, and input/output (I/O) interconnects. Electromagnetic discharge (ESD) is a transfer of electric charge between bodies of different electrostatic potential in proximity or through direct contact [27].

In general, EMI problems can be caused by electromagnetic energy at any frequency in the spectrum, as shown in Fig. 9.5.

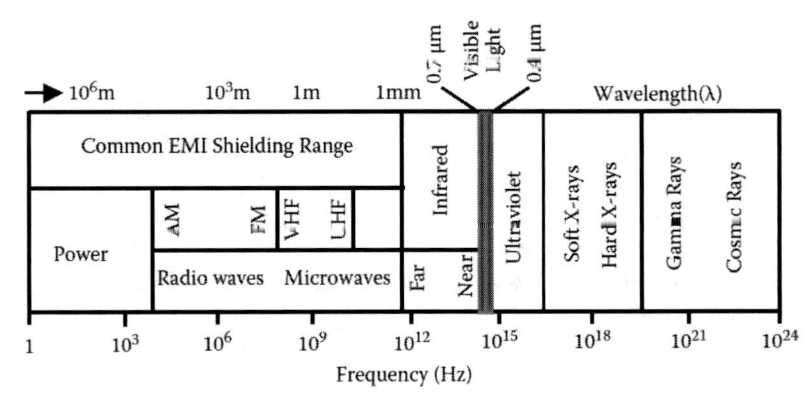

Figure 9.5 Electromagnetic spectrum [27].

Most problems are caused by energy in the radio frequency range, which extends from about 100 kHz to 1 GHz. Radio frequency interference (RFI) is a specific form of EMI. Microwave, light, heat, x-rays, and cosmic rays are other specific types of electromagnetic energy. Other common interference includes EMP, which is a kind of broadband; high-intensity short duration burst of electromagnetic energy, such as lightning or nuclear explosion; and electrostatic discharge (ESD), which is a transient phenomenon involving static electricity friction.

In these analyses, EMI shielding is an important and effective method in EMC design. Certainly, advanced materials and process technology are the key to achieving successful EMI shielding [28]. Electromagnetic interference can be dominated by radiation or conduction, depending upon the type of coupling or propagation path involved. Usually, conduction always accompanies some radiation, and vice versa. Radiated interference happens when a component predominately emits energy, which is transferred to the receptor through the space.

Conducted interference occurs when the source is connected to the receptor by power or signal cables, and the interference is transferred from one unit to the other along the cables or wires. Conducted interference usually affects the main supply to and from the system/device and can be controlled using filters. Either magnetic or electric could be the origin of the EMI. When a circuit with large dV/dt (voltage slew rate) has a significant parasitic capacitance to ground EMI is electrically generated. Magnetically generated EMI appears when a circuit loops with large dI/dt (current slew rate) it has significant mutual coupling to a group of nearby conductors. In addition, there is an important energy exchange between two modes. This effect is known as differential–common mode conversion.

Radiated EMI emission starts from an emitting source, propagates via a radiating path, and reaches a susceptible receiver. The strength of the radiated EMI is determined by the source, the media surrounding the source, and the distance between the source and the susceptive element. The space surrounding a source of radiation can be broken into two regions, as shown in Fig. 9.6. Close to the source is the near or induction field. At a distance greater than the wavelength (λ) divided by 2π (approximately one sixth of a wavelength) is the far or radiation field. The region around $\lambda/2\pi$ is the transition region between the near and far fields.

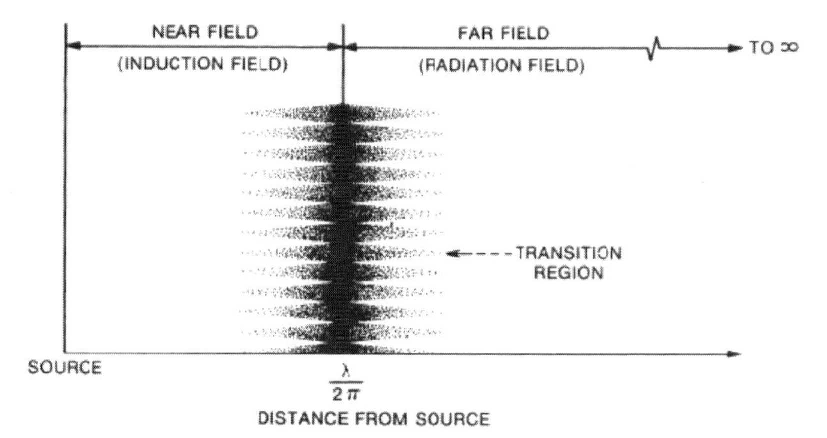

Figure 9.6 The space surrounding a source of radiation can be divided into two regions, the near field and the far field. The transition from near to far field occurs at a distance of $\lambda/2\pi$ [29].

The basic component of EMI is an electromagnetic wave. Any source of the radiated EMI energy generates expanding spherical wave fronts, which travel in all directions from the source. At any point the wave consists of an electric field (E) and a magnetic field (H), which are perpendicular to each other and to the direction of propagation.

The ratio of the electric field (E) to the magnetic field (H) is the wave impedance (Z_0). In the far field, this ratio equals the characteristic impedance of the medium (e.g., $E/H = Z_0 = 377\ \Omega$ for air or free space). In the near field, the ratio is determined by the characteristics of the source and the distance from the source to where the field is observed. If the source has high current and low voltage ($E/H < 377$), the near field is predominantly magnetic. Conversely, if the source has low current and high voltage ($E/H > 377$), the near field is predominantly electric.

The source of the radiated interference may be part of the same system as the receptor or a completely electrically isolated unit. If the conducted emissions are reduced, the relative radiated emissions are also often reduced. However, the dominant radiated interference can affect any signal path within and outside the system/device and is much more difficult to shield [29].

9.5 Electromagnetic Interference Shielding Properties

What is EMI shielding meaning? It means to use a shield (a shaped conducting material) to partially or completely envelop an electronic circuit, that is, an EMI emitter or susceptor. Therefore, it limits the amount of EMI radiation from the external environment that can penetrate the circuit and, conversely, it influences how much EMI energy generated by the circuit can escape into the external environment.

Hence, it is possible to identify two purpose of shied: the first it is to prevent the emissions of the electronics of the product or a portion of those electronics from radiating outside the boundaries of the product, as it shows in Fig. 9.7a; the second it is to prevent radiated emissions external to the product from coupling to the product's electronics, which may cause interference in the product, as it shows in Fig. 9.7b.

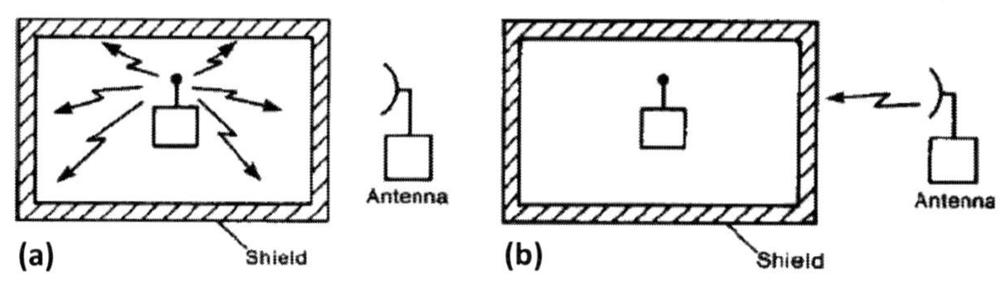

Figure 9.7 Illustration of the use of a shielded enclosure: (a) to contain radiated emissions and (b) to exclude radiated emissions [30].

The basic requirement of an Electromagnetic (EM) shielding material is that a shield fabricated with this material should meet the EMC aspects of achieving a specified extent of shielding under given EMI environmental. Generally, the choice of EM shielding material is decided by

- electromagnetic properties of material to provide a given shielding effectiveness (SE)
- its compatibility for specific shielding applications vis-a-vis interference due to electric, magnetic, or electro-magnetic (radiated) fields
- geometrical considerations, shape and size
- mechanical considerations such as rigidity, flexibility, weight. Structural mating (fastening and joints), and withstand ability against shocks and vibration
- performance under hostile thermal environments
- bandwidth of operation: the effective frequency range with acceptable shielding performance
- ease of fabrication of shields
- cost-effectiveness [31]

Reminding that EM energy could be associated with static and/or time-varying electric and magnetic force fields (known as induction fields) in the vicinity of the source, or it could be a radiated field detached from the source, as it shows in Fig. 9.8, it is easy to understand how suitable materials for EMI shielding mechanism depend on specific and distinct EMI source involved. Several materials could be, and have been, chosen and used for EMI shielding for their own properties: electrical conductivity, magnetic permeability, mainly, with addition of, high mechanical performance, chemical stability, fire retardant, thermal management and geometries [31].

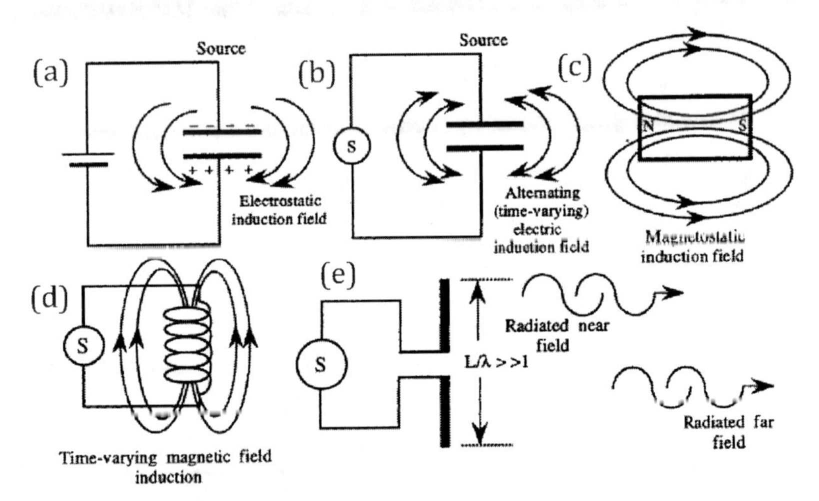

Figure 9.8 Illustration of EM induction fields: (a) electrostatic (dc) induction field; (b) time-varying electric induction field; (c) magnetostatic induction field; (d) time-varying magnetic induction field; (e) radiated near and far EM fields [30].

The mechanism of shielding is that the shielding medium (of appropriate material and geometry) when placed in the region of electromagnetic field offer a barrier impedance Z_m sufficiently large in comparison to Z_0 in case of an electric field dominant ambient so that the waves is impeded sufficiently and shielded off from entering the region beyond the shielding barrier, as shown in Figs. 9.8a,b for magnetic field where value of Y_m, barrier admittance, must be higher than value of Y_0 ($Y_0 = 1/Z_0$).

When designing a shield that works in time-varying fields, it is important to consider the appropriate choice of material and the geometry to provide the necessary barrier-immittance parameter. Also, it is relevant to underline as shields invariably contain apertures or openings used for access and ventilation, as well as a number of joints and seams for practical manufacturing. Shields for practical equipment also allow attachment to wires or pipes used for signaling and services. All these components constitute breaches of the shield integrity and play a decisive part in the overall performance of the shield [30].

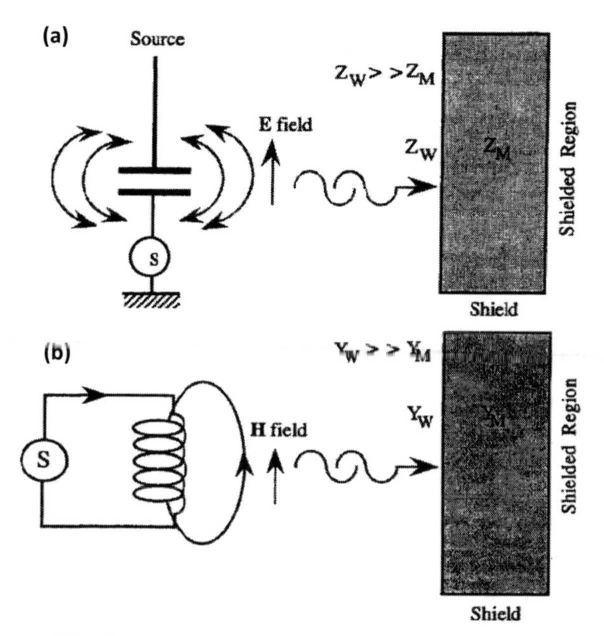

Figure 9.9 (a) Electric field dominant ambient and (b) Magnetic field dominant ambient [30].

Shielding effectiveness

Shielding can be specified in the terms of reduction in magnetic (and electric) field or plane-wave strength caused by shielding. The effectiveness of a shield and its resulting EMI attenuation are based on the frequency, the distance of the shield from the source, the thickness of the shield, and the shield material. For a quantitative description of main aspects of shielding, it is possible to consider three phenomena as responsible for effective shielding (SE):

(1) **Conductive reflection**: The time-varying magnetic field component of incident EM waves induce electric current in the shielding material and these currents in turn provide opposing magnetic field (Faraday–Lenz's law) minimizing the total field beyond the shield.

(2) **Magnetic reflection**: when the shielding material has high magnetic permeability, the magnetic flux lines (time-varying or static) are confined as conductive (low reluctant) paths through the shield and do not link to the region being shielded.

(3) **Conductive energy absorption**: the energy dissipated into conductive shielding medium manifesting as the field attenuation [31]. Shielding effectiveness is normally expressed in decibels (dB) as a function of the logarithm of the ratio of the incident and exit electric (*E*), magnetic (*H*), or plane-wave field intensities (*F*):

$$\mathrm{SE(dB)} = 20 \log (E_0/E_1)$$

$$\mathrm{SE(dB)} = 20 \log (H_0/H_1)$$

$$\mathrm{SE(dB)} = 20 \log (F_0/F_1)$$

With any kind of EMI, there are three mechanisms contributing to the effectiveness of a shield. Part of the incident radiation is reflected from the front surface of the shield, part is absorbed within the shield material, and part is reflected from the shield rear surface to the front where it can aid or hinder the effectiveness of the shield depending on its phase relationship with the incident wave, as shown in Fig. 9.10.

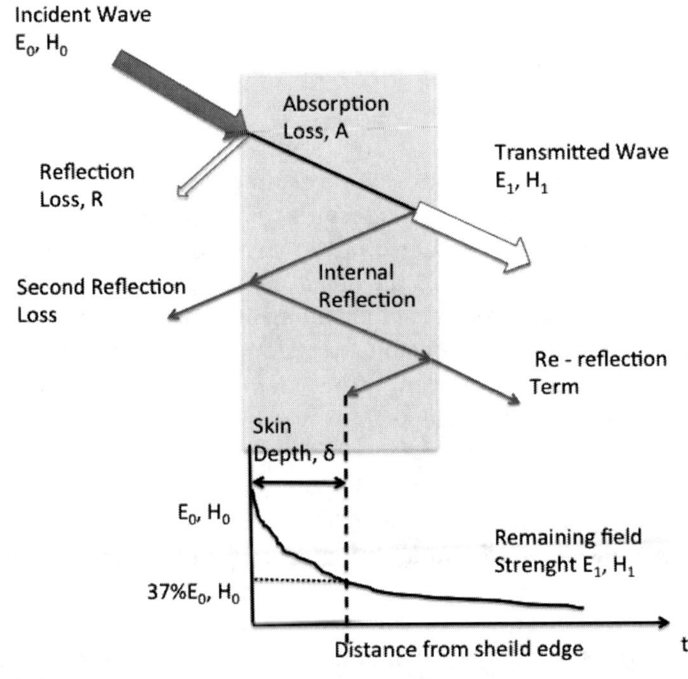

Figure 9.10 Graphical representation of EMI shielding [28].

Total shielding effectiveness (SE) of a shielding material equals the sum of the absorption factor (A), the reflection factor (R), and the correction factor to account for multiple reflections in thin shields:

$$\text{SE (dB)} = R(\text{dB}) + A(\text{dB}) + B(\text{dB}) \tag{9.51}$$

The higher the SE value in decibel (dB) is, the less the energy passes through the sample. It is important to notice how the multiple reflection factor B can be neglected if the absorption loss A is greater than 10 dB. In practical calculation, B can also be neglected for electric fields and plane waves.

Following is the explanation of the two main terms of SE:

Absorption Loss

Absorption losses A are a function of the physical characteristics of the shield and they are independent of the type of source field.

When an electromagnetic wave passes through a medium, used as a shield, its amplitude decreases exponentially. This decay is an effect of currents induced in the medium that produce ohmic losses and heating of the material, and E_1 and H_1 can be expressed as

$$E_1 = E_0 e^{-t/\delta}$$
$$H_1 = H_0 e^{-t/\delta}, \tag{9.52}$$

where t is the thickness of shielding material, expressed in mm, and δ is called the skin depth, and it is defined as the distance required for the wave to be attenuated to $1/e$ or 37%. Therefore, the absorption term A is given by

$$A(\text{dB}) = 20(t/\delta)\log(e) = 8.69(t/\delta) = 131t\sqrt{f\mu_r\sigma_r} \tag{9.53}$$

It should be noted that the absorption term A is the same for all three waves. In fact, in its expression, a particular field (E, H, and F) does not appear.

In Eq. (9.39), f is frequency in MHz; μ_r is relative permeability (1 to copper); σ_r is conductivity relative to copper in IACS (International Annealed Copper Standard) $\sigma_r = \sigma/\sigma_0$, where σ_0 is the conductivity of copper at room temperature ($\sigma_0 = 5.80 \times 10^{-7}$ Siemens/m) [30].

The skin depth δ can be expressed as

$$\delta = 1/\left(\sqrt{\pi f \mu_r \sigma_r}\right) \tag{9.54}$$

The absorption loss of one skin depth in a shield is approximately 9 dB. The skin effect becomes relevant at low frequencies, where usually magnetic field has predominant effects with lower wave impedance than 377 W.

Figure 9.11 shows the skin depth variation with frequency for copper, aluminum, steel, and stainless steel. Copper and aluminum have over five times the conductivity of steel and therefore are very good at stopping electric fields, but they have relative permeability of 1 (the same as air). Typical mild steel has relative permeability of around 300 at low frequencies, falling to 1 as frequencies increase above 100 kHz, and its higher permeability gives it a reduced skin depth, making a reasonable thickness of mild steel better than aluminum for shielding low frequencies. Different grades of steel (especially stainless) have different conductivity and permeability, and their skin depth varies considerably as a result.

To summarize, it is possible to conclude that a good material for a shield will have high conductivity and high permeability and sufficient thickness to achieve the required number of skin depths at the lowest frequency of concern [28], from the absorption loss point of view.

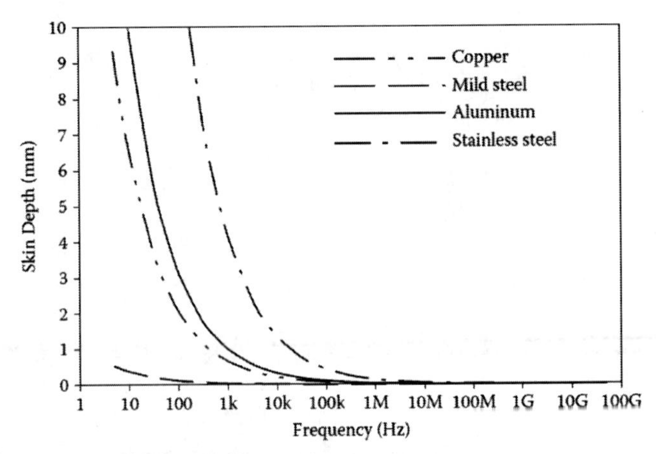

Figure 9.11 Skin-depth variation with frequency for copper, aluminum, steel, and stainless steel [28].

Reflection Loss

The reflection loss, R, at the interface between two media is related to the difference in characteristic impedances between the media, or, in other words, the relative mismatch between the incident wave and the surface impedance of the shield, as shown in Fig. 9.12.

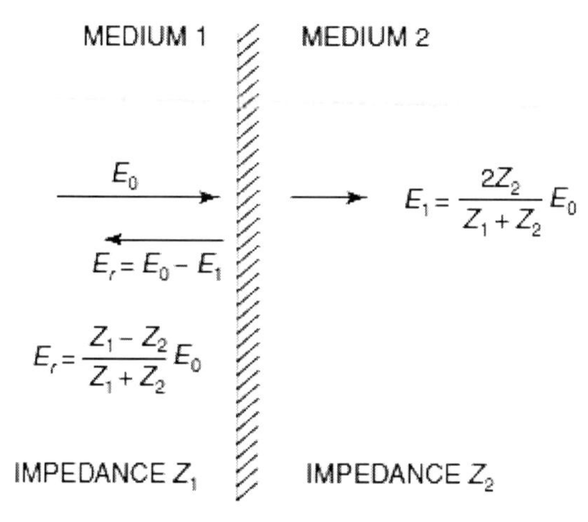

Figure 9.12 An incident wave is partially reflected from, and partially transmitted through, an interface between two media. The transmitted wave is E_1 and the reflected wave is E_r [29].

The computation of reflection losses can be greatly simplified by considering shielding effectiveness for incident electric fields as a separate problem from that of electric, magnetic, or plane waves. The equations for the three principal fields are given by [28,29]

$$R_E = 321.8 + 10 \log (\sigma_r/(f^3 r^2 \mu_{r=0})) \quad (9.55a)$$

$$R_H = 14.6 + 10 \log (fr^2 \sigma_r/\mu_{r=0}) \quad (9.55b)$$

$$R_p = 168 - 10 \log (f\mu_{r=0}/\sigma_r), \quad (9.55c)$$

where R_E, R_H, and R_P are the reflection losses for the electric, magnetic, and plane wave fields, respectively, expressed in dB; σ_r is the relative conductivity referred to copper; f is the frequency in Hz; $\mu_{r=0}$ is the relative permeability referred to free space; and r is the distance from the source to the shielding in m.

Multiple Reflection Correction Factor

The factor B can be mathematically positive or negative (in practice it is always negative), and becomes insignificant when the absorption loss $A > 6$ dB. It is usually only important when metals are thin and at low frequencies (i.e., below approximately 20 kHz). The formulation of factor B can be expressed as [28,32]:

$$B\,(\text{dB}) = 20\log|1 - [(K-1)/(K+1)]\,(10^{-A/10})\,(e^{-i227A})|, \quad (9.56)$$

where A is the absorption loss (dB);

$$K = Z_S/Z_H = 1.3(\mu/fr^2 s)^{1/2}, \quad (9.57)$$

where Z_S is the shield impedance and Z_H is the impedance of the incident magnetic field.

When $Z_H << Z_S$, the multiple reflection factor for magnetic fields in a shield of t and skin depth δ can be simplified as

$$B = 20\log(1 - e^{-2t/\delta}) \quad (9.58)$$

The total shielding effectiveness for electric, magnetic, and plane wave fields can be obtained by Eq. (9.51) with a combination of the related equations of absorption and reflection losses, as well as correction factor B.

Figures 9.13a,b show the plots of reflection loss, absorption loss, and total shielding effectiveness for copper and iron when term B is neglected, comparing between them, iron and copper electromagnetic behavior, for example. These illustrations give a good physical representation of the behavior of the component parts of an electromagnetic wave. From them, it is easy to notice how copper offers more shielding effectiveness than iron in all cases except for absorption loss due to the high permeability of iron. In fact, for a given thickness, magnetic materials such as iron or steel provide higher absorption losses than nonmagnetic materials such as copper.

Therefore, plots also illustrate the reasons because it is so much more difficult to shield magnetic fields than electric fields or plane waves. At low frequencies, the absorption losses are low and rise gradually to high levels at high frequencies. Increasing the thickness of the shield or by using a metal with high permeability, it is possible to improve the absorption performance. However, metals with high permeability usually exhibit low electrical conductivity. Therefore, in that case, it is better to use only magnetic materials that effectively shield against magnetic fields at low frequencies. If magnetic shielding is required, particularly at frequencies below 14 kHz, it is customary to neglect all terms in Eq. (9.36) except the absorption term A. Conversely, if only electric field or plane-wave protection is required, reflection (R) is an important factor to consider in the design. When frequency is around 1 GHz, the value of reflection losses is nothing and the total shield effectiveness of the material in all three fields tends to be the same it is possible to distinguish that from Figs. 9.13a,b.

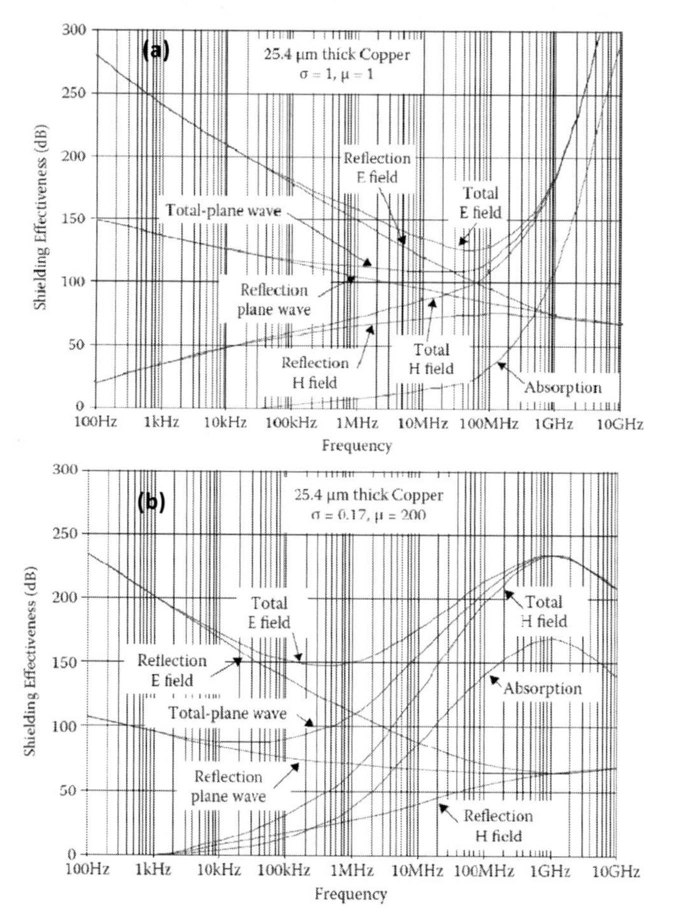

Figure 9.13 (a) Shielding effectiveness of a copper shield and (b) shielding effectiveness of an iron shield [28].

Moreover, higher electrically conductive materials (like copper) seem to give a better electrical and magnetic shielding, at higher frequencies, where both reflection and absorption losses increase.

To summarize, the amount of reflection loss depends on the impedance of the EM wave and the shield. When an EM wave encounters a shield, if the wave's impedance differs significantly from that of the shield, the wave is partially reflected back. Conversely, if the wave's and the shield's impedance are closely matched, the EMI energy will pass through the shield with minimal reflection reminding that when we said before an electrically dominant wave (E-field) in the near field has high impedance (Z_0 > 377 Ω). Higher conductive metals have low impedance and, therefore, are successful at reflecting back electrically dominant waves because of the impedance mismatch. Reflection of EMI is the primary shielding mechanism for electrically dominant waves.

Magnetically dominant waves (H-fields), on the other hand, have low impedance (Z_0 < 377 Ω). With these waves, absorption plays an important role in shielding. Magnetic waves are more difficult to shield; however, their energy generally diminishes as the distance from the source increases. Over greater distances (far field), the electric field component dominates the wave, and it is this electrical component that must be dealt with through EMI shielding.

On the amount of the absorbed wave's energy from shielding material, its relative thickness plays an important role. However, thickness has little effect on the amount of the wave's energy that is reflected. Both the absorption and reflection of EMI are important considerations in selection of a shielding method. Attenuation or reduction of E-fields can be handled effectively by metal sheet, foil, and metal coatings.

Reflection loss is very large relative to absorption loss. Primary reflection occurs at the front surface of the shield, so that very thin sections of conductive material provide good reflection properties. Moreover, multilayer shields provide increased reflection loss due to the additional reflecting surfaces. Conversely, where absorption plays a key role in attenuation, strong magnetic fields are better handled by thick magnetic materials or thick-skin shields, such as those provided by conductive paints (GE Plastics, n.d.) [28].

References

1. Deborah, D., Chung, L. (ed.). *Applied Materials Science Applications of Engineering Materials in Structural, Electronics, Thermal, and Other Industries*, CRC Press 2001, Print ISBN: 978-0-8493-1073-7.
2. Bhushan, B. (ed.). *Springer Handbook of Nanotechnology 3rd Revised and Extended Edition*, © Springer-Verlag Berlin Heidelberg 2010, ISBN: 978-3-642-02524-2 DOI 10.1007/978-3-642-02525-9.
3. Ajayan, P. M., Schadler, L. S., Braun, P. V. (ed). *Nanocomposite Science and Technology*. Copyright © 2003 WILEY-VCH Verlag GmbH Co. KGaA, Weinheim, ISBN: 3-527-30359-6.
4. Alexandre, M., Dubois, P. *Mater Sci Eng R Rep*, **28**, 1–63 (2000).
5. Herron, N., Thorn, D. L. *Adv Mater*, **10**, 1173–1184 (1998).
6. Messersmith, P. B., Giannelis, E. P. *Chem Mater*, **6**, 1719–1725 (1994).
7. Sumita, M., Tsukumo, Y., Miyasaka, K., Ishikawa, K. *J Mater Sci*, **18**, 1758–1764 (1983).
8. Kumara, A. P., Depana, D., Tomerb, N. S., Singha R. P. *Prog Polymer Sci*, **34**, 479–515 (2009).
9. Krishnamoorti, R., Vaia, R. A. (eds.). *Polymer Nanocomposites: Synthesis, Characterization and Modeling*, in ACS Symposium Series, vol. 804, ACS, Washington DC, (2001).
10. Vaia, R. A., Wagner, H. D. *Mater Today*, 7, 32–37 (2004).
11. Zax, D. B., Yang, D.-K., Santos, R. A., Hegemann, H., Giannelis, E. P., Manias, E. *J. Chem. Phys.*, **112**, 2945, (2000).
12. Kim, S. H., Chung, J. W., Kang, T. J., Kwak, S.-Y., Suzuki T. *Polymer*, **48**, 4271–4277 (2007).
13. Ahankari, S. S., Kar, K. K. *Mater Lett*, **62**, 3398–3400 (2008).
14. Krishnamoorti, R., Vaia, R. A., Giannelis, E. P. *Chem Mater*, **8**, 1728–1734 (1996).
15. Díez-Pascual A. M., Naffakh, M., González-Domínguez, J. M., Anson, A., Martinez-Rubi, Y., Simard, B. *Carbon*, **48**(12), 3485–3499 (2010).
16. Díez-Pascual, A. M., Naffakh, M., González-Domínguez, J. M., Anson, A., Martinez-Rubi, Y., Simard, B. *Carbon*, **48**(12), 3500–3511 (2010).
17. Fan, Z., Hsiao, K.-T., Advani, S. G. *Carbon*, **42**(4), 871–876 (2004).
18. Díez-Pascual, A. M., Ashrafi, B., Naffakh, M., González-Domínguez, J. M., Johnston, A., Simard, B., Martínez, M. T., Gómez-Fatou M. A. *Carbon*, **49**, 2817–2833 (2011).

19. Karapappas, P., Tsotra, P., Scobbie, K., *Polym Lett*, **5**(3), 218–227 (2011).

20. Lincoln, D. M., Vaia, R. A., Brown, J. M., Benson Tolle, T. H. *IEEE Aerospace Confer Proc*, **4**, 183–192 (2000).

21. Business Communications Company Inc. RP-234 Polymer nanocomposites: nanoparticles, nanoclays and nanotubes.

22. Stratton, J. A. (ed.). *Electromagnetic Theory*, McGraw-Hill Book Company, Inc, Copyright © 1941.

23. Sette, D., Bertolotti, M. (ed.). Lezioni di Fisica—Elettromagnetismo/Ottica, Masson Milano Copyright © 1998.

24. Buckley, F., Maryott, A. A. *J. Res. Natl. Bur. Std.*, **53**, 229 (1954).

25. Chen, L. F., Ong, C. K., Neo, C. P., Varadan, V. V., Varadan, V. K. (ed.). *Microwave Electronics: Measurement and Materials Characterization*. John Wiley & Sons Copyright © 2004 ISBN: 0470844922.

26. Christos, C. (ed.). *Principles and Techniques of Electromagnetic Compatibility* (Electronic Engineering Systems) Publisher CRC Press © 1995 by Taylor & Francis Group, LLC, ISBN:0849378923.

27. Montrose, M. I. (ed.). *Emc and the Printed Circuit Board, Design, Theory and Layout Made Simple*, Wiley-IEEE Press 1999 ISBN: 0-7803-4703-x.

28. Tong, C., Tong, X. C. (ed.). *Advanced Materials and Design for Electromagnetic Interference Shielding*, CRC Press © 2009 by Taylor & Francis Group, LLC, ISBN: 1420073583.

29. Ott, H. W. (ed.). *Electromagnetic Compatibility Engineering*, Copyright © 2009 by John Wiley & Sons, Inc. ISBN 978-0-470-18930-6.

30. Paul, C. R. *Introduction to Electromagnetic Compatibility* (Wiley Series in Microwave and Optical Engineering), second edition, Copyright © 2006 John Wiley & Sons, Inc ISBN-10: 0-471-75500-1.

31. Neelakanta, P. S. (ed.). *Handbook of Electromagnetic Materials, Monolithic and Composite Version and Their Applications*, CRC Press Inc © 1995 ISBN-10: 9780849325007.

32. Vasaka, C. S. *Theory, Design and Engineering Evaluation of Radio-Frequency Shielded Rooms* (Report NADC-EL-54129). Johnsville, PA: U. S. Naval Development Center (1956).

Chapter 10

Epoxy Nanocomposite Based on Carbon Nanotubes for Electromagnetic Interface Shielding

Stefano Bellucci and Federico Micciulla

Laboratori Nazionali di Frascati, Instituto Nazionale di Fisica Nucleare, Via Enrico Fermi, 40 00044 Frascati, Rome, Italy

bellucci@lnf.infn.it

10.1 Introduction

The conventional electromagnetic interface (EMI) shielding materials include applying metal-based shielding materials, but metals are heavy and have a tendency to corrode. The conducting plastics and coatings have good potential for EMI-shielding materials. There are two types of conductive coatings: intrinsic conductive polymer and composite conductive coating. The intrinsic conductive polymers include polyaniline, polypyrrole, and polypydine. Their poor processing ability and chemical toxicity limit their applications. The composite conductive coatings consist of electromagnetic absorbents (EMA) and coating matrix. Typical EMI absorbents, such as ferromagnetic metal oxides, carbon clusters, silicon carbide like materials, are widely applied to EMI-shielding

Advanced Nanomaterials for Aerospace Applications
Edited by Carlos R. Cabrera and Félix A. Miranda
Copyright © 2014 Pan Stanford Publishing Pte. Ltd.
ISBN 978-981-4463-18-8 (Hardcover), 978-981-4463-19-5 (eBook)
www.panstanford.com

coating. However, there are still numerous drawbacks, such as oxidation and corrosion of metal-like absorbents, inadequate conductive of carbon-like absorbents, inferior dispersing power and bonding strength with coatings, narrow frequency range shielding, and thick and heavy coating layer [1].

In recent years, as EM shielding, conductive polymeric composites filled with carbon fibers have been used. They have many advantages because they combine their good mechanical, chemical stability properties of matrix to significant reinforcing effects of fillers: their low density, which makes them suitable for lightweight shielding materials and their ease of forming by extrusion or injection molding, which makes them suitable to be processed in batches. These materials could have a wide range of applications in military, industrial, and commercial fields. For example, they can be used as casings for computers and television sets to shield electromagnetic waves and thus make them work safely. However, an obvious weakness of the shielding materials filled with carbon fibers is that the conductivity of carbon fibers is much lower than the metal [2]; so a greater volume fraction of carbon fibers is needed in order to provide the same shielding effect. Therefore, in order to decrease the amount of carbon fibers content and still retain a good shielding effect, a layer of metal is usually plated on the surface of fibers. Nickel and copper are the two main metals used as coatings. Nickel is generally preferred because of its stability and copper is preferred because of the readiness with which it is oxidized [3].

With the recent improvements in nanotechnology, it is expected to exploit the promising nanomaterials [4–6] to overcome the aforementioned drawbacks and to develop EMI-shielding coating with high shielding efficiency, broadband frequency, light and thin coating layer, and chemical stability as well as user-friendly attributes [1]. The nanoscale design and engineering of multi-functional materials and devices is based on a new type of carbon composites: carbon nanotubes composites. These are moving a large number of research activities to thriving and remarkable number of industrial products. A large number of these research activities are focused on, but not limited to, the studies of multifunctional composites based on carbon materials as EMI-shielding materials.

10.2 Carbon Nanotubes

Carbon nanotubes are an allotropic structure of carbon. In the solid phase, carbon can exist in three allotropic forms: graphite, diamond and buckminsterfullerene. Graphite is made up of layered planar sheets of sp^2-hybridized carbon atoms bonded together in a hexagonal network. Graphite is soft, slippery, opaque, and electrically conductive and is made up of different geometry of the chemical bonds. Electrons can move freely from an unhybridized p orbital to another, forming an endless delocalized π bond network that gives rise to the electrical conductivity. Diamond has a crystalline structure where each sp^3-hybridized carbon atom is bonded to four others in a tetrahedral arrangement. The crystalline network gives diamond its hardness (it is the hardest substance known) and excellent heat conduction properties (about five times better than copper). The sp^3-hybridized bonds account for its electrically insulating property and optical transparency. Buckminsterfullerenes, or fullerenes, are the third allotrope of carbon and consist of a family of spheroidal or cylindrical molecules with all the carbon atoms sp^2-hybridized. The tubular form of the fullerenes is called nanotubes.

They were discovered by in 1991 by Dr. S. Iijima, a researcher of NEC Corporation, during his analysis by the transmission electron microscope on fullerenes achieved by arc discharge method, in 1991 [7,8]. Fullerenes were discovered in 1985 by Rick Smalley and coworkers [9]. The first fullerene to be discovered was C_{60}. Sixty carbon atoms bonded together in pentagons and hexagons compose an icosahedral-shaped molecule, as a soccer ball ("bucky ball"). The carbon atoms are sp^2 hybridized, but in contrast to graphite, they are not arranged on a plane. The geometry of C_{60} strains the bonds of the sp^2-hybridized carbon atoms, creating new properties for C_{60}. Graphite is a semimetal, whereas C_{60} is a semiconductor.

Studying interstellar dust, the long-chain polyynes formed by red giant stars, Kroto discovered C_{60} fullerene. Its discovery was, like many other scientific breakthroughs, an accident. There are two types of carbon nanotubes (CNTs): CNTs are concentrically rolled graphene sheets with a large number of potential helicities and chiralities rather than a graphene sheet rolled up like a scroll; each ones are separated from one another by approximately 0.34 nm. These types of CNTs are called multiwall carbon nanotubes (MWNTs). Or it is possible to have an only one sheet of graphite to roll up. It can be visualized as a hollow cylinder. It is called single-wall carbon nanotubes (SWNTs) [10] as shown in Fig. 10.1. Figure 10.2 shows the TEM image of three types of MWNTs, with different diameters and number of walls.

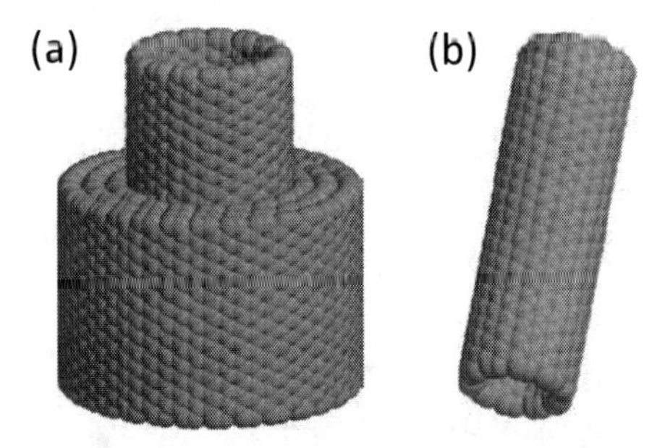

Figure 10.1 (a) Multiwall nanotube. (b) Single-wall nanotube.

Figure 10.2 Transmission electron images (TEMs) of the first observed multiwall carbon nanotubes (MWNTs) reported by Iijima in 1991 [6].

The way the graphene sheet is rolled determines the electronic properties of the tube. In fact, it is possible to have two types of electric behavior of CNTs—metallic or semiconductor—due to the different electronic structures of the tube itself. It is determined by chirality vector $\mathbf{C}_\mathrm{h}$ and 1D translation vector $\mathbf{T}$ of the nanotubes. $\mathbf{T}$ is normal to $\mathbf{C}_\mathrm{h}$ and together both define the elementary cell of the carbon lattice, or unit cell.

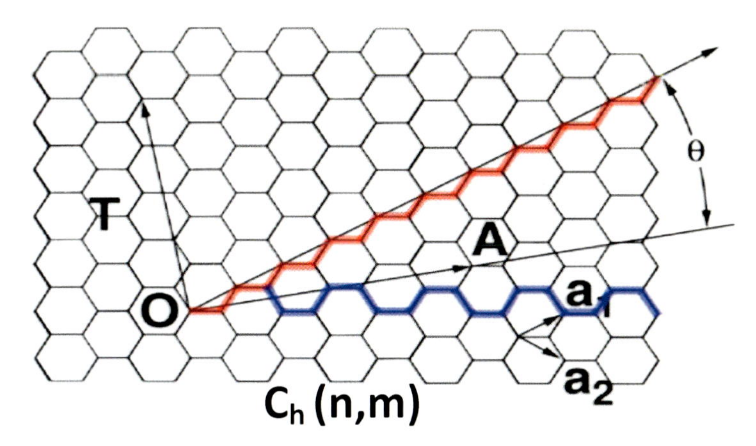

Figure 10.3 Graphite sheet and chirality vector and translation vector.

$\mathbf{C}_\mathrm{h}$ is the vector that defines the circumference on the surface of the tube connecting two equivalent carbon atoms:

$$\mathbf{C}_\mathrm{h} = n\hat{\mathbf{a}}_1 + m\hat{\mathbf{a}}_2, \tag{10.1}$$

where $\hat{\mathbf{a}}_1$ and $\hat{\mathbf{a}}_2$ are the two basis vectors of graphite and n and m are integers; n and m are also called indexes and determine the chiral angle:

$$\theta = \tan^{-1}[\sqrt{3(n/(2m + n))}] \tag{10.2}$$

It is used to separate carbon nanotubes into three classes differentiated by their electronic structure:

The tubes with $m = n$ are commonly referred to as *armchair tubes* and those with $m = 0$ as *zigzag tubes*. Others are called *chiral tubes* in general with the chiral angle, θ, as reported in Eq. (9.46) (Chapter 9).

The angle θ is 0 for zigzag ($m = 0$) and 30° for armchair ($m = n$) tubes. It is a convention to refer to $n \geq m$. Figure 10.4 illustrates a sample of different chirality of tubes.

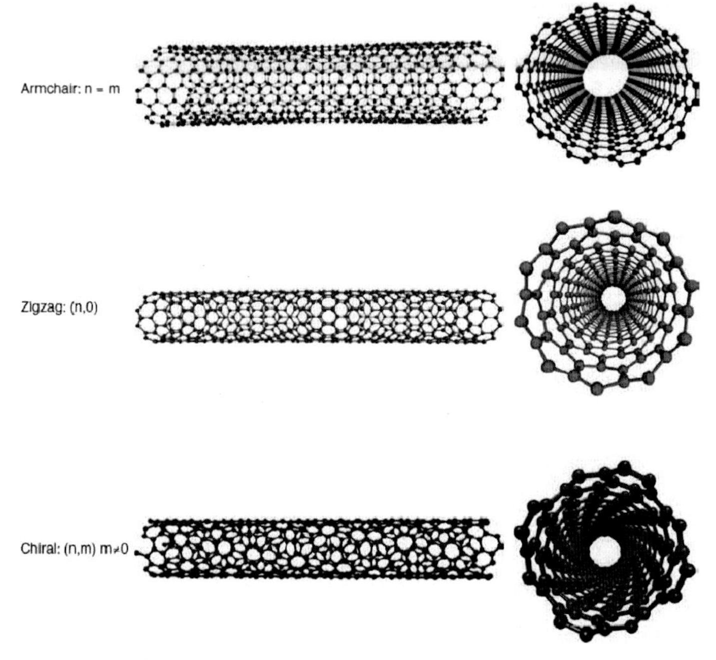

Figure 10.4 Examples of the three types of SWNTs identified by the integers (n, m) [7].

Armchair carbon nanotubes are metallic (a degenerate semi-metal with zero bandgap). Zigzag and chiral nanotubes can be semi-metals with a finite bandgap if

$$(n - m)/3 = q, \tag{10.3}$$

where q is an integer and $m \neq n$; the tube is metallic and semiconductor in all other cases.

Figures 10.5a,b show the different electronic density of states and bandgaps, respectively, relate to different chiral structures.

Electronic properties of nanotubes have received the greatest attention in research fields and applications. Extremely small size and the highly symmetric structure allow for remarkable quantum effects and electronic, magnetic, and lattice properties of the nanotubes.

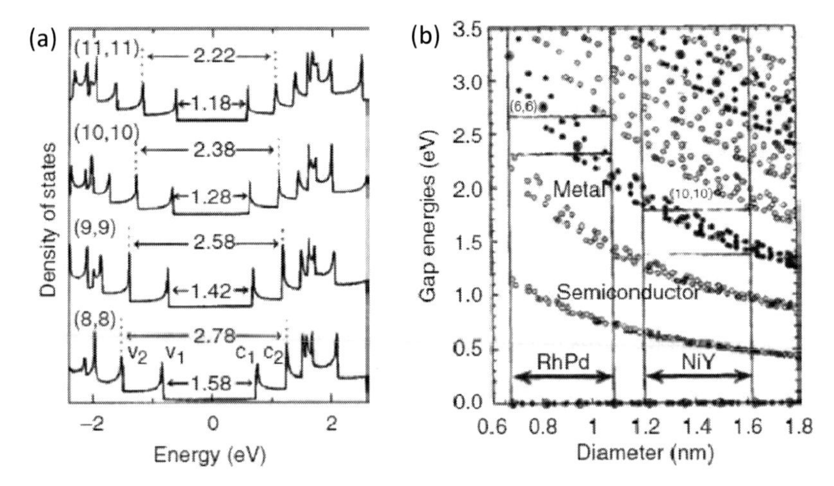

Figure 10.5 (a) Electronic density of states (DOSs) calculated with a tight binding model for (8, 8), (9, 9), (10, 10), and (11, 11) armchair nanotubes [18]; (b) Bandgap energies between mirror-image spikes in DOSs calculated for γ = 2.75 eV. Semiconductor SWNTs are open circles; metallic SWNTs are solid circles with the armchair SWNTs as double circles [19].

It is possible to describe the electronic properties of a nanotube, using the simplest model [11–14], these properties derived from the dispersion relation of a graphite sheet with the wave vectors $(\mathbf{K}_x, \mathbf{K}_y)$

$$E(\mathbf{K}_x, \mathbf{K}_y) = \pm\gamma\{1 + 4\cos((\sqrt{3}\mathbf{K}_x a)/2)\cos(\mathbf{K}_y a/2) + 4\cos^2(\mathbf{K}_y a/2)\},^{1/2} \quad (10.4)$$

where γ is the nearest neighbor-hopping parameter and a is lattice constant. γ = 2.5–3.2 eV from different sources [13] and a = 0.246 nm. When the graphite is rolled up to form a tube, a periodic boundary condition is imposed along the tube circumference or the $\mathbf{C}_h$ direction. This condition quantizes the two-dimensional wave vector $\mathbf{K} = (\mathbf{K}_x, \mathbf{K}_y)$ along this direction. The $\mathbf{K}$ satisfying

$$\mathbf{K}\gamma C_h = 2\pi q \quad (10.5)$$

is allowed where q is an integer. This leads to the following condition at which metallic conductance occurs, as it is shown in Eq. (10.3). All that suggests that one third of the tubes are metallic and two thirds are semiconducting. The bandgap for a semiconducting tube is given by

$$E_g = 2d_{cc}\gamma/D, \tag{10.6}$$

where d_{cc} = 0.142 nm is the C–C bond length and D is the diameter of tube, it is expressed by

$$D = |\mathbf{C}_h|/\pi = a(n^2 + nm + m^2)^{1/2}/\pi \tag{10.7}$$

The bandgap of a 1 nm semiconducting tube is roughly 0.70 to 0.9 eV. It is important to consider the intertube coupling when the results of a SWNT are used for a SWNT rope or a MWNT. Calculations have shown interesting intertube coupling properties. The intertube coupling induces a small bandgap for certain metallic tubes but a reduced bandgap by 40% for semiconducting tubes in a SWNT rope. Similar observations can be expected for a MWNT as well, but the intertube coupling is relatively smaller because of bigger diameter in a MWNT [14–17].

The first discovered CNTs was multiwall and they were made by THE arc discharge method. Alternative methods for preparing CNTs were reported by Smalley's group using laser evaporation [20] and Dai et al. [21] by chemical vapor deposition. To summarize, the most used methods of synthesis are

- arc discharge, using two pure graphite electrodes
- chemical vapor deposition (CVD), filling hydrocarbon gas, or, into a hot furnace, or near hot filament, or system where it could be possible to dissociate hydrogen and carbon atoms
- laser ablation, focusing a high powered laser on a substrate covered by carbon

Each technique has some advantages and drawbacks (Table 10.1). The choice of method of synthesis depends on which type of application CNTs are needed.

It is important to note how the synthesis of MWNTs does not require a catalyst for their growth by either the laser vaporization or the carbon arc method; instead, it is a basic prerequisite to use metal catalyst species, such as the transition metals Fe, Co, and Ni, for the growth of SWNTs. However, there are many challenges to win before carbon nanotechnology reaches its full potential.

For instance, high-quality carbon nanotubes of a reproducible, uniform structure are not yet available in large quantities. Defects, such as the 5–7 rings, kinks, junctions, Stone–Wales's defects, and impurities may be present in CNTs. The closure of the graphene cylinders by the incorporation of topological defects such as pentagons in the hexagonal carbon lattice has interesting structural properties. Complexes and structures can arise, for instance, conical-shaped sharp tips, due to the way pentagons are distributed near the ends for full closure. Theoretical studies have suggested that the ends of the tubes should have different electronic structure due to the presence of topological defects [22]. Pentagon–heptagon pair defects (widely separated defect pairs lead to surface steps) could give as effect of leading to the interesting possibility of changing curvature and helicity without significant bond distortions. CNTs with changing helicity along the axis have been studied [23]. The CVD method is an easier way to produce a large amount of CNTs at low cost, but the obtained tubes are full of defects and have traces of metals, as catalyst, inside themselves. In CNTs obtained by CVD, there may be a lot of defects due to lack of a good set of growing parameters: open-ended MWNTs, bamboo-like dangling bonds.

Table 10.1 Pros and cons of different technique of CNT synthesis

	Pro	Cons
• Arc Discharge	• Quite easy process • Suitable for synthesis without the use of a catalyst • Few topological defects	• Low yield rate • Wide range of dimensions among the synthesized CNTs • Purification needed
• Laser Ablation	• Long CNTs • SWNT diameters controllable	• High concentration of defects • Complexity of the apparatus
• CVD	• High yield • Possibility to grow oriented CNTs and with predefined geometries	• Requires the use of a catalyst for hydrocarbons decomposition • High concentration of defects

The MWNTs grown by the arc discharge method have generally the best structures due to the high temperature of the synthesis process. The CNTs synthesized by arc discharge show high crystallinity and stiffness and few defects, such as pentagons or heptagons existing on the sidewalls of the CNTs.

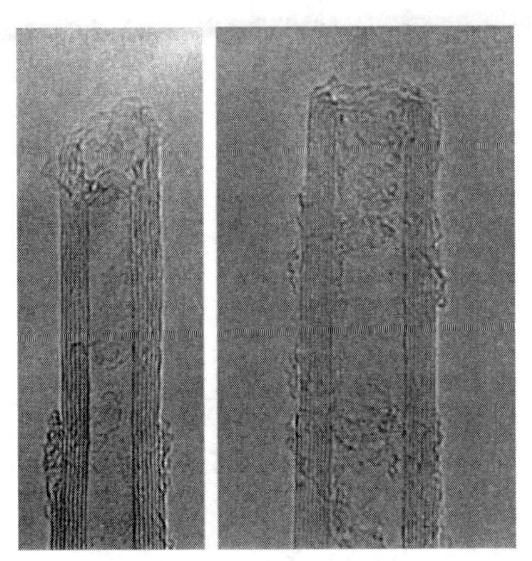

Figure 10.6 Transmission electron micrographs showing open-ended multiwall carbon nanotubes [24].

CNTs morphologies can vary considerably with the synthesis method. This is especially true for catalytically grown CNTs. If the growth conditions are not adjusted properly, then poorly graphitized walls result, and tubes with so-called coffee-cup structures form [24]. To summarize, it is possible to consider two families of defects:

- **topological defects** (pentagon–heptagon)
- **structural defects** (discontinuous or cone-shaped walls or bamboo structure)

Therefore, issues such as controlled syntheses of high-quality carbon nanotubes, producing CNTs with the same chirality, new methods for functionalization of nanomaterials, and novel nanodevice design and integration are being actively investigated. The field of carbon nanotechnology is expanding at a tremendous rate [26].

CNTs show high mechanical, electric, and thermal properties [27,28], as shown in Table 10.2. These properties are driven by their chirality structure and presence of defects. For example, Young's modulus of a cantilevered individual MWNT was measured, and it showed a value of 1.0 to 1.8 TPa from the amplitude of thermally driven vibrations observed in the TEM [29]. At the low end, this is only ~20% better than the best high-modulus graphite fibers. Exceptional resistance to shock loads has also been demonstrated [30]. In other case, the value of tensile strength was measured between 0.8 and 64 GPa for MWNTs and around 200 GPa for SWNTs. It is possible to estimate a value of Young's moduli ranging from 270 GPa to about 1 TPa, consistent with ideal behavior. It is important to remark that both the modulus and strength are strongly dependent on the nanotube growth method and the processing that follows, no doubt due to variable and uncontrolled defects. Also, there always exists some ambiguity regarding the choice of the appropriate cross-sectional area to use in evaluating the stress–strain data [31].

Table 10.2 Properties of CNTs

Electrical	• Can stand current densities as high as 10^9 A/cm^2 (Copper melts at 10^6 A/cm^2); • Can be either metallic ($n - m = 3k, k \in \mathbb{N}$) or semiconducting depending on the chirality (bandgap changes as $1/d$)
Mechanical	• Tensile strength of 64 GPa, as against 2 GPa of a high strength steel • Theoretical Young's modulus of 1 TPa, as compared with 208 GPa for steel • Can bend many times up to 90° without breaking, thanks to the rehybridization of sp^2 orbitals
Thermal	• Thermal conductivity up to 6000 W/mK, depending on the alignment of the tubes (diamante 3300 W/mK) • Stable under vacuum up to 2800°C
Chemical	• Hydrophobic behavior, not wet by many aqueous solutions • Wet by HNO_3 and many organic solvents • Capable of absorbing chemical species such as O_2 and NO_2

Atomic vibrations or phonons dominate the thermal conductivity K of carbon materials. Even in graphite, which is a good electrical conductor, the electronic density of states is so low that thermal transport via "free" electrons is negligible at all temperatures. Pioneering Tomanek group made calculations of the lattice contribution to K, yield values in the range of 2800–6000 W/mK [32].

The electrical behavior of SWNTs is determined by their chirality, either metallic or semiconductor [32]. The longitudinal conductance of a metallic one is quantified, namely, $G = nG_0$, with $G_0 = 2e^2/h = 77.47$ μS and n a natural number. The behavior of MWNTs is metallic if one sheet has a metallic chirality. A theoretical analysis of the conductance of infinitely long, defect-free MWNTs shows that the tunneling current between states on different walls is vanishingly small [33], which leads to the quantization of the conductance. This model shows that in a finite nanotube, the interwall conductance is negligible compared with the intrawall ballistic conductance.

The great interest in CNTs chemical properties is due to their strong sensitivity to chemical or environmental interactions. That includes opening, wetting, filling, adsorption, charge transfer, doping, intercalation, etc. A wide range of applications are possible: chemical and biological separation, purification, sensing and detection, energy storage, and electronics. CNTs are hydrophobic and do not show wetting behavior for most of common aqueous solvents. Using intercalation of the alkali metals with nanotubes it is possible to enhance their metallic conductivity or, for charge- or energy-storage with nanotubes applications, employing halogens intercalations. Experimental observations and theoretical calculations show that these intercalating agents mainly enter intertube spaces or defects on nanotubes for enhanced electrochemical capability for charge transfer and storage.

Indeed, nanotubes as electrode materials show enhanced electrochemical capability. The reduction and oxidation reactions that occur at the electrodes produce a flow of electrons that generate and store energy. The substitutional doping with B and N dopants was pursued to make nanotubes p and n types.

Several computational models have shown that unroped SWNTs are the best choice of CNTs for mechanical enhancement of composites, but a variety of conditions of different CNTs can play an important role with and without chemical functionalization. It is important because it allows a better dispersion into polymer and bounding between the filler and the polymer matrix, but not only. It is important to wet CNTs in aqueous solutions. Initially, nanotube functionalization was made at the various defect sites (MWNTs being higher defective than SWNTs), but functionalizations both at nanotube ends and along the sidewalls without disruption or degradation of the tubes have been demonstrated. A variety of functionalized nanotubes are being developed for composite applications, including fluoronanotubes (f-SWNTs); carboxyl-nanotubes with various end functionalization; and numerous covalently bonded SWNTs such as amino-SWNTs, vinyl-SWNTs, epoxy-SWNTs, and many others to provide for matrix bonding, cross-linking, and initiation of polymerization. Wrapped nanotubes (w-SWNTs) with noncovalent bonding are also another variety of nanotubes and find particular use when the electrical properties of the nanotubes need to be preserved [14].

10.3 Epoxy Nanocomposite Filled with Carbon Nanotubes: DC Measurements

Using CNTs as reinforcement in epoxy nanocomposites can widen their use in potential applications such as coatings, adhesives, potting compounds, encapsulates, structural materials, and liquid crystal displays [34,35].

Electromagnetic shielding applications required highly conductive, strong, easily processed, lightweight, and low-cost materials. High electrical conductivity and high aspect ratio make carbon nanotubes (CNTs) one of the most promising fillers for highly conductive polymer composites, with a large number of studies [36–45] on that. They have prospective applications in aerospace, biomedical, and electronic systems and other fields.

Epoxy resins are categorized as thermosetting resins as well as being phenol resins, urethane resins, etc. The history of epoxy resins goes back to the development of dental materials in 1938. The characteristics of these resins can be summarized as follows [46–56]:

(1) Because their viscosity is low due to comparatively small molecular weight, they can be used for various applications and can be molded under various conditions.
(2) Epoxy resins show smaller curing shrinkage and better dimensional stability than condensation thermosetting resins such as phenol resins and radical polymerization-type resins such as unsaturated polyester resins, because their cross-linking reactions are open-ring addition polymerization.
(3) Various molecule designs are possible because of different phenol compounds as the raw materials.
(4) By selecting suitable hardeners based on the required characteristics, the expected high performance is drawn out.
(5) High heat resistance, electrical properties, adhesion, and mechanical properties.
(6) The curability and moldability of epoxy resins are better than polyimides.

Owing to these excellent characteristics, epoxy resins have been used in a wide range of applications, such as in coating, electronics, adhesives, and engineering work. In Japan, epoxy resins are primarily used in electronics. Twenty-five percent of the total consumption of epoxy resins in the Japanese market is used for electromagnetic components and printed circuit boards (PCB) [48].

Generally, epoxy resin is combined with hardeners, or curing agents, which are chemical substances that have active hydrogen atoms. The high reactivity of the epoxides with amines, anhydrides, and other curing agents provides facile conversion into highly cross-linked materials. During the curing process, the addition reaction between epoxy groups in the epoxy resins and hydroxyl groups in the hardeners is carried out.

After the curing process, epoxy resins provide a low dielectric constant; for this reason, they are useful for several applications such as in semiconductor devices and PCB used for high-performance computers, cellular phones, and satellite communications equipment. Mainly, it is used as coating on PCB, power supply and packaging. Low dielectric constant materials make it possible to improve the operating speed of high frequency equipment. Expression (10.8) gives the relationship between the dielectric constant of an insulating material and the signal propagation speed, when $\mu_r = 1$,

$$V \propto c/\sqrt{\varepsilon}, \tag{10.8}$$

where V is the rate of propagation, C is the velocity of light, and ε is the dielectric constant.

Generally, for electromagnetic components, plastic packages are used as a substitute for ceramic packages because they have shown improvements in semiconductor packages. Because the dielectric constant of plastics is lower than that of ceramics, it will minimize the signal propagation speed under high frequency and disordered wave shape [48].

Here, the idea is to modify the electrical properties of net epoxy resin loading it with carbon nanotubes: decreasing the resistivity of resin and increasing its conductivity to achieve a high-conductivity coating, paint that can reduce the EMI and, hence, work as a good EMI shielding.

Therefore, it is important to quantify the amount of enhancement in the conductivity of resin materials when they are loaded with CNTs and then to evaluate the EM properties of nanocomposites.

In order to evaluate the electric behavior of an epoxy nanocomposite loaded with CNTs, testing was conducted using, as standard specification, a U.S. military standard (MIL-I-4658C), on measurement of resistivity of common epoxy coating on electric circuit for aircraft. A Y-shaped electrical circuit (as shown in Fig. 10.7) with two parallel lines as the tail of the "Y" with a 1 mm gap between them and with a length of about 2.5 cm was used [50–53].

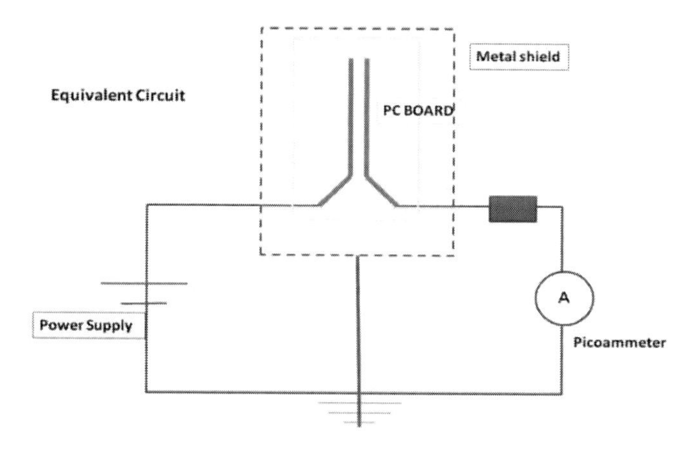

Figure 10.7 Schematic of circuit used for electrical measurements.

The circuits were made on a PC base with silver print and the two arms of the "Y" were connected to the picoammeter and a high voltage supply. The composite mixtures were spread, like thin films, on the circuit and electrical resistance tests were carried out using Keithley 6485 Picoammeter with short circuit protection. The current through the sample was recorded for three different applied dc voltages, namely 200, 500, and 1000 V, and for each value of voltage the measurement was taken at four times intervals: 2, 10, 60, and 120 s.

The EPIKOTE resin that was used is a commercial Hexion product, namely, Epon 828. It is a DGEBA resin. The main properties of the resin are shown in Table 10.3.

Table 10.3 Resin data sheet: Epon 828 DGEBA type

Viscosity	100–150 Pa s at 25°C
Epoxy equivalents	182–194
Density	1.16 g/cm^3

Two types of patented curing agent were used along with the resin: A1 curing agent and PAP8 agent.

A1 is a polyaminoalkylolic curing agent. It was obtained by reacting tetraethylenepentamine (TEPA) with formaldehide; the molecular ratio between the reagents is (1:1); they determine the insertion of alkylolic groups on the amino groups of the polyamine [53].

PAP8 is a polyaminophenolic curing agent. It was obtained by reaction between TEPA, formaldehide, and phenol; the molecular ratio was 1:1:0.8; the CH_2O group was added to the solution of Phenol in TEPA, gradually. After a thermal treatment, CH_2O creates the condensation and insertion of phenolic group onto the amino groups of the TEPA [54].

For preliminary testing, epoxy resin was mixed with graphite. It was bought from Aldrich Company. Table 10.4 shows the data sheet of graphite. The samples were loaded by 20 wt% of graphite.

Table 10.4 Graphite data sheet

Density	2.21 g/m^3
Granulometry	<20 μm
Purity	99.99%

It is important to underline that the first curing agent, A1, possesses polar groups in its chemical composition, whereas the second agent, PAP8, contains benzene groups. As a consequence, the mechanical properties of composites where the PAP8 agent has been used turn out to be improved. To understand the stability of the behavior of the curing agent, in the first step, the experiment was repeated under three different pressures: atmospheric, 10^{-2} mbar, and 10^{-6} mbar.

The low-pressure measurements indirectly gave the effect of moisture on the resistivity values of the samples. The plots in Figs. 10.8 and 10.9 show the resistivity versus applied voltages for various samples under varying voltage and pressure conditions.

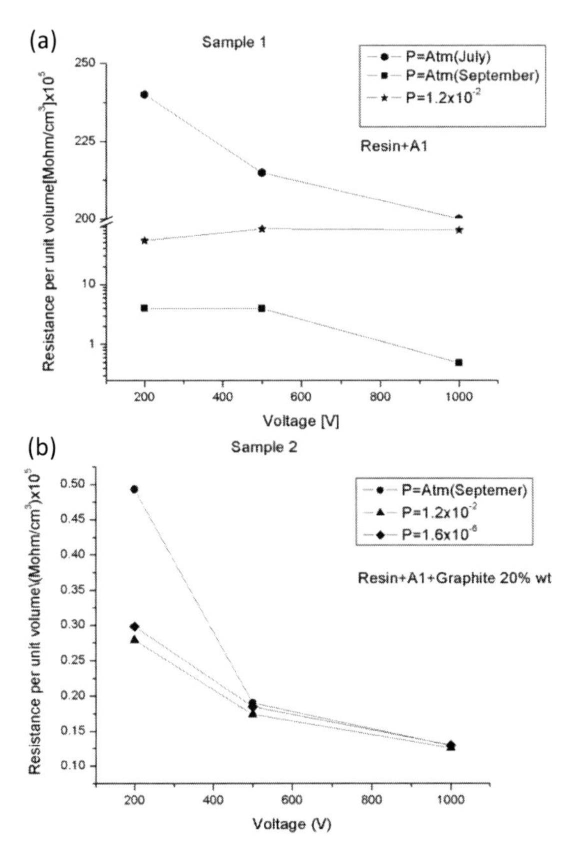

Figure 10.8 (a) Plot of resistivity vs. voltage for the samples. Resin + A1 with no graphite added. (b) Plot of resistivity vs. voltage for the sample. Resin + A1 graphite added.

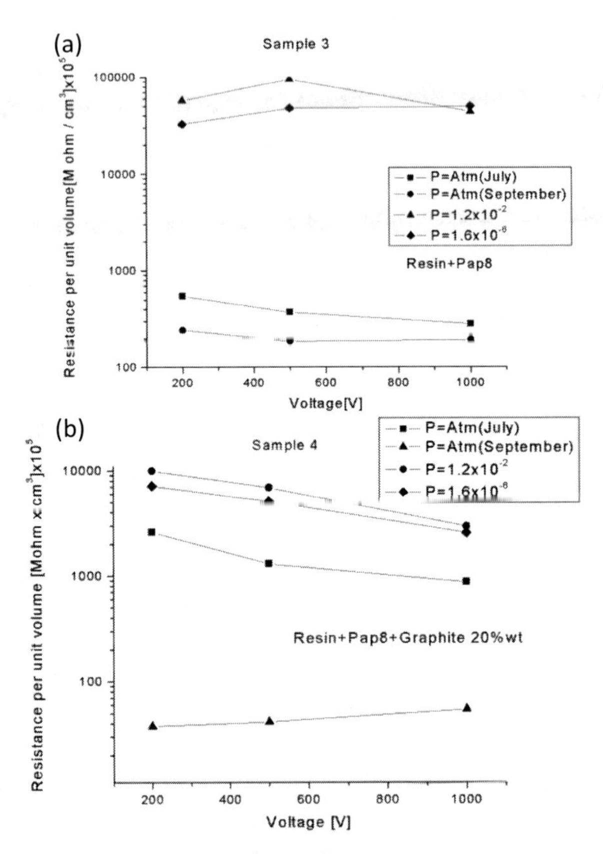

Figure 10.9 (a) Plot of resistivity vs. voltage for the samples. Resin + PAP8 with no graphite added; (b) Plot of resistivity vs. voltage for the sample. Resin + PAP8 graphite added.

It is important to note that the absolute change in resistivity is less over a wide voltage range of 200–1000 V for the sample with A1 curing agent (as seen in Figs. 10.9a,b), whereas for the sample with PAP8 curing agent the resistivity changes marginally more with increasing voltage. Note that the resistivity data were collected with the same samples at two different times of the year (i.e., July 2005 and September 2005) in order to have a rough estimate of the influence of climatic and environmental conditions on the performance. From the preliminary analysis of the data (Figs. 10.9a,b), it appears that the resistivity values of the composites employing PAP8 agent show a large difference between atmospheric and low-pressure conditions, and the composites' behavior was not affected by the addition of graphite. In the case of composites with A1 curing agent, the behavior is quite different (Figs. 10.8a,b), i.e., the stability of the material increases as graphitic additions are included. This seems to favor the use of A1 curing agent from the point of view of the optimization of its use in aerospace applications.

Therefore, it is possible to conclude that the resin + A1 curing agent + graphite combination seems to be an ideal candidate for applications in various pressure ranges as well as voltage ranges. In the second step, for studying resistivity versus concentration, the data were taken under atmospheric pressure conditions only.

The samples were loaded with micro powder of micro-size of graphite from Sigma Aldrich Company, and CNTs obtained by the arc discharge method (at Frascati National Laboratories (LNF)-INFN, "NEXT" Group). Carbon nanotubes were synthesized in a dc arc plasma system in helium atmosphere at a pressure of 700 mbar. The arc was struck between two electrodes of high purity graphite rods (purity 99.97%), bought from Goodfellow.

The discharge is typically carried out at a voltage of 22 V and a current in the range of 90–100 A for 50 min. Some of the evaporated carbon condenses on the tip of the cathode, forming a slag-like hard deposit, as shown in Fig. 10.10. The deposit, essentially on the cathode, consists of bundles of carbon nanotubes mixed with a small quantity of amorphous carbon [56,57].

Figure 10.10 Scanning electronic microscope images of cathode deposit of carbon nanotubes taken using ISI ABT-DS 130S Microscope by A. Grilli [49].

Solution processing of CNT–epoxy resin was the employed method. Samples were prepared as follows: In the beginning, the fillers were mixed in isopropylic alcohol and ultrasonicated for 30 min. Then this solution was mixed with a known quantity (weight) of resin (around 1 g) and heated in an oven for 2 h at 80°C. The alcohol evaporated off and the resin with CNT was again sonicated for 15 min. Immediately after this, the hardener A1 is mixed and the mixture is applied on the surface of the electrical circuit and allowed to set.

Composites of resin A1 mixed with both graphite as well as CNTs were studied separately with three compositions of 0.1, 0.5, and 20 wt% and 0.1, 0.25 and 0.5 wt%, respectively. Mean resistivity measurements gave the data shown in Table 10.5.

Table 10.5 Graphite and CNT resistivity

Matrix + CNTs	**0.1%**	**0.25%**	**0.5%**	**wt**
Resistivity	1.63E + 02	9.19E + 00	5.24E-02	MΩ × cm
Matrix + Graphite	**0.1%**	**0.50%**	**20%**	**wt**
Resistivity	7.95E + 05	1.17E + 05	1.79E + 00	MΩ × cm

As can be seen in the plot given in Fig. 10.11, the resistivity decreases drastically from thousands of megaohms to hundred megaohms with even a small (0.1 wt%) addition of CNTs. Further increase (0.25 wt%) in CNTs concentration results in the resistivity value of 9.19 MΩ (a decrease of two orders of magnitude). An increase of CNTs to 0.5 wt% results in the decrease of resistivity to 0.05 MΩ, i.e., again a change by two orders of magnitude. The concentration of 0.25 wt% was considered only for the CNTs sample but not for graphite.

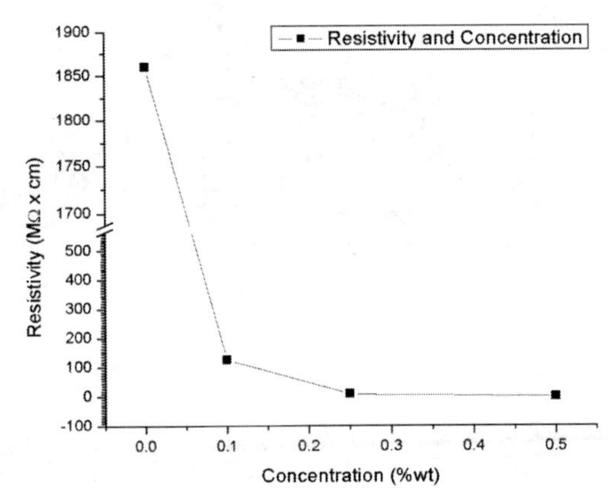

Figure 10.11 Resistivity vs. concentration of CNTs [47].

The sharp fall in resistivity (from few thousand MΩ to few hundred MΩ) for even a small change in the concentration of

0.1 wt% can be seen clearly. The change in the resistivity values for CNT-based composites turns out to be significant, even for small changes in the added CNT percentage. These results might be important for determining the most suitable "recipe" for the realization of composite materials for high-fidelity circuits in aerospace applications or even in devices exposed to disturbances predominantly electromagnetic in their nature.

It is possible to carry out the measurement of electric properties of composite using other testing method. According to the specific standards ASTM D257 and ASTM D4496-04, the evaluation of the electric behavior of an epoxy nanocomposite loaded with carbon material was done by the "two-probe method." The morphology of CNTs is an important factor that affects the SE of CNTs composites. The used epoxy resin was EPON 828 and carbon materials employed were

- MWNTs produced by Heji, using CVD following some of their characteristics and image shown in Fig. 10.12.
- SWNTs produced by Heji with the characteristics and morphology shown in the Fig. 10.13.
- CNTS produced by the arc discharge method (at Frascati National Laboratories (LNF)—INFN, "NEXT" Group), based on the electric arc primer in the presence of an inert gas (helium gas was used). The equipment used has been described earlier in the text. The CNTs made by Arc discharge owing to its chemical purity and low defectiveness.

Carbon black is a particulate form of industrial carbon that has a microstructure "nearly graphitic," but unlike graphite, the layers' orientation is random [57]. Carbon black used here is Printex 90 (mean particle diameter 14 nm, BET surface area 300 m^2/g, volatile matter at 950°C (%) ~1), generously provided to our group by Evonik Degussa GmbH.

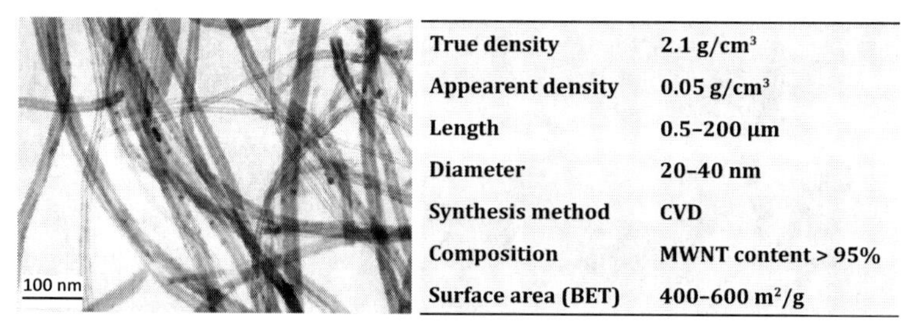

True density	2.1 g/cm^3
Appearent density	0.05 g/cm^3
Length	0.5–200 μm
Diameter	20–40 nm
Synthesis method	CVD
Composition	MWNT content > 95%
Surface area (BET)	400–600 m^2/g

Figure 10.12 TEM micrograph of multiwall nanotubes.

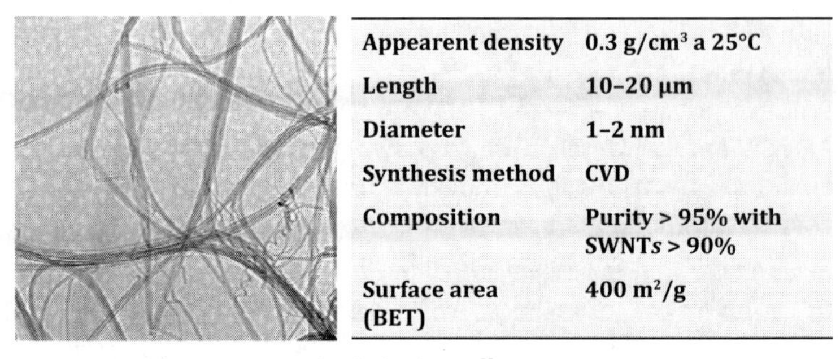

Appearent density	**0.3 g/cm³ a 25°C**
Length	**10–20 μm**
Diameter	**1–2 nm**
Synthesis method	**CVD**
Composition	**Purity > 95% with SWNT*s* > 90%**
Surface area (BET)	**400 m²/g**

Figure 10.13 TEM micrograph of single-wall nanotubes.

Figure 10.14 TEM micrographs of arc discharge multiwall nanotubes performed by CIGS.

Samples were made in the "bar" shape, length 12 cm and width and depth 1 cm. Then the bars, once removed from the mould were cut into many "cubes" (dimension 10 mm × 10 mm × 10 mm). Figure 10.15 shows the samples. On these, the electrical measurements were carried out: the current intensity, in order to obtain the value of the resistance, and consequently the resistivity (or conductivity).

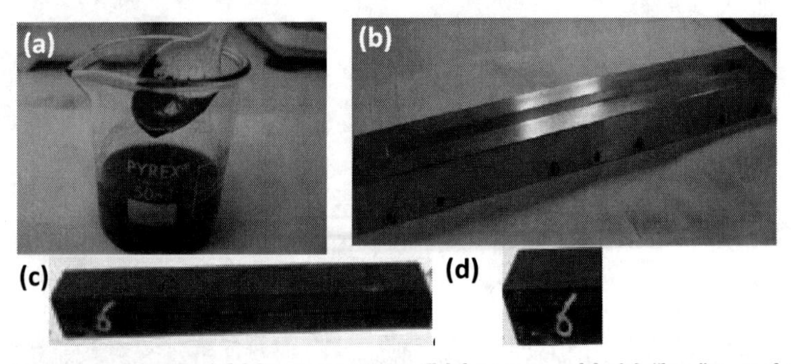

Figure 10.15 Images of (a) preparation; (b) brass mould; (c) "bar" samples; (d) "cube" samples.

Table 10.6 shows the procedure of the preparation of "bar" samples. Solution processing of CNTs–epoxy resin was the employed method. It is based on mixing of two components into a certain solvent and, then, evaporate the latter to form a composite material (film, bar). This one is the most common method for preparing CNT–polymer composites. Ultrasonication is an effective method to disperse CNTs in liquids having a low viscosity, such as water, acetone, propanol, and ethanol. However, most polymers are either in a solid or viscous liquid state, which requires the polymer to be dissolved or diluted using a solvent to reduce the viscosity before dispersion of CNTs, or heating of polymer. Other sintering processes widely employed are melt shear mixing, bulk mixing, calendaring mixing, roll mixing, polymerization in situ, and also a combination of these methods.

Table 10.6 Procedure for preparing "bar" samples

- Degassing of the resin under vacuum (1–3 mbar), t = 12–14 h
- CNTs' dispersion in propanol through ultrasonication for 1 h 30 min
- Mixing of the obtained suspension (CNT–propanol) with epoxy resin followed by alcohol
- Evaporation at 130–150°C × 2 h
- Further sonication of CNTs into the epoxy resin, 1 h 30 min
- Slow manual mixing with the curing agent (A1) $x \sim 7$ min
- The mix is poured into the moulds
- 20 min in air
- 1 h in oven at 40°C for further degassing
- 20 h in air
- 6 h at 80°C

Standard laboratory sonicators (in a water bath) run at 20–23 kHz with a power less than 100 W. Commercial probe sonicators have an adjustable amplitude ranging from 20% to 70% and a power of 100–1500 W. The probe is usually made of an inert metal such as titanium. Most probes are attached with a base unit and then tapered down to a tip with a diameter from 1.6 to 12.7 mm [59,61].

The effective utilization of the CNT material in composite applications depends strongly on their ability to be dispersed individually and homogeneously within a matrix. An optimized interfacial interaction between the CNT sidewalls and the matrix should result in an efficient load transfer to the "hard" component of the composite.

Using a dc power supply, tests were performed on each "cube" at different voltages: 50, 100, 200, 300, 500, 750, and 1000 V. The current intensity was measured at four time intervals: 2, 10, 60, and 120 s. Each type of mixture (i.e., the mixture containing carbon black, the one with MWNTs, SWNTs, and with INFN-CNTs) was loaded at different concentrations of fillers, such as 0.1%, 0.25%, and 0.5 wt% (estimated percentage on the weight of the resin). From each mixture, we obtained 12 "cubes": Each one was tested at different voltages for different times. The *I*–*V* graphs (where *I* denotes the average of the current's values measured for the different cubes obtained from each bar of a given mixture and filler concentration) were realized for each type of filler at different concentrations for different times; the *I*–*V* graphs related to the time 120 s and to the concentration of 0.5 wt% are shown in Figs. 10.16–10.18 [62,63].

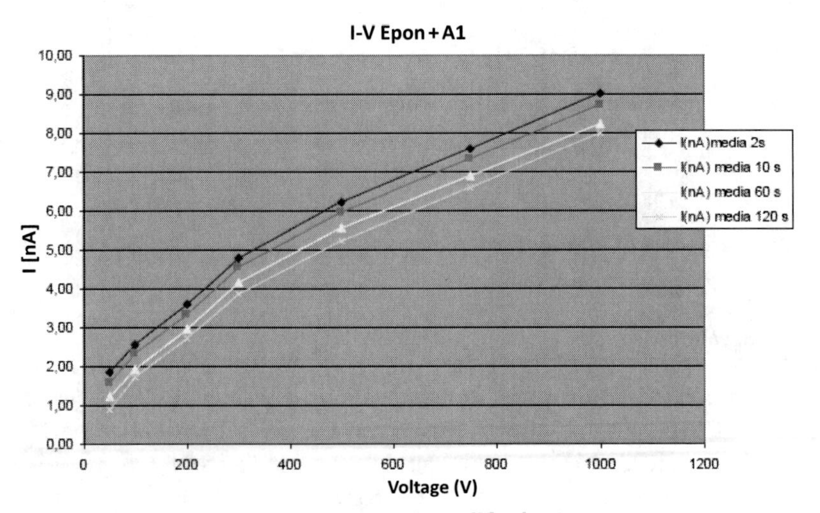

Figure 10.16 Plot of current (*I*) vs. voltage (*V*) of only resin composites.

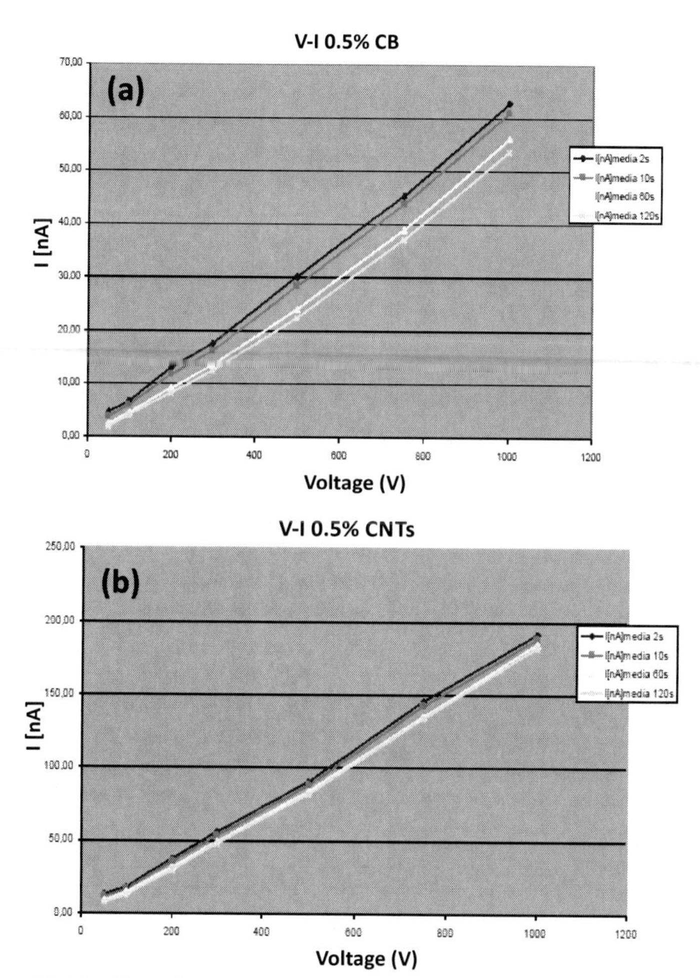

Figure 10.17 Plot of current (I) vs. voltage (V): (a) Resin + CB composites; (b) resin + INFN CNT composites.

When the resin is loaded with filler, at the same percentage in weight, it is easy to identify a linear trend of (I, V) plots, as an ohmic behavior, specially, when the matrix is loaded with 0.5 wt% of carbon nanotubes. The trend of the net resin at different times of measurement shows a quite different behavior, not ohmic behavior, due to the absence of a conductivity path and the presence of tunneling resistance. The resistance to electron flow in conductive nanocomposites comprises three resistances: the intrinsic resistivity of the filler material, the particle–particle contact resistance, and the tunneling resistance [64]. The particle–particle contact resistance is the resistance due to forcing of an electron through a small distance between two conduct particles. The tunneling resistance is a result of electrons passing through a very thin insulating film, which is possible when the film thickness is less than 100 Å. Any modifications to the composite that affect at least one of the three resistances can change the composite conductivity.

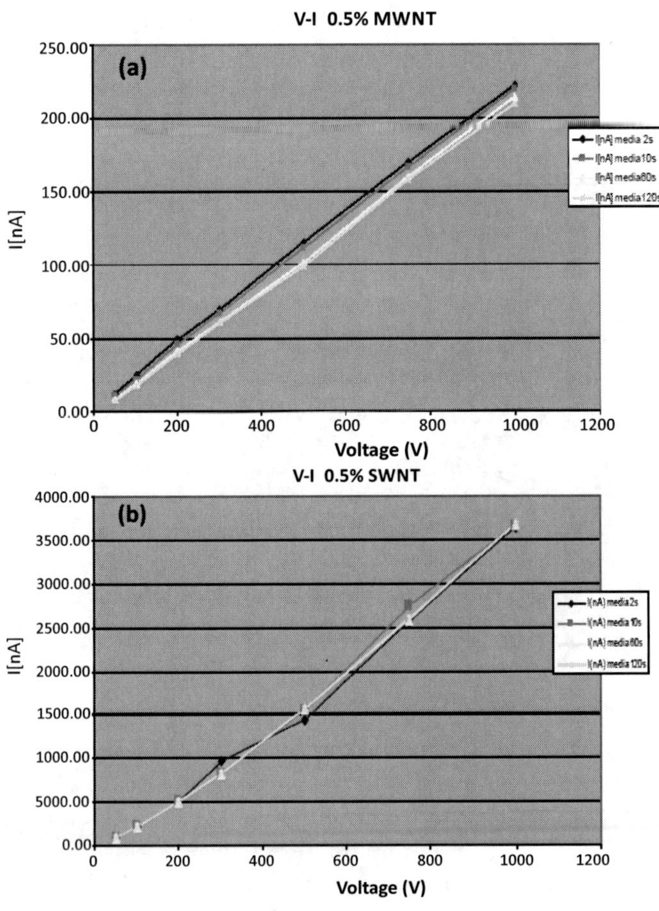

Figure 10.18 Plot of current (I) vs. voltage (V): (a) Resin + HeJi MWNTs composites; (b) resin + Heiji SWNTs composites.

In order to achieve conduction in filled polymer systems, conductive pathways of filler particles are required throughout the polymer matrix so as to allow electrons to move freely through the material. CNTs possess high aspect ratios and have the added advantage of achieving percolation at lower compositions than spherical fillers. Percolation describes the range of compositions where a three-dimensional network of the filler is formed. When percolation is reached, the conductivity increases dramatically because the network has created a conductive path. The percolation behavior can be expressed by a power law equation [65–67]

$$\sigma_{\mathrm{v}} \approx (v - v_{\mathrm{c}})^{t}, \tag{10.9}$$

where σ_{v} is the conductivity as a function of the filler concentration, v is the volume fraction of filler, v_{c} is the volume fraction where percolation occurs, and t is the critical exponent, which appears to depend on sample dimensionality.

However, numerous studies show that the percolation threshold and conductivity depend strongly on the polymer type and synthesis method, geometric parameters of CNTs (length and diameter), disentanglements of CNTs, agglomeration, uniform spatial distribution of individual CNTs and degree of alignment [68,70]. Table 10.7 presents a comparison between the different types of polymeric matrices, CNTs type and additional treatment (purification/functionalization), processing method, and electrical characteristics such as room temperature. Also, it clearly reveals a large variation of the electrical properties values as a function of the polymer matrix, processing method, alignment or not of tubes and CNT type (amount of defects, impurity inside the tubes). Correlations and general dependencies between the above-mentioned parameters are difficult to establish [71]. During engineering and designing process, it is better to take into account all these parameters to obtain a high-performance material.

Analyzing the trend of resistivity versus concentration, in Fig. 10.20, it is easy to see that while the values of resistivity of the matrix with arc discharge and the one with MWNTs by CVD are comparable to each other, the value of resistivity of the matrix with SWNTs is approximately three orders of magnitude lower.

Table 10.7 Electrical properties of CNT-based polymer composites [71–87]

Matrix	CNT type	CNTs wt%	Processing method	Matrix—electric conductivity (S/m)	Composite—electric conductivity (S/m)
Epoxy	MWNTs	≤0.18	Solution mixing	$\approx 4 \times 10^{-9}$	$\approx 10^{-2}$
Epoxy	MWNTs	≤4	Bulk mixing	$\approx 10^{-8}$	$\approx 4 \times 10^{-1}$
Epoxy	Aligned MWNTs	≤1	Shear mixing	$\approx 10^{-9}$	$\approx 2 \times 10^{-1}$
Epoxy	CNTs	≤2.5	Solution mixing	2.4×10^{-14}	$\approx 1.3 \times 10^{-2}$
Epoxy	Palmitic acid-modified CNTs	≤0.8	Mechanical mixing	7.9×10^{-14}	6.9×10^{-3}
Epoxy	MWNTs with and without non-ionic surfactant (Tergitol)	≤12	Solution mixing	$\approx 2 \times 10^{-14}$	$\approx 5 \times 10^{-5}$
Epoxy	SWNTs	≤0.21	High frequency sonication Method	—	$\approx 1.25 \times 10^{-3}$
Epoxy	MWNTs	≤1.5	Solution mixing	$\approx 10^{-7}$	≈0.5
Epoxy	SWNTs\DWNTs	≤0.4	Solution mixing	10^{-13}	10^{-2}
Epoxy	MWNTs SWNTs treated in EtOH/NaOH mixture Ball milled SWNTs	≤0.5	High-shear mixing	$\approx 10^{-8}$	≈0.8
Epoxy	SWNTs	≤15	Solution mixing	$\approx 2 \times 10^{-11}$	≈10
Epoxy	MWNTs	≤0.5	Calendering process	$\approx 10^{-8}$	≈0.01
Epoxy	MWNTs	≤5	Calendering process	$\approx 10^{-15}$	≈50
Epoxy	MWNTs	≤10	Solution mixing	1.2×10^{-14}	$\approx 3 \times 10^{-3}$
Epoxy	SDS suspended MWNTs	≤0.5	Bulk mixing	5.1×10^{-12}	2.5×10^{-7}
Epoxy	Oxidized MWNTs	≤1	Solution mixing	$\approx 10^{-13}$	$\approx 10^{-2}$
Epoxy	Pristine MWNTs	≤1	Solution mixing	—	$\approx 10^{-2}$
Epoxy	MWNTs	≤0.75	Roll milling	$\approx 10^{-15}$	$\approx 10^{-1}$

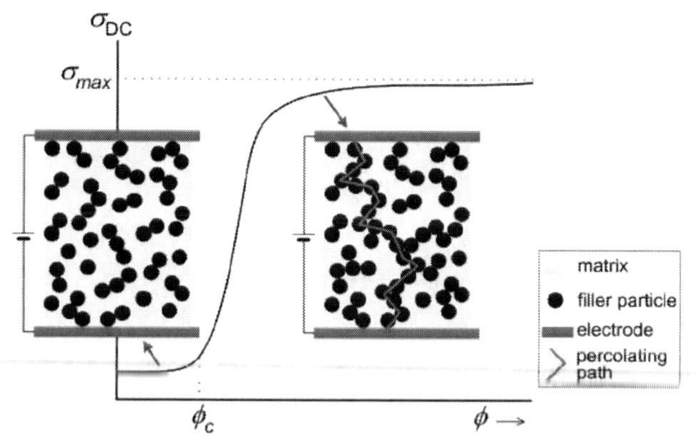

Figure 10.19 σ_{DC} as a function of filler fraction v. A steep increase in conductivity is observed at a critical filler fraction v_c. For $v < v_c$, the filler fraction is too small to form a continuous network through the sample, as shown in the left drawing of the nanocomposite, whereas for $v > v_c$, such a continuous network exists.

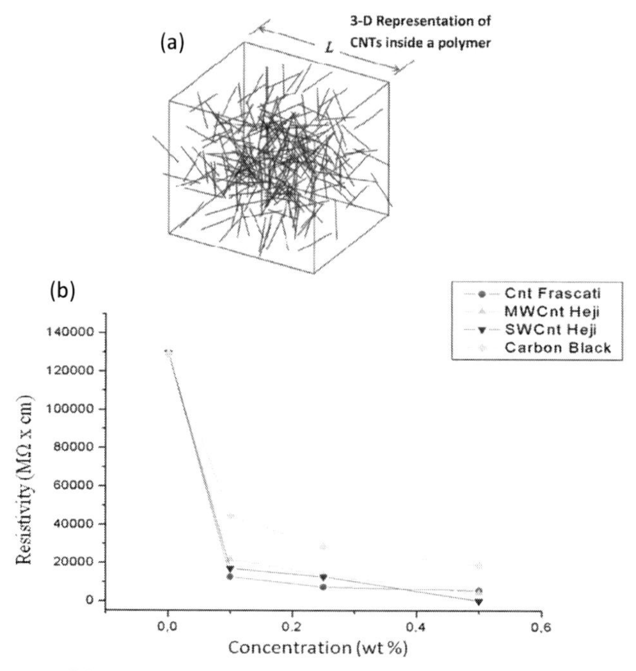

Figure 10.20 (a) Resistivity vs. concentration of carbonaceous fillers and (b) 3D representative [96] CNT volume elements inside the polymer.

This means that doubling the CNTs concentration (from 0.25% to 0.5%) shifts the relevance of the parameter affecting the most the performance in conductivity of the nanocomposite, from the density of topological defects (or equivalently the CNTs deposition method) to the aspect ratio of the filler. Remember that the value of the aspect ratio of SWNTs is higher than the one of MWNTs, which means that the extension's value of the matrix-filler interface, at equal weight concentration, is higher in the first ones than the latter. Therefore, increasing the amount of nanotubes, a mechanism of percolation is activated, in which an interconnected network is formed (so we are beyond the percolation threshold), in order to create a preferential way for the current passage inside the loaded matrix [88], as it is possible to see in Fig. 10.19, which shows a schematic representation of interconnection of CNTs inside a cubic volume, as used sample for this analysis. The addition of conductive particles into an insulating polymer can result in an electrically conductive composite if the particle concentration exceeds the percolation threshold. CNTs are very effective fillers, with an electric current-carrying capacity cited to be as large as 1000 times higher than copper wires [89]. A good dispersion and correct alignment, less defects and high aspect ratio of CNTs into the composite give quite different conductivity behavior of nanocomposites. There are clearly differences in electrical conductivity due to CNT features, especially distribution, degree of alignment, CNT–CNT and CNT–polymer interactions and hopping or tunneling behavior [90–92], but also differences due to the polymer and test methods (particularly surface resistivity testing of films vs. bulk resistivity, for example) [93]. Formation of CNTs networks can be induced by the application of an external electric field (dc field of 100 V/cm) [93]. Therefore, electric fields not only align CNT but also enhance the attractive forces between neighboring CNT. Conduction may occur by electron hopping from a nanotube to an adjacent one when they are close enough. Above percolation, conductive paths are formed through the composites due to quantum tunneling effects where the distance between the conductive inclusions is such that electron hopping can occur. Percolation theory assumes that paths are made up of conductive inclusions in direct contact [94].

For polymer nanocomposites, the electrical conductivity is mostly determined by the tunneling resistance among neighboring nanotubes, under condition of uniform dispersion of conductive CNTs into insulating matrix materials. The tunneling resistance between a CNT–polymer–CNT is very sensitive to polymer separation layer thickness and also depends on the relative orientation between the CNTs.

At low concentration in volume, an average value more than one order larger in magnitude than the diameter of nanotubes is possible to take account as the average value of distance of separation between nanotubes, thus the CNT curvature effect can be neglected in theoretical analysis and simulations. In a typical insulating polymer matrix, the value of average effective junction resistance due to tunneling among CNTs was estimated in (R_0 =)10^{13} Ω by Foygel et al. [95], by fitting their formula to the experimental results reported by Ounaies et al. [94]. They demonstrated that the high value of resistance is driven by contacts between the nanotubes belonging to the backbone percolation cluster. To understand and know better, the real mechanism of percolation into the polymer, a lot of simulation modeling were tested.

In their simulation modeling, Yu et al. showed how the conductivity of CNTs inside the insulator polymer material is determined by tunneling resistance, which can be explained by the following two factors [96]:

- the large difference in the tunneling resistance
- the resistance of individual nanotubes and the unique structure of resistor networks consisting of alternating these two types of resistors

Yu et al. considered that the simulation the parameters characterizing the nanotube properties are assumed to be constant or follow simple distributions: The position and the orientation of CNTs are assumed to follow the uniformly random distribution and tubes are assumed to be identical in shape and size. If there is a good dispersion of CNTs inside the resin, it will be possible to ignore the resistance of nanotubes, during the calculation. In this simulation, the "soft core" model was used in the current Monte Carlo simulation with the expectation that CNTs bend at the junctions [97]. Using Simmons formula, the tunneling resistance between two neighboring CNTs can be approximately estimated as follows [96]:

$$R_{\text{tunnel}} = V/AJ = [(h^2 d/Ae^2\sqrt{2m\lambda})]\exp[(4\pi d/h)(\sqrt{2m\lambda})] \quad (10.10)$$

where J is the tunneling current density, V the electrical potential difference, e the quantum of electricity, the mass, h Planck's constant, d the distance between CNTs, k the height of barrier (for epoxy, 3.98 eV), and A the cross-sectional area of tunnel (the cross-sectional area of the CNT is used here as an approximation).

They have shown that the tunneling resistance, between two crossed CNTs separated by a 1 nm-thick epoxy layer, ranges from around 3×10^{10} to 5×10^{11} Ω as the diameter of CNTs changes from 8 to 2 nm, respectively. The resistance increases to 10^{19} Ω for a separation of 1.8 nm and the tube diameter of 2 nm intrinsic resistance. Also, the resistance of the CNTs was estimated by following formula

$$R_{\text{tube}} = \rho \text{ X } (L/\pi R^2), \tag{10.11}$$

where ρ is the nanotube resistivity, L and R are the length and the radius of the tube, respectively. The resistivity of conductive CNTs was reported to be as low as 5.1×10^{-8} Ω m. The resistivity of a typical metallic CNTs is usually taken as 10^{-6} Ω m, as confirmed by Thostenson et al. [88]. Given this resistivity, the intrinsic tube resistance is around 10^6 Ω for a nanotube with length of 3 μm and a diameter of 2 nm. However, it is important to underline how two tunneling resistors have to be connected by a tube segment, and similarly, two tube resistors are always bridged by one tunneling resistor. This unique configuration of effective networks, and the fact that the tunneling resistors are much larger in value as compared to the tube intrinsic resistances, justifies the simplification of only taking into account the tunneling resistance during the simulation; therefore R_{tube} can be negligible. Under these considerations, the simulation results showed a good agreement with the classic percolation theory on estimation of the critical behavior of the conductivity of nanocomposite [97].

Notably, in real material systems, physical parameters, such as the tunneling resistance, can have large variations as a function of the junction configuration and materials properties of nanotubes and polymer. It is important also to consider that "the soft core" model seems to contradict the physical principle that tubes can never penetrate through each other. On the other hand, the "hard core" model has not the above-mentioned contradiction, but its implicit assumption that the tube have to be infinitely stiff leads to high rejection rates of the generation of tubes during the simulation. In reality, the tubes with high aspect ratios can be easily bent at junctions to avoid penetrating each other [97].

Using a Monte Carlo simulation, Ning Hu et al. decided to consider CNTs as "soft-core" cylinders of length L and diameter D, and they are allowed to penetrate each other, also it was considered the distance between two noncontact CNTs (J), where J there is the tunneling current (Fig. 10.21). They want to create a "3D-cell" model, more exactly, a combination of 3D resistor network and 3D fiber reorientation model, to predict the electrical conductivity of the nanocomposite. Most of these models have been based on the RC (resistance–capacitance) of a simulated microstructure. For convenience, in their research, the electrically conductive paths in the matrix phase are completely neglected. There is no obvious aggregation in their various specimens using different fabrication processes for this nanocomposite.

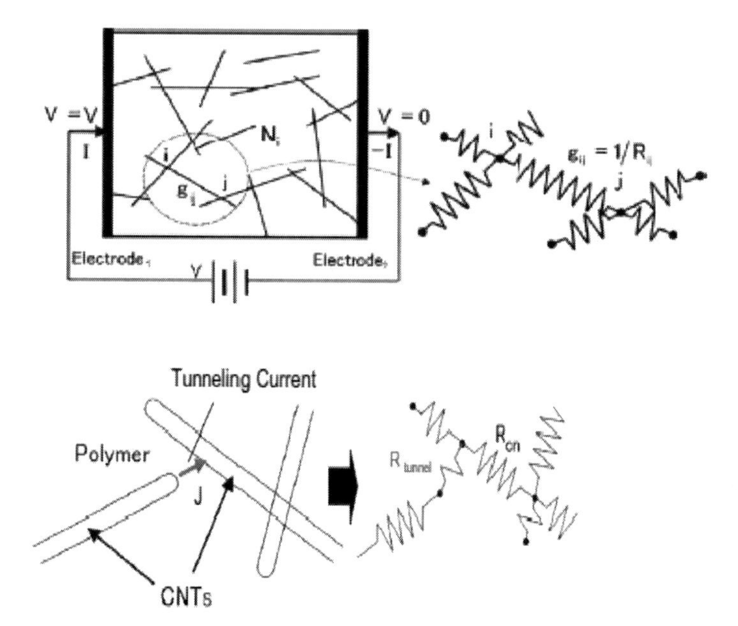

Figure 10.21 Schematic of a resistor model with a random distribution of CNTs [96].

For a CNT with two contacting points i and j with neighboring CNTs, the conductance g_{ij} between i and j (the inverse of resistance R_{ij}) can be evaluated as

$$g_{ij} = \sigma_{\mathrm{CNT}} (S_{\mathrm{CNT}}/l_{ij}), \tag{10.12}$$

where l_{ij} is the length between the points i and j, σ_{CNT}, and S_{CNT} are electrical conductivity and cross-sectional area of the CNT, respectively, as shown in Fig. 10.21. The tube–tube contacts among CNTs are assumed here to be perfect with zero resistance. Using Eq. (10.12), the tunneling resistance effect on electrical conductivity of nanocomposite was estimated as shown in Fig. 10.22 (for epoxy, $K = 0.5$–2.5 eV).

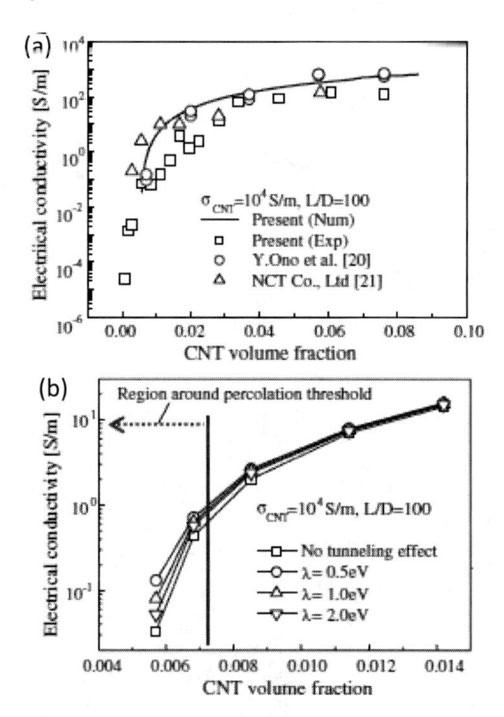

Figure 10.22 Results of electrical conductivity of nanocomposites: (a) comparison of experimental and numerical conductivities and (b) effect of tunneling on conductivity [98].

The tunneling effect was identified by the increase in electrical conductivity when the volume fractions of CNTs are near the percolation threshold of the composite [98]. The tunneling effect disappears gradually with increasing additions of CNTs.

Comparing their results, Ning Hu et al., found a good agreement between numerical simulation and experimental measurement, as shown in Fig. 10.22. Bauhofer et al. noted that this type of modeling on statistic percolation theory has some limitations. It is derived for ideal systems, and they do not consider that CNTs/polymer composites are far away from being ideal systems; in fact, homogeneous dispersion of identical particles is very difficult to achieve. Because CNTs have a wide range of properties, i.e., length, diameter, chirality, entanglement and waviness, it is also possible to have different effects of CNTs in bundles and ropes with regard to alignment and interfacial bonding. Also, they try to identify main parameters leading to conductivity in polymer composite. They suggest that the shear mixing technology and the presence of solvent during the processing of the composite seem to improve the conductivity of the composite.

The choice of polymer is more important than that of fillers because each matrix has an own tunneling resistance, the extreme distance dependence of tunneling through polymer barriers between CNT. It seems that some polymer types and processing methods favor the formation of insulating polymer coatings of different thicknesses around the CNTs. According to Connor et al., tunneling between CNT separated by a thin isolating layer should be described by following the relationship:

$$\ln\sigma_{DC} \sim -v^{1/3} \quad (10.13)$$

Connor et al. suggest a relation between dc conductivity (σ_{DC}) and filler load (v) [99].

After these considerations, it is important to remark that, the mechanism of charge transport in CNT/polymer composites has also been addressed by temperature dependent conductivity measurements [100]. Charge transport via phonon-induced tunneling of electrons between randomly distributed localized states (variable range hopping [VRH]) often follows a temperature dependence of the type

$$\sigma_{dc} \propto \exp[-(T_0/T)^{\gamma}], \quad (10.14)$$

where T_0 is a parameter depending on the density of states at the Fermi level on the distance of the wave functions decay, the value of γ and the interpretation of T_0 depend on the details of the model. The original Mott theory for 3D variable range hopping with a constant density of states (DOS) at the Fermi energy predicts $\gamma = 1/4$, while several modifications of the model have been proposed to describe the frequently observed value $\gamma = ½$ [100].

Bauhofer et al. identified the concept of excluded volume as a powerful theoretical method to estimate the percolation threshold of composites containing statistically dispersed nonspherical particles. The concept is based on the idea that the percolation threshold is not linked to the true volume of the filler particles but rather to their excluded volume (V_{ex}). The excluded volume is defined as the volume around an object in which the center of another similarly shaped object is not allowed to penetrate [101,103].

For randomly oriented cylinders with volume $V = \pi D^2 L/4$ and high aspect ratio $\eta = L/D$, excluded volume is given by

$$(V_{ex}) \approx DL^2\pi/2 \tag{10.15}$$

$$v_c \sim V/V_{ex} = 1/(2\eta). \tag{10.16}$$

This formula relates v_c (it is the volume fraction where percolation occurs) to the aspect ratio (η) of the tubes [71].

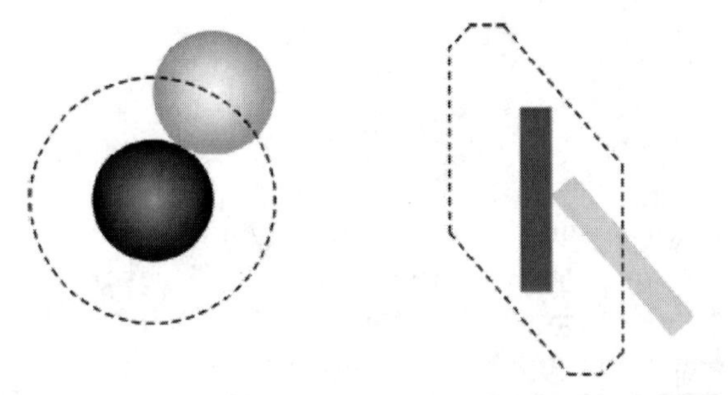

Figure 10.23 Excluded volume (or area) of objects given by the dashed lines for a sphere (left) and a stick (right), where for the latter the angle between the stick and a second stick is fixed and given by the angle between de dark grey and the light grey stick. The excluded volumes of the dark grey objects are shown, while the light grey objects are a guide to the eye [104].

Li et al. indicated, through results of their Monte Carlo simulations, that the electrical conductivity of composites with wavy nanotubes is lower than that of composites with straight nanotubes. The critical exponent of the power-law dependence of electrical conductivity on nanotube concentration decreases with increasing nanotube curl ratio. They showed that the electrical conductivity exhibits an inverse power-law dependence on the curl ratio with a critical exponent in the range of 2.2–2.6 [105].

Recently, Kovacs et al. reported the observation of two percolation thresholds in the same MWNT/epoxy system. The lower one attributed to a kinetic and the higher one to a static network formation process. Statistical percolation refers to a situation where randomly distributed filler particles form percolating paths. In kinetic percolation, the particles are free to move and thereby can form a conducting network at much lower particle concentrations. Particle movement can be caused by diffusion, convection, shearing, or external fields [106].

Rosca and Hoa demonstrated that the MWNT aspect ratio increased 5.5 times the conductivities of the corresponding composites increased by almost 10 times. Low percolation thresholds and optical micrographs evidenced a presence of kinetic percolation rather than a statistic one [135,139].

Rosca and Hoa found how temperature plays an important role to influence the conductivity as it determines the re-aggregation of the nanotubes, at low loadings, close to the percolation threshold. At high loadings, due to the high viscosity and extended entanglement of the nanotubes, the composite conductivity increases slightly with temperature [107].

Comparing several types of epoxy nanocomposites, loaded with different CNTs, Gojny et al. found that any kind of treatment, leading to a reduction of the aspect ratio (functionalization, ultrasonication, etc.) corresponds to the increases of the percolation threshold [108]. Figure 10.24 shows the plots of electrical conductivity versus CNTs concentration of several types of CNTs with different high aspect ratios.

One of the most important parameters that determine the composite conductivity and the percolation behavior of the CNTs—mainly of MWNTs—is the aspect ratio. Higher aspect ratio resulted in lower percolation threshold and higher conductivity.

Kim et al. investigated the electrical conductivity of oxidized MWNT–epoxy composites [109]. Two groups of specimens were made: The MWNTs were oxidized under both mild and strong conditions. Strong oxidation conditions produced partially damaged nanotubes. Consequently, their conductivity decreased and the percolation threshold increased. On other hands, the second group of MWNTs, which were oxidized under mild conditions, presented a high conductivity, independent of oxidation conditions.

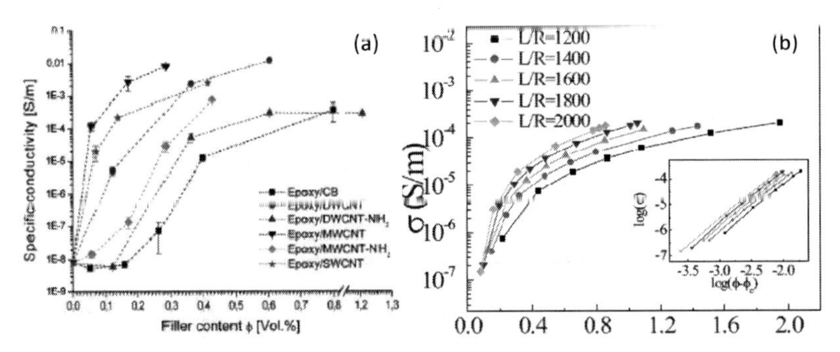

Figure 10.24 (a) Electrical conductivity of the nanocomposites as function of filler content in volume percent [106]. (b) Critical behavior of the electrical conductivity σ as a function of CNT volume fraction V in cases of high aspect ratios. The data are replotted in the inset [96].

Fu et al. studied electrical conductivity as a function of film thickness of nanocomposites and nanotubes volume fraction. They showed that decreasing the thickness of film to a value comparable with the MWNT length, the percolation threshold significantly diminishes. The authors explain this considering that different conductive paths appear with different probabilities in a film of MWNT embedded in polymer [110].

10.4 AC Conductivity Measurements

Macroscopic parameters of material depend on frequency, ω, as ε, μ, σ, they identify the kind of material. In the domain of the frequency, it is necessary to understand what happens when the homogeneous material interacts with applied varying time electromagnetic field.

Various types of charge within the material lead the interaction between the electromagnetic field and the material. They may generate two main effects: **resonance** or **relaxation** phenomenon.

In all materials, there are various types of charges:

(a) The "inner" electrons (i.e., those of the inner electronic shells) tightly bound to the nuclei: Although little affected by the applied field, they "resonate" with high energy (≈104 eV), short-wavelength (≈10–10 m) electromagnetic fields corresponding to the X-ray range.

(b) The "outer" electrons (i.e., those of the outer electronic shells): These are the valence electrons, which contribute to atomic and/or molecular polarizabilities, and also, in the case of elongated molecular structures, to their orientation with respect to the applied field.

(c) The electron-free electrons or conduction electrons, which contribute to the "in phase" conduction. When there is an applied electric field E, these electrons move with a velocity $v = \mu E$, where μ is the mobility of electrons; it is a characteristic of the given material at given temperature and accounts for all the inelastic collisions that confer to the electrons a velocity μE in field E.

(d) The bound ions, which represent ionic molecular dipoles (for example, Cl^- H^+) or a dipole associated in a lattice ($[Li]^-$ Mg^+ in LiF where $[Li]^-$ is a vacancy and Mg^+ is a substitutional cation). These permanent dipoles experience an orientational torque in a uniform field; in non-uniform field, a net force also acts, in addition to the torque.

(e) The free ions, as in electrolytes and nonstoichiometric ionic crystal (e.g. excess of K^+ in KCl), which move in the applied field with low mobility.

(f) The multipoles, and mainly the quadrupoles, or an antiparallel association of two dipoles that undergo a configurational strain under a uniform field and experience a torque in a divergent field [111].

Applying an alternative electric field, this puts into oscillation one or more types of charges or charge associations among listened above. Each configuration having its own critical frequency, above which the interaction with the field becomes vanishingly small, the lower the frequency, the more configurations are excited.

It is important to remark that the critical frequency of a given configuration depends on the relevant masses and the elastic restoring and frictional forces: elastic forces as Columbic attraction inside the molecules give the character of a resonance to the interaction, damped to a lesser or greater degree: radiation friction or Brehmstrahlung. Figure 10.25 shows both real ε' and imaginary ε'' part of a typical complex permittivity spectrum of a polar material containing space charges. Starting from the high-frequency side, electrons of the inner atomic shell have critical frequencies of order of 10^{19} Hz (X-ray range). So an electromagnetic field of frequency greater than 10^{19} Hz (or wavelength shorter than 1 Å) cannot excite any vibration in the atoms; hence, it has no polarizing effect on material, which has the same permittivity of vacuum, ε_0, for this frequency (zone 1).

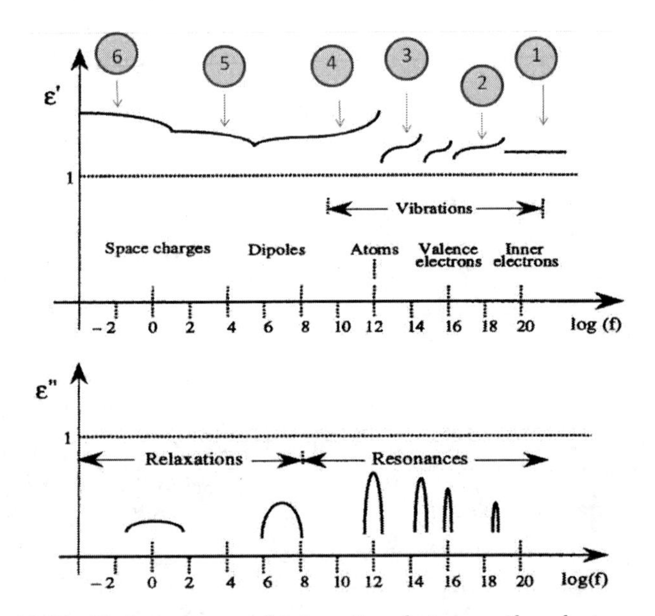

Figure 10.25 Various types of interaction between the electromagnetic field and matter, and the relative permittivity [112].

When the frequency is lower than the resonant frequency of the inner electrons, they can "feel" the electric component of electromagnetic field and they vibrate with field. So they polarize the material, which raises the relative permittivity above unity (zone 2).

If the frequency of electromagnetic field is lower than the resonant frequency of the valence electrons, which is in range 3×10^{14} to 3×10^{15} Hz (i.e., in the optical range from ultraviolet [0.1 μm] to the near infrared [1 μm]), these electrons also take part in the dielectric polarization, and their contribution again raises the permittivity (zone 3).

At the frequencies of atomics vibrations in molecules and crystals, in the range 10^{12} to 3×10^{13} Hz, the same type of "resonance" process occurs (zone 4).

In all these reported processes, the charges affected by field can be considered to be attracted toward their central position by forces that are proportional to their displacements, i.e., by linear elastic forces. This mechanical approach of an electronic resonance is only approximate, since electrons cannot be treatments of these properly by classic mechanics. Quantitative treatments of these processes require the formalism of quantum mechanics. The quantum numbers of these systems are generally so large that a classical resonance model including a friction term (to account for radiation damping) gives a fair description of these interactions.

Lazy collective orientation of dipoles or the accumulation of an ionic space charge near the electrodes when a field is applied or cutoff is an example of such process, which is known as "relaxation."

The subsets of the relaxation phenomenon are as follows:

- **Dipole oriental relaxation**, which is the time-dependent polarization due to the orientation of dipoles.
- **Interface polarization and relaxation**, in heterogeneous materials (Maxwell–Wagner effect) correspond to the evolution from the "capacitive" voltage distribution of the short time (or high frequency), heterogeneous permittivity to the "resistive" distribution of the long time (or low frequency) heterogeneous resistivity.
- **Space charge polarization and relaxation**: It occurs in materials containing carriers that do not recombine at the electrodes and therefore behave in a low-frequency ac field, as macroscopic dipoles that reverse their direction each half period.

Dielectric relaxation in changing electric fields could be considered analogous to hysteresis in changing magnetic fields (for inductors or transformers). Relaxation, in general, is a delay or lag in the response of a linear system, and therefore dielectric relaxation is measured relative to the expected linear steady state (equilibrium) dielectric values. The time lag between the electrical field and polarization implies an irreversible degradation of free energy (G).

Debye relaxation is the dielectric relaxation response of an ideal, non-interacting population of dipoles to an alternating external electric field. It is usually expressed in the complex permittivity ε^* of a medium as a function of the field's frequency ω:

$$\varepsilon^* = \varepsilon + (\varepsilon_s - \varepsilon_\infty)/(1 + j\omega\tau), \tag{10.17}$$

where ε_s is the static (dc) dielectric constant, ε_∞ dielectric constant at infinite frequencies (optical wavelength) and τ is the relaxation time.

This formula is derived under the following assumptions:

(1) At any point in the material, local electric field is the same as the applied field.
(2) The dc conductivity of the material can be neglected.
(3) All dipoles have the same relaxation time.

These assumptions are very strong and they are not so precise. The first assumption does not consider that local electric field is not the same as the applied electric field, but there exists a Lorentz field, as the first approximation. The second is right for materials where conductivity is equal to zero ($\sigma = 0$), but not for conducting polar dielectric where conductivity is present ($\sigma \neq 0$). The third assumption is limited at fixed time into material, only one type of dipoles, different types of dipoles could have different relaxation time, τ. Taking into account all these corrections to Debye relation, it is possible to write the following:

$$\tau' = (\varepsilon_s - 2\varepsilon_0)/(\varepsilon_\infty + \varepsilon_0) \tag{10.18}$$

$$\varepsilon^* = \varepsilon_\infty + (\varepsilon_s - \varepsilon_\infty)/(1 + j\omega\tau) - j\sigma/\omega \tag{10.19}$$

$$\varepsilon^* = \varepsilon_\infty + \textstyle\sum_{i = 1,N} [\varepsilon_i/(1 + j\omega\tau_i)], \tag{10.20}$$

where N is the number of relaxational processes involved. In Eq. (10.18), τ' replaces the relaxation time τ. This value is greater than the real relaxation time. In some anisotropic materials, τ' could be up to three orders of magnitude longer than the relaxation time measured in a dilute and nonpolar material. When the conductivity is taken account, it is possible to obtain Eq. (10.19). It has a trend, as shown in Fig. 10.26, where it is possible to identify three regions, named I, II, and III.

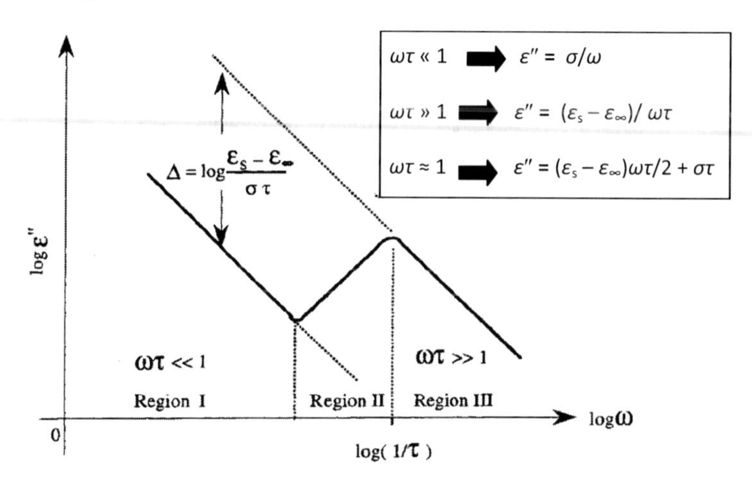

Figure 10.26 Loss dielectric response versus frequency [112].

It is necessary remark that defects, imperfections, contaminants, temperature, etc., influence the Debye relation. The previous description was made taking into account the homogeneous material: non-inclusions inside. Determining the effective electromagnetic properties of heterogeneous mixtures has been a continuing interest in physics and other scientific fields for over 150 years since the initial work of Faraday and Maxwell.

The Clausius–Mossotti formula,

$$(\varepsilon - \varepsilon_0)/(\varepsilon + 2\varepsilon_0) = \mathbf{N}\alpha/3\varepsilon_0, \qquad (10.21)$$

It is a first-order approximation for the dielectric properties of mixtures but contains parameters more relevant to describing molecular effects (e.g., in dilute gases) than heterogeneous mixtures with inclusions. In 1904, Maxwell Garnett (MG) re-derived a macroscopic version of the Clausius–Mossotti formula to describe the effective refractive index and/or dielectric properties of a suspension of spheres, and the Maxwell Garnett formula is still commonly used for two-phase mixtures. The MG mixing rule has been successfully applied to engineering microwave absorbing materials containing carbon particles [113], using the formula for the effective relative permittivity. It has been generalized for a multiphase mixture [114] (see also below).

The MG model is valid for dielectric composites with dilute conductive phases (*below the percolation threshold*). This model implies the quasi-static approximation. Its main features are as follows:

- The mixture is electrodynamically isotropic.
- The mixture is linear, that is, none of its constitutive parameters depends on the intensity of electromagnetic field.
- The mixture is nonparametric, that is, its parameters do not change in time according to some law as a result of external forces: electrical, mechanical, etc.
- Inclusions are separated by distances greater than their characteristic size.
- The characteristic size of inclusions is small compared to the wavelength in the effective medium.
- Inclusions are arbitrary randomly oriented ellipsoids.
- If there are conducting inclusions, their concentration should be lower than the percolation threshold.

The MG model involves an exact calculation of the field induced in the uniform host by a single spherical or ellipsoidal inclusion and an approximate treatment of its distortion by the electrostatic interaction between the different inclusions. This distortion is caused by the charge dipoles and higher multipoles induced in the other inclusions. The induced dipole moments cause the longest range distortions and their average effect is included in the MG approximation, which results in a uniform field inside all the inclusions.

$$\frac{\varepsilon_{\mathrm{MG}}(\omega) - \varepsilon_{\mathrm{h}}(\omega)}{\varepsilon_{\mathrm{MG}}(\omega) + 2\varepsilon_{\mathrm{h}}(\omega)} = f\frac{\varepsilon_{\mathrm{i}}(\omega) - \varepsilon_{\mathrm{h}}(\omega)}{\varepsilon_{\mathrm{i}}(\omega) + 2\varepsilon_{\mathrm{h}}(\omega)}, \tag{10.22}$$

where $\varepsilon_{\mathrm{MG}}$ is the effective permittivity by Maxwell Garnet method, ε_{i} and ε_{h} are the permittivity of the inclusions and the host medium, and f is the volume fraction of the inclusions ($f \ll 1$).

When a mixture is placed in an electric field, the electrical conduction and polarization would depend on the random spatial dispersion (and relative orientation) of the shape inclusions in the medium; therefore, the dielectric or permittivity characteristics of mixture would be essentially statistical in nature as determined by random particle dispersion. So, the calculation of the inducted electric field becomes a probabilistic calculation, statistic problem.

So the statistic value of effective permittivity is given by

$$\varepsilon_{\text{eff}} = F_1\,(\varepsilon_1, \varepsilon_2, f, q), \tag{10.23}$$

where q is the form factor, it takes into account the shape of the inclusions.

Calculation of the effective permittivity is difficult, because the calculation of F_1 is difficult. F_1 must be a homogeneous function of the first degree extracted from the set of independent variables ε_1 and ε_2. Several models exist for the calculation of F_1 [31]. Also, different models on dielectric material emerge from the MG model. Kuzhir et al. presented a comparative study of the electromagnetic shielding effectiveness in X-band (8–12 GHz), Ka-band (26–37 GHz), and W-band (78–118 GHz) provided by carbon black (CB), and commercially available CVD single-wall and multiwall CNT (SWNT and MWNT) dispersed in epoxy resin in low concentration (0.5 wt%). The data collected for microwave frequencies were found to be well correlated with the theoretical simulations on the basis of the generalized Maxwell Garnett theory. They consider the CNT-based resin composite as a collection of small conductors of different lengths randomly dispersed in the dielectric matrix and electrically characterized by a frequency dependent polarizability $\alpha(\omega)$ [115]:

$$\alpha(\omega) = (R^2L/6) \times (\omega_p^2/(\omega^2 - \omega_p^2 n_z + i\omega\nu))\tilde{\sigma}_{\text{axial}}/\omega, \tag{10.24}$$

where L and R are length and radius of the CNT, respectively, $n_z \approx 4(R/L)^2\,[\ln(L/R) - 1]$, $\omega_p = (4e/R)(3\gamma_0 a/2\pi h^2)^{1/2}$ is plasmonic frequency, γ_0 = 2.7 eV is the overlap integral, ν is the relaxation frequency, e is the electron charge, a = 0.142 nm is the interatomic distance in graphene, σ_{axial} is CNT axial conductivity. The CNT polarizability is calculated rigorously using the ideas of CNT as transmission line developed within the linear nano-electrodynamics in Ref. [116] for zigzag, armchair, and chiral SWNT in the microwave and the infrared frequency regimes, and in Ref. [117] for MWNT.

In order to take into account the size distribution of CNT particles, we consider a composite as k-component material; each components enumerated by the index $j = 1,2,..., k$, and N_j is a density of j-component, $N_j \ll 1$ for arbitrary j. As the next step, we model the composite as a single-component one comprising particles of the $j = 1$ type; all other $N - 1$ components are tentatively included into the host with the dielectric function ε_{hl}. In accordance with the Maxwell Garnet model, an effective dielectric function is given by

$$\varepsilon_{\text{eff}}(\omega) = \varepsilon_{\text{h1}} + 4\pi N_1\alpha_1(\omega)\left[1/\left(\frac{1}{3}4\pi N_1\alpha_1(\omega)\right)\right]\left[1/\left(1-\frac{1}{3}4\pi N_1\alpha_1(\omega)\right)\right] \tag{10.25}$$

where $\alpha_1(\omega)$ is dynamic polarizability of j-filler with $j = 1$. Finally, recurrence transformation leads to the effective dielectric function expressed in terms of the dielectric function of the host medium ε_h and polarizabilities and the densities of inclusions N_j as follows:

$$\varepsilon_{\text{eff}} = \varepsilon_{\text{h}} + \left[4\pi\Sigma_{j=1,k} N_j \alpha j(\omega)/\left(1-\frac{1}{3}4\pi\Sigma_{j=1,K} Nj\alpha j(\omega)\right)\right] \tag{10.26}$$

P. Kuzhir et al. (Fig. 10.27a) that the theoretical simulation of SWNT/resin permittivity is in pretty nice agreement with the experimental results collected in microwave frequencies. SWNT (6, 0), (10, 10), (20, 0), (24, 0) with diameter 1.1, 1.36, 1.6, and 2 nm, respectively, and the lengths 10, 15, and 20 μm were used for numerical estimations.

Results showed that SWNTs and MWNTs produced via large-scale technology (commercial CVD) do not demonstrate a high EM shielding effectiveness in X-, Ka-, and W-bands for such a small CNT concentration as 0.5 wt%. The easiest way to improve the EM shielding ability is to increase the CNT concentration up to at least 1.5 wt%. See the results of theoretical modeling for SWNT-based composites presented in Fig. 10.27b. One can see that the increase of the nanocarbon concentration from 0.5 to 1.5 wt% shall lead to the decrease of the EM transmittance through the sample by two times [118].

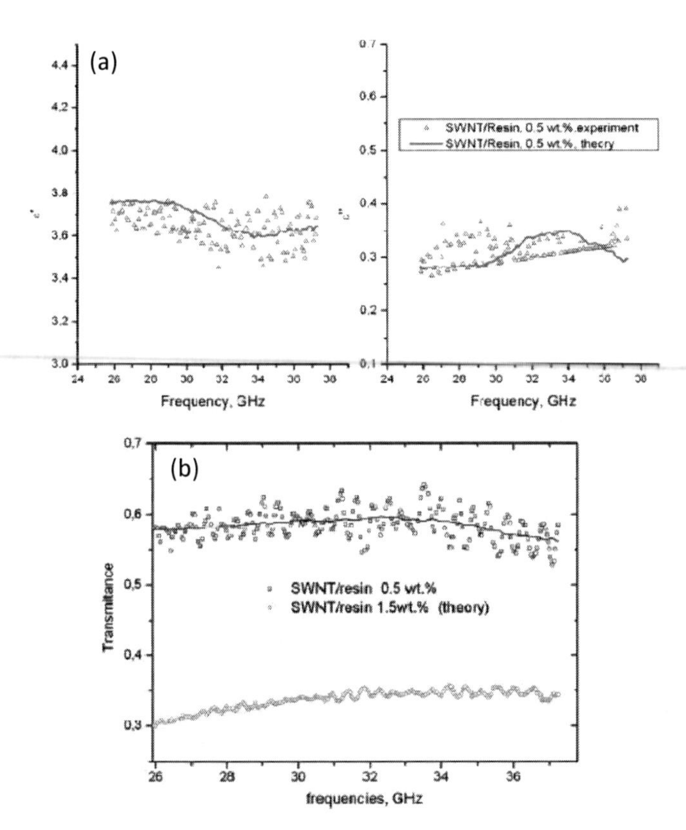

Figure 10.27 (a) Frequency dependence of permittivity of SWNT/resin composites, 0.5 wt% of SWNT inclusions (theoretical simulation and experimental data, color online); (b) SWNT/resin EM transmittance versus frequency, theoretical simulation for 0.5 and 1.5 wt% of nanocarbon inclusions (color online) [118].

The measured ac conductivity $\sigma_{\text{ac}}(\omega)$ of polymers is characterized by the transition above a critical (angular) frequency ω_{c} from a low-frequency dc plateau to a dispersive high-frequency region. The ac conductivity for all systems has been calculated from the dielectric loss factor, ε', expressed by Eqs. (10.25) and (10.27):

$$\sigma_{\text{ac}} = \varepsilon_0 \omega \varepsilon', \tag{10.27}$$

where $\varepsilon_0 = 8.85 \times 10^{-12}$ F m^{-1} is the permittivity of free space and ω the angular frequency ($\omega = 2\pi f$).

Equation (10.28) has been used in the literature to model either the ac effects of the free charge and partially bound free charge in hopping and tunneling conduction

$$\sigma_{\text{ac tot}}(\omega, T) = \sigma_0(0, T) + j\omega(\varepsilon(\omega) - \varepsilon_0) \qquad (10.28)$$

It is important to underline how most models of ac conductivity are based on charged particles in potential wells, yielding a percolation threshold, where energy fluctuations determine whether the particle can surmount the barrier and thereby contribute to the conductivity.

Vavouliotis et al. examined the ac conductivity of epoxy resin loaded with CNTs in the frequency range from 101 to 106 HZ at ambient temperature. According to the percolation theory, an enhancement of conductivity was evidenced increasing the weight content of the conductive nanofiller. The ac conductivity was found to be frequency dependent beyond a critical frequency that increased with the nanofiller content. They guessed that the critical frequency also follows a percolation-type law with the weight fraction of the nanotubes. With increasing frequency and CNT weight content, the ac conductivity increases up to around 10 orders of magnitude due to its dependency on both frequency and CNT content. The ac conductivity increases with CNT weight content in the whole frequency range.

The ac spectra of neat epoxy and the specimen with the lowest concentration of CNTs display curve similarities and values proximity. A tendency to acquire constant values in a sort frequency range at the low-frequency edge is recorded in tandem with extensive frequency dispersion [119].

In the low-frequency edge, A. Vavouliotis et al. have observed an abrupt increase of conductivity (up to eight orders of magnitude), between the specimens loaded with 0.1 and 0.3 wt% of CNTs. For the nanocomposites with CNTs content higher than 0.3 wt%, conductivity spectra exhibit both a wider plateau, called "apparent dc conductivity," and a narrower range where conductivity values are frequency dependent, as it is shown in Fig. 10.28. Using ac conductivity values for a frequency of 0.1 Hz, they calculated a percolation threshold of 0.098 wt% and a critical exponent of 3.204 (Fig. 10.28). The dispersion of ac conductivity with frequency is a common characteristic in heterogeneous and disordered solids (see Table 10.8) [120,121].

Table 10.8 Summary of calculated parameters and critical frequency according the best fit of Eqs. (10.27) and (10.28) on AC experimental data of all nanocomposites [120]

CNT[%]	σ_0 [S/m]	A	s	ω_c [Hz]
Neat epoxy	6.09E − 12	1.40E − 11	1.218	5.03E − 01
0.1%	2.54E − 10	2.66E − 10	1.136	9.63E − 01
0.3%	3.58E − 05	5.76E − 08	0.717	7.91E + 03
0.5%	4.11E − 05	1.36E − 10	1.239	2.56E + 04
0.6%	9.49E − 04	2.78E − 09	1.134	7.56E + 04
0.8%	1.25E − 03	1.57E − 08	0.822	9.17E + 05
1.0%	3.27E − 03	8.52E − 12	1.393	1.45E + 06

Note: Experimental data of Vavouliotis et al.'s work confirm the dependence of the critical frequency upon the filler content, which in this case are the carbon nanotubes.

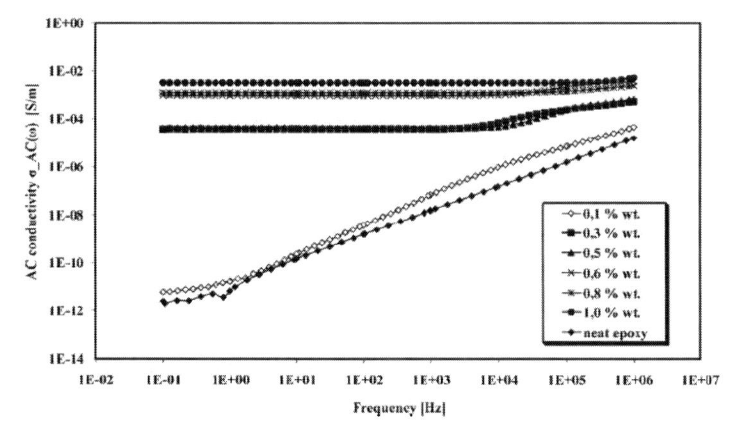

Figure 10.28 Log–log plot of the ac conductivity of all the examined nanocomposites as a function of frequency [121].

The total measured conductivity $\sigma_{ac\ tot}(\omega, T)$ for a wide class of dielectric materials (semiconductors, organic, inorganic, glass, etc.) shows dispersion behavior through its dependence on the frequency (ω) according to Jonsher relation:

$$\sigma_{ac\ tot}(\omega, T) = \sigma_0(0, T) + \sigma_{ac}(\omega, T) \tag{10.29}$$

$$\sigma_{ac}(\omega, T) = A(T) \cdot \omega^{s(T)} \quad (10.30)$$

At constant temperature, the response of ac conductivity with frequency could be in general expressed mathematically with the following power law expression:

$$\sigma_{ac\ tot}(\omega) = \sigma_0 + A \cdot \omega^s, \quad (10.31)$$

where σ_0 is the $\omega \to 0$ limiting value of $\sigma_{ac\ tot}$ and A and s are parameters which, according to the literature, are dependent on temperature, morphology, and composition. Exactly, the frequency exponent(s) value lies in the range $0 < s < 1$. The exponent measures the degree of interaction of mobile ions with the environment [122]. $A(T)$ determines the polarizability [123]. For Debye-type interaction, $s \approx 1$ and for pure ionic crystal, $s \approx 0$, i.e., s decreases with the increase of interaction. The temperature dependence of s determines, to a large extent, the type of conduction mechanism [123]. Equation (10.29) is often called "the ac universality law" [124]; it is used to analyze the ac conductivity in all disordered solids [125]. It is an experimental fact that the ac conductivity of conducting polymers and, more generally, of many disordered solids, as a function of frequency, consists of a frequency-independent low-frequency region and, above a critical frequency value ω_c, a high-frequency one, which is dispersive. Equation (10.31) shows a combination of two parts; the first one is a constant term, and the second one is a power law. Switching to the power law trace occurs at a certain critical frequency (ω_c), which is common in both branches; ω_c depends on temperature and particle concentration [126] and can be calculated from the equilibrium point of the two branches of Eq. (10.31), via the following relation:

$$\omega_c = [\sigma_0/A]^{1/s} \quad (10.32)$$

For a given angular frequency $\omega < \omega_c$, the measured conductivity $\sigma_{ac\ tot}(\omega)$ results from the macroscopic conductivity (along paths connecting the opposite surfaces of the specimen, where electrodes are attached) and from the charge flow along paths, which are larger than v/ω, where v is some typical value for the mean velocity of the transferring charge carriers. In the dispersive region $\omega > \omega_c$, the measured ac conductivity is the sum of the macroscopic conductivity and the conductivity along paths with lengths equal or larger than v/ω (i.e., the lengths corresponding to frequencies from to ω) [127].

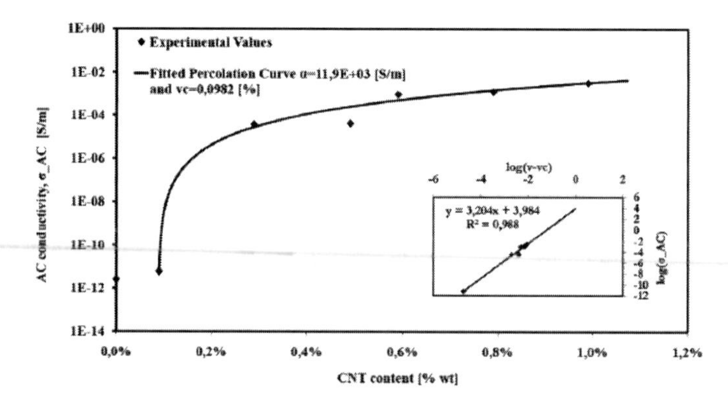

Figure 10.29 The ac conductivity for a frequency of 0.1 Hz as a function of CNT content [120].

Vavouliotis et al. found that ac conductivity exhibits strong dispersion with frequency and below a critical value becomes constant tending to dc conductivity, while above the critical value it varies as a power of frequency. They were able to eliminate and/or reduce effects of contact resistances, between CNTs, probably on all scales of interaction levels from ac conductivity measurements.

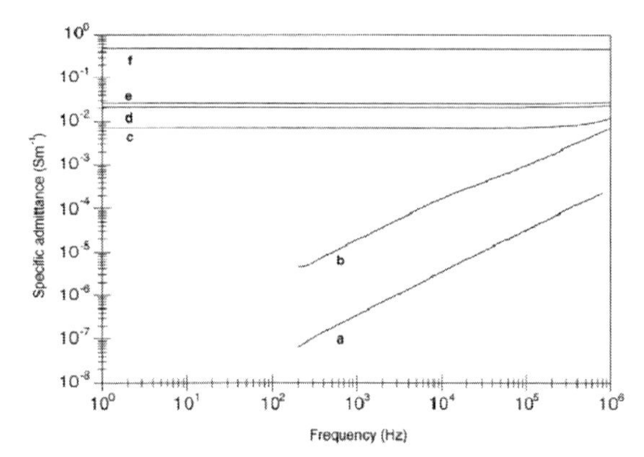

Figure 10.30 Log–log plot of the specific admittance of the nanocomposites containing untreated catalytically grown nanotubes as a function of the frequency. (a) Pure epoxy; (b) 0.0225 wt% nanotubes; (c) 0.04 wt%; (d) 0.10 wt%; (e) 0.13 wt%; (f) 0.15 wt% [36].

Finally, critical frequency (ω_c), which was found to follow a percolation-type dependency on nanotubes concentration, diminishes with CNT content indicating the facilitation of the conduction process [127]. In studies conducted by Sandler et al., catalytically grown carbon nanotubes were dispersed as conductive fillers in an epoxy matrix, and the resulting electrical properties were comparable to those obtained using an optimized process for carbon black. They chose to measure the ac conductivity using the following formula:

$$\sigma_{ac} = Y'x(d/A), \tag{10.33}$$

where Y' is the real part of the complex admittance, d is the sample length or distance between the electrodes, and A is the contact area. From the ac impedance spectroscopy performed at room temperature, the real and imaginary parts of the complex impedance (Z^*) were obtained as a function of the frequency. The complex admittance ($Y^* = 1/Z^*$) of the nanocomposites can be modeled as a parallel resistor (R) and capacitor (C) and written as a function of angular frequency (ω).

$$Y^*(\omega) = Y' + jY' = 1/R + j\omega C, \tag{10.34}$$

where the specific admittance (y) of the nanocomposite is the module of complex admittance; it was calculated to understand the behavior of admittance with the frequency (Fig. 10.29), as the following formula shows:

$$y(\omega) = |Y^*(\omega)|d/A \tag{10.35}$$

From their analysis, Sandler et al. achieved the same values for dc conductivity and ac conductivity at low frequencies, as in literature. Using carbon nanotubes they could both reduce the percolation threshold to below 0.04 wt%, and increase the overall conductivity achieved. The results indicated that nanocomposites would be a good coating against electrostatic charges [36].

Garrett et al. analyzed the resistances of the junctions between CNTs using a modeling ac impedance spectroscopy (IS). The data obtained are generally analyzed in terms of an equivalent circuit model. The analyst tries to find a model whose impedance matches the measured data. Through this model data, the resistance of the junctions and bundles is separated to an equivalent circuit of two parallel resistance–capacitance elements in series, as shown in Fig. 10.31b: Voigt elements. In IS measurements, each frequency probes a certain length scale within the sample. The characteristic length scale of the conductive network can be associated with the distance between junctions in the SWNTs networks. This correlation length is probed in the impedance measurements when it is comparable with the ac modulation period, thus allowing charges to travel from one junction to another along the SWNTs bundle. At frequencies lower than the correlation frequency, the charge carriers move over much larger distances, which are comparable to the scale of the entire nanotube network, and the impedance approaches the dc resistance of the network. At higher frequencies, the carriers move a shorter distance within each ac cycle, and many of the carriers only move within bundles, with the junctions making little contribution to the resistance. A transition from the long-range transport to localized carrier confinement is characterized by a falloff frequency in the real impedance and is correlated with the maximum in the imaginary impedance. As the concentration of SWNT increases, the falloff frequency shifts to higher values, while real impedance is seen to decrease simultaneously [128].

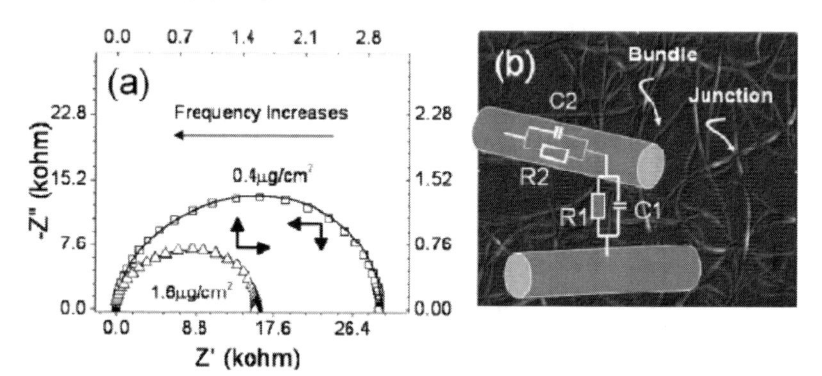

Figure 10.31 (a) Complex impedance of 0.4 and 1.6 μg/cm^2 SWNTs networks prepared from SDS solutions. (b) Equivalent circuit model of the SWNT networks with parallel R1, C1 elements describing electrical properties of the junctions, and R2, C2 elements modeling the nanotube bundles. The SEM image of a percolating SWNT network with the bundle size around 20 nm is shown in the background [128].

Generally, for an RLC circuit, the complex impedance (Z^*) is given by three contributions (resistor, inductor, and capacitor):

$$Z^* = R + j(\omega L - 1/\omega C), \tag{10.36}$$

where R is the ohmic resistance, L is the inductance, and C is the capacity.

For the RC circuit, the complex impedance is given by

$$Z^* = R - j(1/\omega C) \tag{10.37}$$

It is possible to write

$$Xc = 1/\omega C \tag{10.38}$$

so Eq. (10.38) becomes

$$Z^* = R - jXc, \tag{10.39}$$

where $Z' = R$ is the real part of the complex impedance and $Z'' = Xc$ is the imaginary part of complex impedance.

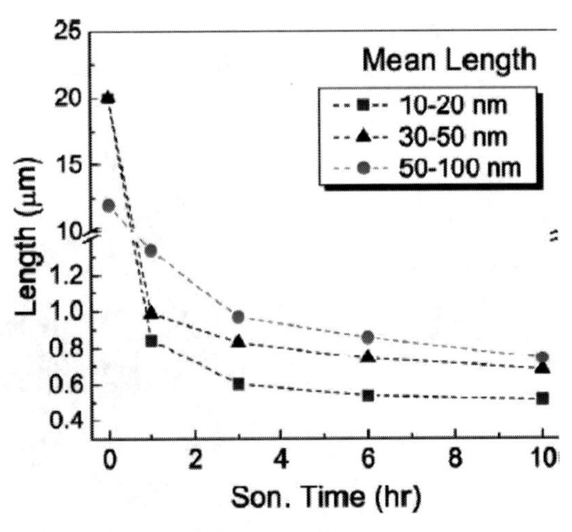

Figure 10.32 Plot of mean length against sonication time for the three diameter MWNTs investigated in the present study: 10–20, 30–50, 50–100 nm [133].

During the modeling impedance properties of the SWNT films measured, they have found that the equivalent circuit model, consisting of two Voigt elements in series, showed high correspondence with the nature of conduction through junctions and bundles. A Voigt element is a circuit with a resistor and a capacitor connected in parallel. This is a common model used to model the properties of materials with more than one conduction mechanism such as the case of polycrystalline solids, where one Voigt element is correlated with the bulk grain properties, and the other to the grain boundaries [128]. They have assigned the lower resistance element to the nanotube bundles (Fig. 10.31b) where the charges are strongly localized. The higher resistance Voigt element was assigned to the junctions, where conduction is dominated by granular-metal type tunneling. Previous studies on thick nanotube mats have suggested that variable range hopping through nanotube defects is a greater contributor to resistance than transport between bundles [128].

At a much lower frequency, films showed dependence from frequency, suggesting that interbundle transport is the dominating contributor. Due to much higher quality nanotubes that are available now than when the earlier study was done, and their work involved much sparser networks, where continuous, branching "super-bundles" are not prominent. The junction resistance was found to be 3.3 ± 0.3 times the bundle resistance and was independent of nanotube loading in all samples. This constant ratio of junction resistance to bundle resistance supports theoretical predictions that state that the number of junctions increases linearly with the number of bundles [129–132].

The junction resistance was found to be 3 to 3.5 times higher than the bundle resistance [130]. Using the ac IS method of analysis, Kerr et al. discovered that the tube length minimally affects the dielectric response of these composites when the doping level is below the percolation threshold. They used MWNTs with three different lengths and diameters. For the composite systems, ac impedance measurement indicated that polydimethylsiloxane (PDMS) doped with large-diameter tubes (30–50 nm and 50–100 nm) were predominantly capacitative in the frequency range of 100 Hz to 3 MHz, with voltage–current phase offsets close to 90°. In other case, no change was observed with the addition of MWNTs (10–20 nm) inside PDMS. Kerr et al. analyzed the effect of ultrasonication at different time, on modification of length of MWNTs, as shown in Fig. 10.29. They found that when the aspect ratio decreases, the effect of MWNTs on the matrix is reduced.

At 1 wt% loading, the resulting MWNTs-PDMS composites show percolation behavior only for the CNTs with high aspect ratios [133].

Jianfei Chang et al. analyzed the electrical properties of nanocomposite when the same filler (MWNTs) is enclosed into the resin with different treatment for curing processes. Based on the same components (MWNTs and epoxy resin), composites prepared by thermal or microwave technique exhibited significantly different structures and thus properties.

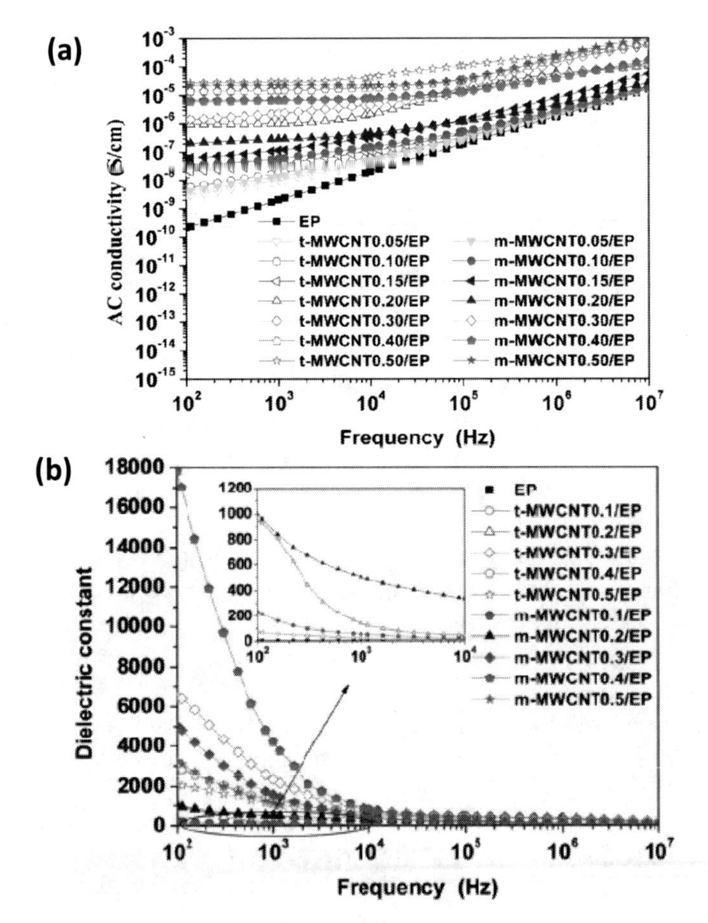

Figure 10.33 (a) Dependence of electrical conductivity on frequency for t-MWNT/EP and m-MWNT/EP composites at room temperature. (b) Dependence of dielectric constant on frequency for composites with different contents of MWNTs [134].

As shown in Fig. 10.33, the electrical conductivities of both composites increase as the volume fraction of MWNTs and frequency increase. With regard to the magnitude of the conductivity, it is closely related with the volume fraction of MWNTs in the corresponding composite. Specifically, the dielectric constant and loss at 100 Hz of m-MWNT/epoxy (microwave treatment) composite with 0.04 vol% MWNTs are about 2.5 and 0.05 times the corresponding values of t-MWNT/epoxy (thermal treatment) composites, respectively, because of their different structures. They discovered that using the microwave curing technique, MWNTs cannot only be uniformly dispersed in the matrix but also align in a direction. This unique structure of MWNTs allows the development of novel composites with a very higher dielectric constant and extremely low dielectric loss. This investigation suggests a new method to develop high-performance electric conductor/polymer composites through controlling the dispersion and spatial distribution of conductors using different curing techniques [134].

It is useful to remark that the application of both ac and dc electric fields during nanocomposite curing was used to induce the formation of aligned conductive nanotube networks between the electrodes.

The network structure formed in ac fields was found to be more uniform and more aligned compared to that in dc fields. In this case, it is possible to obtain a good path where the carriers can move easily.

10.5 Epoxy Nanocomposite Filled with Carbon Nanotubes: Shielding Effectiveness Measurements

All EMC regulations and testing standards are under the International Electrotechnical Commission (IEC). Within the IEC, three technical committees are related to EMC: CISPR (International Special Committee on Radio Interference), TC77 (concerned mainly with EMC in electrical equipment and networks), and TC65 (concerned mainly with immunity standards). The CISPR committee gives outputs in the form of standards. These cover a wide range of EMC-related topics with the designation CISPR10 to CISPR23.

One of the most useful designations is CISPR 22 for conducted and radiated emissions of information technology equipment; it has formed the foundation for most of the major national standards. It is important to underline that there has been incompatibility between the requirements of different countries because

(a) the recommendations proposed by CISPR cannot become law unless the individual member countries take the appropriate action themselves, and
(b) individual countries have made adjustments to the recommendations before adopting them as national standards.

In Europe, the European Standards Committee (ESC), through the European Committee for Electrotechnical Standards (CENELEC), is responsible for the adoption of suitable standards. These standards cover broad terms and broad classes of equipment:

- Class 1 (residential, commercial, and light industry)
- Class 2 (heavy industry)
- basic standards (setting out test procedures)
- product standards (covering specific types of apparatus, e.g., information technology equipment)

In the United States, civilian radio, communications, and interference are regulated by the Federal Communications Commission (FCC). An individual or organization must comply with all applicable FCC standards and equipment authorization requirements to legally import or market a product in the United States. FCC regulations specify EMC standards, test methods (ANSI C63.4), equipment authorization procedures, and marking requirements.

For the military market, electronic systems generally have to comply with MIL-STD-461, which imposes limits on the susceptibility of a system to conduct and radiate EMI, as well as specifying its maximum conducted and radiated emissions. The accompanying document, MIL-STD-462, specifies the testing methods. Mil-STD-461 has been recognized and applied by many defense organizations outside the United States, as well as some nonmilitary agencies [134].

For commercial applications, the target value of the EMI-shielding effectiveness needed is around 20 dB (i.e., equal to or less than 1% transmission of the electromagnetic wave). Commercially conducted emission limits of the FCC, ESC, and CISRR 22 are more severe than those for military equipment because the ambient noise in which commercial systems operate must be kept to a minimum since there are generally no susceptibility requirements for this type of equipment. On the other hand, the more stringent military standard for radiated emissions takes into account the closer proximity of electronic equipment in aircraft, ships, tanks, and so forth, compared with commercial installations, which are typically more widely dispersed.

EMI shielding is a rapidly growing application of carbon materials, particularly discontinuous carbon fibers. It is useful to remark some characteristics of the parameters that combine to give the Eq. (10.37) of SE.

- *Reflection parameter:* The shield is required to have mobile charge carriers (electrons or holes) that interact with irradiated EM waves. As a result, the shield has good conductivity—although high conductivity is not specifically required—and it has a volume resistivity of 1 Ω cm, typically [133,135].
- *Absorption of the radiation:* The shield should have electric and/or magnetic dipoles that interact with the electromagnetic fields in the radiation. The electric dipoles may be provided by $BaTiO_3$ or other materials that have a high value of dielectric constant. The magnetic dipoles may be provided by FeO or other materials having a high value of the permeability (μ) [136], which may be enhanced by reducing the number of magnetic domain walls through the use of a multilayer of magnetic films [137,138].

In order to increase the shielding effectiveness through multiple internal reflections, it is important to have a conductive polymer nanocomposite [139]. The advantage of internal reflections is that the material absorbs and attenuates instead of reflecting the electromagnetic radiation. For addressing the requirements of a new generation of high-performance materials to use as a light shield against EMI, several studies on carbon materials have been and are being conducted. A wide range of polymers, more generally plastic materials, were used as the matrix. Taking into account the first experiments on the polymer filled with carbon nanotubes, mainly on foamy materials, Yang et al. developed carbon nanofiber– and carbon nanotube–filled polystyrene conductive foams for lightweight EMI shielding [140]. They obtained a composite with a density about 0.56 g/cm^3 and obtained an EMI-shielding effectiveness of approximately 20 dB. Compared to carbon nanofiber–filled composite foams, carbon nanotube–filled foams had higher EMI-shielding effectiveness at lower concentrations [141]. Thomassin et al. developed MWNT/polycaprolactone composite foams by using supercritical CO_2 foaming technology and achieved EMI-shielding effectiveness as high as 60/80 dB together with low reflectivity at very low volume concentrations of MWNT (0.25 vol%). The shielding effectiveness that Thomassin et al. achieved is comparable to that of the most efficient metallic EMI-shielding materials [142]. They also found that increasing the carbon nanotube concentration, the shielding effectiveness increases, possibly due to increased conductivity [143]. The shielding effectiveness in conductive carbon nanotube/polymer composite foams results mostly from the loss due to internal multiple reflections; increasing the bubble density (and hence increasing internal area) should increase the shielding effectiveness [144]. Increasing the bubble density, the alignment of the nanotubes changes; this can decrease the conductivity and thus the shielding effectiveness [145]. Controlling the properties of the MWNTs and the bubble density inside the nanocomposite foams to get high shielding effectiveness is of great importance for such EMI-shielding applications [146–150].

Loading MWNTs in polystyrene (PS) matrix, Yang et al. tested the achieved nanocomposites, in the frequency range of 8.2–12.4 GHz (X band). They found a value of SE of 20 dB for 7% MWNTs [146,147]. Kim et al. studied the electrical conductivity and EMI properties of MWNTs in poly(methyl methacrylate) (PMMA) containing Fe. They obtained the value of 27 dB (SE) for 40% MWNT loading. MWNT composites with encapsulated Fe have also been studied for their microwave absorption, but with a different phases and shapes of included Fe [150].

Grimes et al. reported that SWNT-polymer composites possess high real permittivity (polarization, ε') as well as imaginary permittivity (adsorption or electric loss, ε'') in the 0.5 to 5.5 GHz range. They found that the permittivity decreases rapidly with increasing frequency [151].

Ning Li et al. measured SE of epoxy nanocomposite with 10 wt% SWNT-long loading. They found the values of the reflectivity (*R*), absorptivity (*A*), and transmissivity (*T*) to be 0.90, 0.08, and 0.02, respectively, at 1.0 GHz. Thus, the contribution of reflection to the total EMI-shielding effectiveness is much larger than that from absorption. Similar results were observed at other frequencies and with other loadings above the percolation threshold [152,153].

For studying the shielding mechanism of MWCNTs, grown by Fe nanoparticles as catalyst, with Fe catalyst, into different polymers—for polystyrene, for the PMMA-MWNTs Fe, and epoxy-MWNTs with crystalline Fe—it was observed that the contribution from absorption to total EMI–SE was larger than that from reflection. However, the authors identified this with the ferromagnetic Fe in the system [154].

Nanocomposites typically contain 2% to 15% loadings on a weight basis. Yet property improvements can equal and sometimes exceed traditional polymer composites, even containing 20% to 35% mineral or glass. The shielding effectiveness of 20 to 30 dB has been obtained in the X-band (8.2 to 12.4 GHz) range for 15% SWNT-loaded epoxy composites [155]. The EMI-shielding effectiveness of a composite material depends mainly on the filler's intrinsic conductivity, permittivity, and aspect ratio.

10.6 Conclusions

This chapter yields a review and an outlook for a new generation of nanomaterials for electromagnetic shielding. It provides a description of epoxy composites loaded with carbon nanotubes. Carbon nanotubes are an allotropic form of graphite with amazing properties. The combination of the matrix and the filler gives a light, strong, highly conducting material, which could be used as an electromagnetic shield in several ways, including coating, paint, and glue.

The chapter shows the principal physical parameters, especially the dielectric properties. The dc and ac studies of the behavior of the nanocomposites and their modeling help in obtaining a better understanding of the physical properties and the ways to obtain high-performance nanocomposites. Lastly, some results of the measurement of shielding effectiveness have been illustrated. Quantifying nanofiller dispersion, distribution, orientations, and aspect ratio is crucial to the understanding of the property–structure relationship.

The aspect ratio is the most important parameter. Its high value is essential for good conductivity, mechanical property, and high shielding effectiveness of nanocomposites. It has a strong influence on several CNT properties [96]. It is possible to confirm that CNTs could be used for the fabrication of effective electromagnetic materials on the basis of the epoxy resin. However, before these materials can find widespread commercialization, studies and modeling are needed to improve the performance of CNT-based nanocomposites and to solve difficulties such as higher interfacial interactions with polymers, optimization of fabrication processes, and well dispersion into the matrix, [59]. This chapter takes into account only carbon nanotubes, but it is important to remark that a large number of investigations are being carried out currently on graphene [152] and magnetic nanoparticles [153]. They could become a novel potential filler to use in the matrix for nanocomposites for electromagnetic shielding in the coming years.

Acknowledgments

The research has been supported by EU FP7 project FP7-266529 BY-NanoERA and by the Italian Ministry PRIN 2008 research program Development and Electromagnetic Characterization of Nano Structured Carbon Based Polymer CompositEs (DENSE). Printex 90 was generously provided to our group by Evonik Degussa GmbH. Epikote 828 resin was generously provided from Hexion Specialty Chemicals Inc. (now known as Momentive Specialty Chemicals Inc.).

References

1. P. Li, Y. Shan, J. Deng, Y. Xijiang, *Electromagnetic Interference Shielding Effectiveness of Carbon-Nanotubes Based Coatings*, Asia Pacific International Symposium on Electromagnetic Compatibility, IEEE (2010).

2. G. J. Vasaka. *Theory, Design and Engineering Evaluation of Radio-Frequency Shielded Rooms* (Report NADC-EL-54129). Johnsville, PA: U.S. Naval Development Center (1956).
3. D. M. Bigg. Mechanical, thermal and electrical properties of metal fiber filled polymer composites. *Polym. Eng. Sci.*, **19**, 1188–1192 (1979).
4. G. Lu, X. Li, H. Jiang. Electrical and shielding properties of ABS resin filled with nickel-coated carbon fibers. *Compos. Sci. Technol.*, **56**, 193–200 (1996).
5. T. S. Wang, G. H. Chen, C. L. Wu, D. J. Wu. Study on the graphite nano-sheets/resin shielding coatings. *Prog. Org. Coatings*, **59**, 101–105 (2007).
6. L. L. Wang, B. K. Tay, K. Y. See, Z. Sun, Li. K. Tan, D. Lua. Electromagnetic interference shielding effectiveness of carbon-based materials prepared by screen pinging. *Carbon*, **47**, 1905–1910 (2009).
7. S. Iijima. Helical microtubules of graphitic carbon. *Nature*, **354**, 56 (1991).
8. M. J. O'Connell (ed). *Carbon Nanotubes: Properties and Applications*, CRC Press, Taylor & Francis Group, LLC (2006).
9. H. W. Kroto, J. R. Heath, S. C. Obrien, R. F. Curl, R. E. Smalley. C60 Buckminsterfullerene. *Nature*, **318**, 162–163 (1985).
10. M. Endo, S. Iijima, S. Mildred (ed.). *Carbon Nanotubes*, Dresselhaus Pergamon. Elsevier Science Limited (1996).
11. N. Hamada, S. Sawada, A. Oshiyama. New one-dimensional conductors: graphitic microtubules, *Phys. Rev. Lett.*, **68**, 1579–1581 (1992).
12. J. W. Mintmire, B. I. Dunlap, C. T. White. Are fullerene tubules metallic?, *Phys. Rev. Lett.*, **68**, 631–634 (1992).
13. A. M. Rao, E. Richter, S. Bandow, B. Chase, P. C. Eklund, K. A. Williams, S. Fang, K. R. Subbaswamy, M. Menon, A. Tess, R. E. Smalley, G. Dresselhaus, M. S. Dresselhaus. Diameter-selective Raman scattering from vibrational modes in carbon nanotubes, *Science*, **275**, 187, (1997).
14. L. Kelly, M. Meyyappan. In *Carbon Nanotubes: Science and Applications* (M. Meyyappan, ed.). CRC Press LLC (2005).
15. P. Delaney, H. J. Choi, J. Ihm, S. G. Louie, M. L. Cohen. Broken symmetry and pseudo-gaps in ropes of carbon nanotubes, *Nature*, **391**, 466–468 (1998).
16. P. Delaney, H. J. Choi, J. Ihm, S. G. Louie, M. L. Cohen. Broken symmetry and pseudogaps in ropes of carbon nanotubes, *Phys. Rev. B.*, **60**, 7899–7904 (1999).

17. P. Lambin, L. Phillippe, J. C. Charlier, J. P. Michenaud. Electronic band structure of multilayered carbon tubules, *Comp. Mater. Sci.*, **2**, 350 (1994).

18. R. Saito, M. Fujita, G. Dresselhaus, M. S. Dresselhaus. Electronic structure of graphene tubules based on C_{60}, *Phys. Rev. B.*, **46**, 1804 (1992).

19. H. Kataura, Y. Kumazawa, Y. Maniwa, I. Umezu, S. Suzuki, Y. Ohtsuka, Y. Achiba. Optical properties of single-wall carbon nanotubes, *Synth Metals*, **103**, 2555–2558 (1999).

20. A. Thess, R. Lee, P. Nikolaev, H. Dai, P. Petit, J. Robert, C. Xu, Y. H. Lee, S. G. Kim, A. G. Rinzler, D. T. Colbert, G. E. Scuseria, D. Tománek, J. E. Fischer, R. E. Smalley. Crystalline ropes of metallic carbon nanotubes, *Science*, **273**, 483 (1996).

21. J. Kong, A. M. Cassel, H. Dai. Chemical vapour deposition of methane for single-walled carbon nanotubes, *Chem. Phys. Lett.*, **292**, 567–574 (1998).

22. F. F. Komarov, A. M. Mironov. Carbon nanotubes: present and future, *Phys Chem Solid State*, **3**, 411–429 (2004).

23. P. M. Ajayan, T. W. Ebbesen. Nanometre-size tubes of carbon, *Rep. Prog. Phys.*, **60**, 1025–1062 (1997).

24. S. Iijima. Growth of carbon nanotubes, *Mater. Sci. Eng. B.*, **19**, 172 (1993).

25. M. S. Dresselhaus, G. Dresselhaus, P. Avouris (eds.). *Carbon Nanotubes, Synthesis, Structure, Properties and Applications*, Series: Topic in Applied Physics, Vol. 80, Springer (2001).

26. L. Dai (ed.). *Carbon Nanotechnology*, Elsevier B.Y. (2006).

27. S. Bellucci. Carbon nanotubes: physics and applications, *Phys. Stat. Sol. (c)*, **2**, 34 (2005).

28. S. Bellucci, Nanotubes for particle channeling, radiation and electron sources, *Nucl. Instr. Meth. B.*, **234**, 57 (2005).

29. M. Treacy, T. W. Ebbesen, J. M. Gibson. Exceptionally high Young's modulus observed for individual carbon nanotubes, *Nature*, **381**, 678–680 (1996).

30. Y. Q. Zhu, T. Sekine, T. Kobayashi, E. Takazawa, M. Terrones, H. Terrones. Collapsing carbon nanotubes and diamond formation under shock waves, *Chem. Phys. Lett.*, **287**, 689–693 (1998).

31. *Nanotubes and Nanofibers*, Advanced Materials Series, series editor: Yury Gogotsi, CRC Press. Taylor & Francis Group, LLC ISBN 0-8493-9387-6 (978-0-8493-9387-7).

32. S. Berber, Y. K. Kwon, D. Tomanek. Unusually high thermal conductivity of carbon nanotubes, *Phys. Rev. Lett.*, **84**, 4613 (2000).

33. C. Dekker. Carbon nanotubes as molecular quantum wires, *Phys Today*, **52**, 22–28 (1999).

34. Y. G. Yoon, P. Delaney, S. G. Louie. Quantum conductance of multiwall carbon nanotubes, *Phys. Rev. B.*, **66**, 073407, (1–4) (2002).

35. E. Camponeschi, B. Florkowski, R. Vance, G. Garrett, H. Garmestani, R. Tannenbaum. Uniform directional alignment of single-wall carbon nanotubes in viscous polymer flow, *Langmuir*, **22**(4), 1858–1862 (2006).

36. S. Ganguli, M. Bhuyan, L. Allie, H. Aglan. Effect of multi wall carbon nanotube reinforcement on fracture behavior of a tetrafunctional epoxy, *J. Mater. Sci.*, **40**(13), 3593–3595 (2005).

37. P. M. Ajayan, O. Stephan, C. Colliex, D. Trauth. Aligned carbon nanotube arrays formed by cutting a polymer resin—nanotube composite, *Science*, **265**, 5176, 1212–1214 (1994).

38. J. Sandler, M. S. P. Shaffera, T. Prasseb, W. Bauhoferb, K. Schulte, A. H. Windle. Development of a dispersion process for carbon nanotubes in an epoxy matrix and the resulting electrical properties, *Polymer*, **40**(21), 5967–5971, (1999).

39. K.-T. Lau, D. Hui. The revolutionary creation of new advanced materials: carbon nanotube composites, *Compos Part B*, **33**, 263 (2002).

40. M. J. Biercuk, M. Liaguno, M. C. Radosavljevic, J. K. Hyun, J. E. Fischer, A. T. Johnson. Carbon nanotube composites for thermal management, *Appl. Phys. Lett.*, **80**, 2767 (2002).

41. J. K. W. Sandler, J. E. Kirk, I. A. Kinloch, M. S. P. Shaffer, A. H. Windle. Ultra-low electrical percolation threshold in carbon-nanotube-epoxy composites, *Polymer*, **44**(19), 5893–5899 (2003).

42. M. B. Bryning, M. F. Islam, J. M. Kikkawa, A. G. Yodh. Very low conductivity threshold in bulk isotropic single-walled carbon nanotube–epoxy composites, *Adv. Mater.*, **17**(9), 1186–1191 (2005).

43. F. H. Gojny, M. H. G. Wichmann, B. Fiedler, I. A. Kinloch, W. Bauhofer, A. H. Windle, K. Schulte. Evaluation and identification of electrical and thermal conduction mechanisms in carbon nanotube/epoxy composites, *Polymer*, **47**(6), 2036–2045 (2006).

44. A. Vavouliotis, E. Fiamegou, P. Karapappas, G. C. Psarras, V. Kostopoulos. DC and AC conductivity in epoxy resin/multiwall carbon nanotubes percolative system, *Polym. Compos.*, **31**(11), 1874–1880 (2010).

45. P. F. Beuins. *Epoxy Resin Technology*, Wiley, New York (1968).

46. W. G. Potter. *Epoxide Resin*, Butterworth, London, (1970).
47. C. A. May. *Epoxy Resins*, Dekker, New York (1988).
48. H. S. Nalwa (ed.). *Handbook of Low and High Dielectric Constant Materials and Their Applications, Vol. 1: Materials and Processing*, Academic Press (1999).
49. S. Bellucci, C. Balasubramanian, F. Micciulla, G. Rinaldi. CNT composites for aerospace applications, *J. Exp. Nanosci.*, **2**(3) 193–206 (2007).
50. S. Bellucci, F. Micciulla, I. Sacco, G. De Bellis, G. Rinaldi. *Mechanical and electrical characterization of epoxy nanocomposites for electromagnetic shielding devices in aerospace applications*, International Symposium on Electromagnetic Compatibility, 2009 (EMC 2009) IEEE (2009).
51. S. Bellucci, C. Balasubramanian, G. De Bellis, F. Micciulla, G. Rinaldi. *Screening electromagnetic interference effect using nanocomposites*, Macromolecular Symposia, vol. 263, issue 1, 21–29, Special Issue: From Polymer Structure and Rheology to Process Modeling EUPOC 2007, (2007).
52. S. Bellucci, C. Balasubramanian, P. Borin, F. Micciulla, G. Rinaldi. *CNT composites for aerospace applications*, Clean Technology 2007 Technical Proceedings: Cleantech Conference and Trade Show, Chapter 8: Nanoparticle Processes & Applications, pp. 246–250 (2007) ISBN: 1-4200-6382-0.
53. G. Rinaldi, D. Rossi. Particulate composites from epoxy resin and fly-ash for the confinement of medium and low level radwastes, *Polym Int*, **31**(3), 227–233 (1993).
54. G. Rinaldi, G. Maura. Durable glass–epoxy composites cured at low temperatures: effects of thermal cycling, UV irradiation and wet environmental, *Polym Int*, **31**(3), 339–345 (1993).
55. S. Bellucci, C. Balasubramanian, F. Micciulla, A. Tiberia. Study of field emission of multiwalled C nanotubes synthesized by arc discharge, *J Phys Condensed Matter*, **19**, 395014 (7 pp.), (2007).
56. S. Bellucci, A. Tiberia, G. Di Paolo, F. Micciulla, C. Balasubramanian. Emission characteristics of carbon nanotubes at large electrode distances, *J Nanophoton*, **4**, 043501 (2010).
57. J. V. Accorsi, *Impact of morphology and dispersion of carbon black on the weather resistance of polyethylene*, Paper given at the International Wire & Cable Symposium, Atlantic City, November 18, (1999) KGK Kautschuk Gummi Kunststoffe 54. Jahrgang, Nr. 6/(2001).
58. <http://www.sonifier.com>, sonifier products, Branson ultrasonics corp (accessed June 2010).

59. P.-C. Ma, N. A. Siddiqui, G. Marom, J.-K. Kim. Dispersion and functionalization of carbon nanotubes for polymer-based nanocomposites*: a review, Composites: Part A*, **41** (2010) 1345–1367.

60. M. Dresselhaus, G. Dresselhaus, P. C. Eklund. *Science of Fullerenes and Carbon Nanotubes*, Academic Press, San Diego (1996).

61. S. Kirkpatrick. Percolation and conduction, *Rev Mod Phys*, **45**, 574–588 (1973).

62. S. Bellucci, L. Coderoni, F. Micciulla, G. Rinaldi, I. Sacco. The electrical properties of epoxy resin composites filled with CNTs and Carbon Black, *J Nanosci Nanotechnol J Nanosci Nanotechnol.*, 2011 Oct, **11**(10): 9110–7.

63. S. Bellucci, L. Coderoni, F. Micciulla, G. Rinaldi, I. Sacco. Mechanical and electrical characterization of polymer nanocomposites with carbon nanotubes, *Nanoscience and Nanotechnology Letters,* **3**(6), 826–834.

64. C. A. Martina, J. K. W. Sandler, M. S. P. Shaffer, M.-K. Schwarz, W. Bauhofer, K. Schulte, A. H. Windle. Formation of percolating networks in multiwall carbon-nanotube–epoxy composites, *Compos. Sci. Technol.*, **64**(15), 2309–2316 (2004).

65. L. Valentini, D. Puglia, E. Frulloni, I. Armentano, J. M. Kenny, S. Santucci. Dielectric behavior of epoxy matrix/single-walled carbon nanotube composites, *Compos. Sci. Technol.*, **64**(1), 23–33 (2004).

66. M. Moniruzzaman, K. I. Winey. Polymer nanocomposites containing carbon nanotubes, *Macromolecules*, **39**, 5194–5205 (2006).

67. W. Bauhofer, J. Z. Kovacs. A review and analysis of electrical percolation in carbon nanotube polymer composites, *Compos. Sci. Technol.*, 1, **69**, 1486–1498 (2009).

68. J. Sandler, M. S. P. Shaffer, T. Prasse, W. Bauhofer, K. Schulte, A. H. Windle. Development of a dispersion process for carbon nanotubes in an epoxy matrix and the resulting electrical properties, *Polymer*, **40**, 5967–5971 (1999).

69. J. K. W. Sandler, J. E. Kirk, I. A. Kinloch, M. S. P. Shaffer, A. H. Windle. Ultra-low electrical percolation threshold in carbon–nanotube-epoxy composites, *Polymer*, **44**, 5893–5899 (2003).

70. Z. Spitalsky, D. Tasis, K. Papagelis, C. Galiotis. Carbon nanotube–polymer composites: chemistry, processing, mechanical and electrical properties, *Prog. Polym. Sci.*, **35**, 357–401 (2010).

71. S. Barrau, P. Demont, A. Peigney, C. Laurent, C. Lacabanne. DC and AC conductivity of carbon nanotubes–polyepoxy composites, *Macromolecules*, **36**, 5187–5194 (2003).

72. S. Barrau, P. Demont, E. Perez, A. Peigney, C. Laurent, C. Lacabanne. Effect of palmitic acid on the electrical conductivity of carbon nanotubes–epoxy resin composites, *Macromolecules*, **36**, 9678–9680 (2003).
73. S. Cui, R. Canet, A. Derre, M. Couzi, P. Delhaes. Characterization of multiwall carbon nanotubes and influence of surfactant in the nanocomposite processing, *Carbon*, **41**, 797–809 (2003).
74. B. Kim, J. Lee, I. Yu. Electrical properties of single-wall carbon nanotubi and epoxy composites, *J. Appl. Phys.*, **94**, 6724–6728 (2003).
75. Moh. Abdalla, D. Dean, D. Adibempe, E. Nyairo, P. Robinson, and G. Thompson. The effect of interfacial chemistry on molecular mobility and morphology of multiwalled carbon nanotubes epoxy nanocomposite, *Polymer*, **48**, 5662–5670 (2007).
76. Y. S. Song, J. R. Youn. Influence of dispersion states of carbon nanotubes on physical properties of epoxy nanocomposites, *Carbon*, **43**, 1378–1385 (2005).
77. S. Barrau, P. Demont, C. Maraval, A. Bernes, C. Lacabanne. Glass transition temperature depression at the percolation threshold in carbon nanotube–epoxy resin and polypyrrole–epoxy resin composites, *Macromol Rapid Commun*, **26**, 390–394 (2005).
78. A. Moisala, Q. Li, I. A. Kinloch, A. H. Windle. Thermal and electrical conductivity of single- and multi-walled carbon nanotube–epoxy composites, *Compos. Sci. Technol.*, **66**, 1285–1288 (2006).
79. N. Li, Y. Huang, F. Du, X. He, X. Lin, H. Gao, et al. Electromagnetic interference (EMI) shielding of single-walled carbon nanotube epoxy composites, *Nano. Lett.*, **6**, 1141–1145 (2006).
80. F. H. Gojny, M. H. G. Wichmann, B. Fiedler, I. A. Kinloch, W. Bauhofer, A. H. Windle, et al. Evaluation and identification of electrical and thermal conduction mechanisms in carbon nanotube/epoxy composite, *Polymer*, **47**, 2036–2045 (2006).
81. E. T. Thostenson, T. W. Chou. Processing–structure–multi-functional property relationship in carbon nanotube/epoxy composites, *Carbon*, **44**, 3022–3029 (2006).
82. S. M. Yuen, C. C. Ma, H. H. Wu, H. C. Kuan, W. J. Chen, S. H. Liao, et al. Preparation and thermal, electrical, and morphological properties of multiwalled carbon nanotube and epoxy composites, *J. Appl. Polym. Sci.*, **103**, 1272–1278 (2007).
83. A. S. Santos, T. Leite, C. A. Furtado, C. Welter, L. C. Pardini, G. G. Silva. Morphology, thermal expansion, and electrical conductivity of

multiwalled carbon nanotube/epoxy composites, *J. Appl. Polym. Sci.*, **108**, 979–986 (2008).

84. B. J. Landi, R. P. Raffaelle, M. J. Heben, J. L. Alleman, W. VanDerveer, T. Gennett. Single wall carbon nanotube–nafion composite actuators, *Nano. Lett.*, **2**, 1329–1332 (2002).

85. Z. Spitalsky, C. A. Krontiras, S. N. Georga, C. Galiotis, Effect of oxidation treatment of multiwalled carbon nanotubes on the mechanical and electrical properties of their epoxy composites, *Compos. A.*, **40**, 778–783 (2009).

86. L. Liu, K. C. Etika, K. S. Liao, L. A. Hess, D. E. Bergbreiter, J. C. Grunlan. Comparison of covalently and noncovalently functionalised carbon nanotubes in epoxy, *Macromol Rapid Commun*, **30**, 627–632 (2009).

87. A. Allaoui, S. Bai, H. M. Cheng, J. Bai. Mechanical and electrical properties of a MWNT/epoxy composite. *Compos. Sci. Technol.*, **62**, 1993–1998 (2002).

88. N. Hu. The electrical properties of polymer nanocomposites with carbon nanotube fillers, *Nanotechnology*, **19**, 215701 (2008).

89. G. P. Collins. Different stripes: physicists struggle to explain high-temperature superconductivity, *Sci. Am.*, **283**, 3 (2000).

90. C. Li, E. T. Thostenson, T. W. Chou. Effect of nanotube waviness on the electrical conductivity of carbon nanotube-based composites, *Compos. Sci. Technol.*, **68**, 1445–52 (2008).

91. C. Y. Li, E. T. Thostenson, T. W. Chou. Dominant role of tunneling resistance in the electrical conductivity of carbon nanotube-based composites, *Appl. Phys. Lett.*, **91** (2007).

92. L. B. Valdes. Effect of electrode spacing on the equivalent base resistance of point-contact transistors, In: *Proceedings of the IRE*, pp. 1429–1434 (1952).

93. C. A. Martin, J. K. W. Sandler, A. H. Windle, M.-K. Schwarz, W. Bauhofer, K. Schulte, M. S. P. Shaffer. Electric field-induced aligned multi-wall carbon nanotube networks in epoxy composites. *Polymer*, **46**(3), 877–886 (2005).

94. Z. Ounaies, C. Park, K. E. Wise, E. J. Siochi, J. S. Harrison. Electrical properties of single wall carbon nanotube reinforced polyimide composites, *Compos. Sci. Technol.*, **63**, 1637–1646 (2003).

95. M. Foygel, R. D. Morris, D. Anez, S. French and V. L. Sobolev. Theoretical and computational studies of carbon nanotube composites and suspensions: Electrical and thermal conductivity, *Phys. Rev. B.*, **71**, 104201 (2005).

96. Y. Yu, G. Song, and L. Sun. Determinant role of tunneling resistance in electrical conductivity of polymer composites reinforced by well dispersed carbon nanotubes, *J. Appl. Phys.*, **108**, 084319 (2010).

97. J. G. Simmons. Generalized formula for the electric tunnel effect between similar electrodes separated by a thin insulating film, *J. Appl. Phys.*, **34,** 6 (1963).

98. N. Hu, Y. Karube, C. Yan, Z. Masuda, H. Fukunaga, Tunneling effect in a polymer/carbon nanotube nanocomposite strain sensor, *Acta. Mater.*, **56**, 2929–2936 (2008).

99. M. T. Connor, S. Roy, T. A. Ezquerra, and F. J. B. Calleja. Broadband ac conductivity of conductor–polymer composites, *Phys. Rev. B*, **57**, 2286–2294 (1998).

100. N. F. Mott. Conduction in non-crystalline materials. III. Localized states in a pseudogap and near extremities of conduction and valence bands, *Philosophical Magazine*, **19**, 835–852 (1969).

101. W. Bauhofer, J. Kovacs. *A review and analysis of electrical percolation in carbon nanotube polymer composites, Compos. Sci. Technol.*, **69**(10), 1486–1498, CNT-NET 07 Special Issue with regular papers (2009).

102. L. Onsager. The Effects of shape on the interactions of colloidal particles, *Ann. NY Acad. Sci.*, 1949, **51**, 627–659.

103. I. Balberg, C. H. Anderson, S. Alexander, N. Wagner, Excluded volume and its relation to the onset of percolation, *Phys. Rev. B.*, **30**(7), 3933–3943 (1984).

104. L. J., Huijbregts. *Charge transport and morphology in nanofillers and polymer nanocomposites*, Technische Universiteit Eindhoven, Eindhoven, the Netherlands (2008). ISBN 978-90-386-1195-2.

105. C. Li, E. T. Thostenson, T.-W. Chou. Effect of nanotube waviness on the electrical conductivity of carbon nanotube-based composites, *Compos. Sci. Technol.*, **68**(6), 1445–1452 (2008).

106. J. Z. Kovacs, B. S. Velagala, K. Schulte, W. Bauhofer. Two percolation thresholds in carbon nanotube epoxy composites. *Compos. Sci. Technol.*, **67**(5), 922–928 (2007).

107. I. D. Rosca, S. V. Hoa, Highly conductive multiwall carbon nanotube and epoxy composites produced by three-roll milling, *Carbon*, **47**, 1958–1968 (2009).

108. F. H. Gojny, M. H. G. Wichmann, B. Fiedler, I. A. Kinloch, W. Bauhofer, A. H. Windle, K. Schulte. Evaluation and identification of electrical and thermal conduction mechanisms in carbon nanotube/epoxy composites, *Polymer*, **47**, 2036–2045 (2006).

109. Y. J. Kim, T. S. Shin, H. D. Choi, J. H. Kwon, Y. C. Chung, H. G. Yoon. Electrical conductivity of chemically modified multiwalled carbon nanotube/ epoxy composites, *Carbon*, **43**, 23–30 (2005).

110. M. Fu, Y. Yu, J. J. Xie, L. P. Wang, M. Y. Fan, S. L. Jiang, Y. K. Zeng. Significant influence of film thickness on the percolation threshold of multiwall carbon nanotube/low density polyethylene composite films, *Appl. Phys. Lett.*, **94**, 012904, (1–3)(2009).

111. R. Coelho. *Fundamental Studies in Engineering 1: Physics of Dielectrics for the Engineer*, Elsevier Scientific Publishing Company (1979).

112. P. S. Neelakanta (ed.). *Handbook of Electromagnetic Materials, Monolitic and Composite Version and Their Applications*, CRC Press Inc. ISBN-10: 9780849325007.

113. J. C. Maxwell Garnett (1904). Colours in metal glasses and in metallic films, *Phil. Trans. R. Soc. London A*, **203**, 385–420.

114. M. Y. Koledintseva, R. E. DuBroff, R. W. Schartz. A Maxwell–Garnett model for dielectric mixtures containing conducting particles at optical frequencies, *Prog. Electromagn. Res., PIER*, **63**, 223–242 (2006).

115. A. Lakhtakia, G. Ya. Slepyan, S. A. Maksimenko, A. V. Gusakov and O.M. Yevtushenko. Effective medium theory of the microwave and the infrared properties of composites with carbon nanotube inclusions, *Carbon*, **36**, 1833–1839 (1998).

116. G. Ya. Slepyan, S. A. Maksimenko, A. Lakhtakia, O. Yevtushenko and A.V. Gusakov. Electrodynamics of carbon nanotubes: Dynamic conductivity, impedance boundary conditions, and surface wave propagation, *Physical Review B*, **60**, 17136–17149 (1999).

117. M. Y. Shuba, G. Ya. Slepyan, S. A. Maksimenko, C. Thomsen and A. Lakhtakia. Multiwalled carbon nanotubes as waveguides and antennas in the infrared and the visible regimes, *Physical Review B*, **79**, 155403 (2009).

118. P. Kuzhir, A. Paddubskaya, D. Bychanok, A. Nemilentsau, M. Shuba, A. Plusch, S. Maksimenko, S. Bellucci, L. Coderoni, F. Micciulla, I. Sacco, G. Rinaldi, J. Macutkevic, D. Seliuta, G. Valusis, J. Banys. Microwave probing of nanocarbon based epoxy resin composite films: toward electromagnetic shielding, *Thin Solid Films*, **519**, 4114–4118 (2011).

119. A. Vavouliotis, E. Fiamegou, P. Karapappas, G. C. Psarras, V. Kostopoulos. DC and AC conductivity in epoxy resin/multiwall carbon nanotubes percolative system, *Polymer Compos*, **31**(11), 1874–1880 (2010).

120. A. Vavouliotis, E. Fiamegou, P. Karapappas, G. C. Psarras, V. Kostopoulos. Electrically conductive epoxies using multiwall carbon nanotubes, *Plastic Research Online*. Society of Plastics Engineers (SPE) (2010).

121. G. C. Psarras. Hopping conductivity in polymer matrix-metal particles composites, *Compos. A.*, **37**, 1545–1553 (2006).

122. R. H. Chen, R. Y. Chang, C. S. Shern, and T. Fukami. Structural phase transition, ionic conductivity, and dielectric investigations in $K_3H(SO_4)_2$ single crystals, *J. Phys. Chem. Solids*, **64**(4), 553–563 (2003).

123. C. Karthik, K. B. R. Varma. Dielectric and AC conductivity behavior of $BaBi_2Nb_2O_9$ ceramics, *J. Phys. Chem. Solids*, **67**(12), 2437–2441 (2006).

124. G. C. Psarras. Hopping conductivity in polymer matrix–metal particles composite, *Composites: Part A*, **37**, 1545–1553 (2006).

125. H. M. Abdelmoneim. Dielectric and AC conductivity of potassium perchlorate, $KCLO_4$, *Acta. Phys. Pol. A.*, **117**, 936–940 (2010).

126. A. K. Jonscher. *Universal Relaxation Law*, Chelsea Dielectrics Press, London, (1992).

127. A. N. Papathanassiou, I. Sakellis, J. Grammatikakis. Universal frequency-dependent ac conductivity of conducting polymer networks, *Appl. Phys. Lett.*, **91**, 122911 (2007).

128. M. P. Garrett, I. N. Ivanov, R. A. Gerhardt, A. A. Puretzky, D. B. Geohegan. Separation of junction and bundle resistance in single wall carbon nanotube percolation networks by impedance spectroscopy, *Appl. Phys. Lett.*, **97**, 163105 (2010).

129. R. A. Gerhardt, in *Encyclopedia of Condensed Matter Physics* (G. Bassani, G. Liedl, P. Wyder, eds.), Elsevier, New York, **350** (2005).

130. M. S. Fuhrera, M. L. Cohena, A. Zettla, V. Crespi. Localization in single-walled carbon nanotubes, *Solid State Commun*, **109**, 105–109 (1999).

131. D. Stauffer, A. Aharony. *Introduction to Percolation Theory*, 2nd ed., Taylor & Francis, London, (1992).

132. I. Balberg. Scaling theory of the low-field Hall effect and magnetoresistance near a percolation threshold, *Philos. Mag. B*, **56**, 991 (1987).

133. S. Kirkpatrick, Percolation and conduction, *Rev. Mod. Phys.*, **45**, 4 (1973).

134. C. Tong, X. C. Tong (ed.). *Advanced Materials and Design for Electromagnetic Interference Shielding*, CRC Press, Taylor & Francis Group, LLC (2009), ISBN: 1420073583.

135. C. J. Kerr, Y. Y. Huang, J. E. Marshall, E. M. Terentjev. Effect of filament aspect ratio on the dielectric response of multiwalled carbon nanotube composites, *J. Appl. Phys.*, **109**, 094109 (2011).

136. D. D. L. Chung. Electromagnetic interference shielding effectiveness of carbon materials, *Carbon*, **39**, 279–285 (2000).

137. G. Lu, X. Li, H. Jiang. Electrical and shielding properties of ABS resin filled with nickel-coated carbon fibers, *Compos. Sci. Technol.*, **56**, 193–200 (1996).

138. V. V. Sadchikov, Z. G. Prudnikova. Amorphous materials in electromagnetic shields, *Stal'*, 4, 66–69 (1997).

139. C. A. Grimes, EMI shielding characteristics of permalloy multi layer thin film. In: *IEEE Aerospace Applications Conf Proc., IEEE*, Computer Society Press Los Alamitos, California, USA: IEEE, 1994, pp. 211–21. *Plastics Eng.*, 1492–1498 (1997).

140. W. J. Biter, P. J. Jamnicky, W. Coburn. In: *7th International SAMPE Electronics Conference, Shielding Improvement Use of Thin Multilayer Films*, vol. 7, Covina, California, USA: SAMPE, 234–242 (1994).

141. Y. L. Yang, M. C. Gupta, K. L. Dudley, R. W. Lawrence. A comparative study of EMI shielding properties of carbon nanofiber and multi-walled carbon nanotubi filled polymer composites, *J. Nanosci. Nanotechnol.*, **5**, 927–931 (2005).

142. D. D. L. Chung. Materials for electromagnetic interference shielding, *J. Mater. Eng. Perform.*, **9**, 350–354 (2000).

143. J.-M. Thomassin, C. Pagnoulle, L. Bednarz, I. Huynen, R. Jerome, C. Detrembleur. Foams of polycaprolactone/MWNT nanocomposites for efficient EMI reduction, *J. Mater. Chem.*, **18**, 792–796 (2008).

144. N. Li, Y. Huang, F. Du, X. He, X. Lin, H. Gao, et al. Electromagnetic interference (EMI) shielding of single-walled carbon nanotube epoxy composites, *Nano. Lett.*, **6**, 1141–1145 (2006).

145. L. Chen, R. Ozisik, L. S. Schadler. The influence of carbon nanotube aspect ratio on the foam morphology of MWNT/PMMA nanocomposite foams, *Polymer*, **51**, 2368–2375 (2010).

146. W.-S. Jou, H.-Z. Cheng, C.-F. Hsu, The electromagnetic shielding effectiveness of carbon nanotubes polymer composites, *J. Alloys. Compounds*, 434–435, pages 641–645, (2007).

147. Y. L. Yang, M. C. Gupta, K. L. Dudley, R. W. Lawrence. Novel carbon nanotube–polystyrene foam composites for electromagnetic interference shielding, *Nano. Lett.*, **5**, 2131 (2005).

148. Y. L. Yang, M. C. Gupta, K. L. Dudley, R. W. Lawrence, Electromagnetic interference (EMI) shielding of single-walled carbon nanotube epoxy composites, *J. Nanosci. Nanotechnol.*, **5**, 927 (2005).

149. H. M. Kim, K. Kim, C. Y. Lee, J. Joo, S. J. Cho, H. S. Yoon, D. A. Pejakovic, J. W. Yoo, A. Epstein. Electrical conductivity and electromagnetic interference shielding of multiwalled carbon nanotube composites containing Fe catalyst. *J. Appl. Phys. Lett.*, **84**, 589 (2004).

150. R. C. Che, L. M. Peng, X. F. Duan, Q. Chen, X. L. Liang, Microwave absorption enhancement and complex permittivity and permeability of Fe encapsulated within carbon nanotubes. *Adv. Mater.*, **16**(5), 401–405 (2004).

151. C. A. Grimes, C. Mungle, D. Kouzoudis, S. Fang, P. C. Eklund, The 500 MHz to 5.50 GHz complex permittivity spectra of single-wall carbon nanotube-loaded polymer composites. *Chem. Phys. Lett.*, **319**, 5–6, 460–464 (2000).

152. N. Li, Y. Huang, F. Du, X. He, X. Lin, H. Gao, Y. Ma, F. Li, Y. Chen, P. C. Eklund. Electromagnetic interference (EMI) shielding of single-walled carbon nanotube epoxy composites. *Nano. Lett.*, **6**, 6 (2006).

153. N. Li, Y. Huang, F. Du, X. He, X. Lin, H. Gao, Y. Ma, F. Li, Y. Chen, P. C. Eklund. The influence of single-walled carbon nanotube structure on the electromagnetic interference shielding efficiency of its epoxy composites. *Carbon*, **45**(8), 1614–1621 (2007).

154. J. R. Potts, D. R. Dreyer, C. W. Bielawski, R. S. Ruoff. Graphene-based polymer nanocomposites. *Polymer*, **52**, 5–25 (2011).

155. J. Azadmanjiri, P. Hojati-Talemi, G. P. Simon, K. Suzuki, C. Selomulya. Synthesis and electromagnetic interference shielding properties of iron oxide/polypyrrole nanocomposites. *Polymer Eng. Sci.*, **51**, 247–253 (2011).

Chapter 11

Bringing NASA-Relevant Nanotechnology Research into the Classroom

Ana-Rita Mayol,[a] Christian L. Menéndez,[a] Melissa Dávila,[b] and Liz M. Díaz-Vázquez[a]

[a]*NASA-URC Center for Advanced Nanoscale Materials, University of Puerto Rico, Rio Piedras Campus, San Juan, PR 00931, USA*
[b]*College of Education, University of Puerto Rico, Río Piedras Campus, San Juan, PR 00931, USA*

armayol@hotmail.com

11.1 Introduction

The researchers and fellows at the NASA-URC Center for Advanced Nanoscale Materials (NASA-URC-CANM) contribute to promoting the diversity of the next-generation workforce by providing research-based education to its participants. NASA-URC-CANM directly supports an average of 23 undergraduate and graduate fellows each semester. Graduate and undergraduate fellows are required to advance the NASA mission by completing NASA relevant research and education and outreach activities. Fellows participate in the development and implementation of educational materials using their research in nanoscale science and application as motivators and correlating it to the K-16 curriculum. This requirement has been instrumental in improving fellows' pedagogical and communication skills. Fellows adapt simplified versions of their research and, under the mentorship of an education expert,

Advanced Nanomaterials for Aerospace Applications
Edited by Carlos R. Cabrera and Félix A. Miranda
Copyright © 2014 Pan Stanford Publishing Pte. Ltd.
ISBN 978-981-4463-18-8 (Hardcover), 978-981-4463-19-5 (eBook)
www.panstanford.com

correlate these to the science curriculum. To this effect, fellows have developed lessons for the K-12 and undergraduate curriculum. These interdisciplinary, inquiry-based lessons are designed to promote active learning and the development of problem solving and analytical thinking skills in the students impacted. Moreover, these lessons provide everyday life application and pertinence to science concepts discussed in the classroom. It is also an excellent strategy to bring the NASA mission into the classroom and to demystify the important contributions that basic research activities in the area of nanoscale science contribute to make space exploration possible. In this chapter the rationale, methodology and overview of the materials developed and implemented will be discussed.

11.2 Bringing Research into the Classroom: Rationale

The most common teaching style used in science and engineering at the K-16 level is traditional instruction, which does not foster interdisciplinary thinking. Many efforts have been done to promote conceptual learning [1–8]. However, within each discipline, limited efforts have been focused on encouraging students to think across disciplines [9,10]. Science instruction at all levels is done typically in the following way:

(1) The material presented illustrates the topics in the curriculum in an outdated fashion.
(2) The curriculum was planned to be taught discipline specific, thus promoting learning of concepts in a fragmented fashion with no links to other disciplines.
(3) Lack of relevance to current research findings detracts students to pursue advanced degrees in STEM fields.
(4) Resistance from faculty that are not familiar with interdisciplinary learning often hinders the implementation of innovative and interdisciplinary educational modules and laboratories.

Nanoscience and technology is an intrinsic interdisciplinary science that brings together chemistry, physics, biology, engineering, mathematics, and ethics [5,11–14]. Incorporating nanoscience concepts into the curriculum creates the perfect scenario to modernize the curriculum, promote interdisciplinary thinking, bring current research finding to the classroom, and encourage societal and ethical implications in the classroom. Having identified the need for engaging and relevant educational materials, the leadership of the center has devised a two prong approach to involve researchers and fellows in contributing to the efforts of educating the next-generation workforce by (1) fostering the development of pedagogical and communication skills of undergraduate and graduate follows, by disseminating interdisciplinary inquiry based educational material developed by fellows inspired by NASA research and (2) infusing NASA mission in the K-16 curriculum by implementing educational materials developed by fellows.

Implementing nanoscience educational materials in the K-16 curriculum present challenges in: the way of inserting the lessons, identifying the level necessary for the students, identifying the nanoscience topics that have relevance in the curriculum, identifying classical topics that must be retained in the curriculum, training educators in the this area, technical language inherent to each discipline, and developing and implementing appropriate activities where students can derive appropriate conclusions from the material presented [9,10,15].

11.3 Methodology: Conceptualization and Designing the Appropriate Hands-On Activities

The result of this initiative bridges the gap between innovative, state-of-the art research relevant to NASA and science education by developing interdisciplinary and relevant educational modules to impact the next generation of future scientists. The research scientist, graduate and undergraduate fellows use their research as an inspiration to develop engaging and hands-on activities in STEM for the K-16 classroom. Researchers and educators work as a team in conceptualizing, developing, implementing and evaluating educational modules that bring research into the classroom and that are relevant to the science curriculum, at the appropriate pedagogical level.

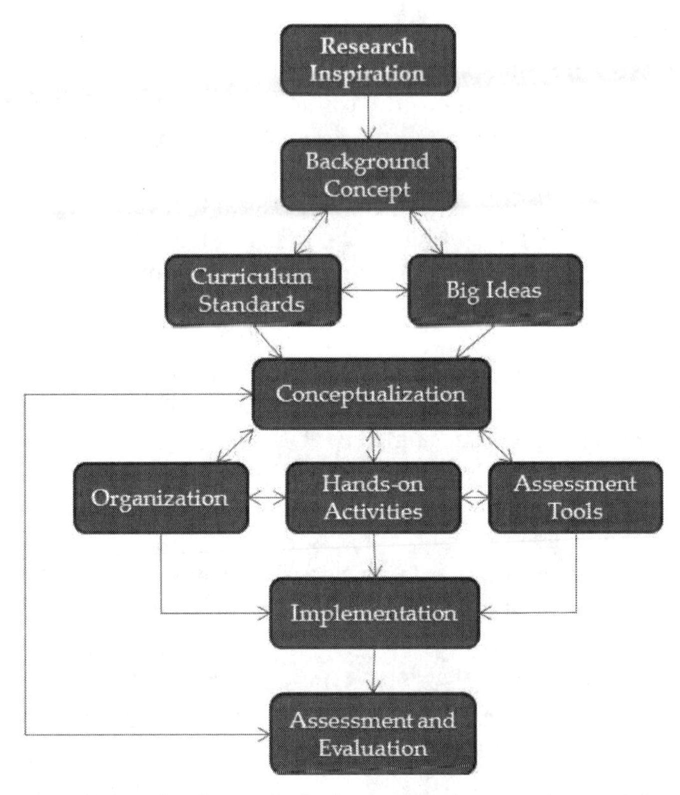

Figure 11.1 Educational module design and implementation model.

The model used by NASA-URC-CANM fellows to develop and implement research based educational modules is presented on Fig. 11.1. Fellows work under the mentorship of their research advisor and an education expert. The first step is to identify the research inspiration of the module. Upon discussion with their mentors, they identify the science concepts that they want to explore within the module. Very often the initial concept is either too broad or has very little relevance to the curriculum. Further discussion with mentor and education expert helps narrowing down the scope of the module. Fellows also narrow the scope of the module by identifying the "big ideas" in nanoscience and technology permeating the research topic they want to explore [6,8,16] and the standards and expectation of the K-16 curriculum [17]—the big ideas as size and scale; structure of matter; forces and interactions; quantum effects; size-dependent properties; self-assembly; tools and instrumentation; models and simulations; science, technology, and society; surface area to volume ratio; and applications [16]. All modules developed are inquiry based and provide hands-on activities for students to discover knowledge.

The conceptualization phase is an iterative process in which the fellow establishes the educational level of the module based on the science concepts and target skills. The science concepts selected mainly determine the appropriate pedagogical level although there are some topics that can be discussed at different levels. Knowing the expected laboratory skills, graphic skills, and numeracy of the K-12 students is fundamental when planning the experimental design. The mathematical skills needed to complete the activities play an important role during all stages of the development and should not be overlooked. An appropriate organization ensures the development of the knowledge built within the inquiry-based hands-on activities. Concepts must be presented one at a time and questions should guide students in the discovery of new knowledge. Fellows use Webb's pedagogical levels to define and align the activities [18,19]. Assessment tools are also incorporated during the conceptualization phase. A first draft of the module is tested among peers that were not involved in its development. The input of this pilot group is taken into account to incorporate changes prior to implement to the target audience. In some instances, more than one pilot group is used.

Teachers use lessons that contain materials they are required to teach in the classroom, as mandated by the science curriculum standards. The activities are designed using the NSTA science standards [18,19]. Effective modules have alignment between goals, objectives, activities, and standards. This alignment is achieved by using the appropriate pedagogical level within the experiment design. Table 11.1 presents a summary of the educational modules developed. The table includes a brief explanation of the module, the pedagogical level, science disciplines that are encompassed and the "Big Ideas" discussed in the module. Section 11.4 includes a case study of an electrochemical module. This module explores the phenomenon of surface area to volume ratio, which is central in energy production. Section 11.5 includes a summary of the development of a water quality module. This module is an example of how to present NASA-relevant research in an accessible way to the K-12 community.

Table 11.1 Educational modules developed or adapted by NASA-URC-CANM researchers and fellows

Module name	Objective	Grade level	Science discipline	Nanoscience concepts
Nanoscience expo	Students will participate in interactive demonstrations, learning key concepts and applications of nanoscience and technology.	K-12	General science	All the big ideas
Size and scale	Students will explore the size of different objects, explore the scale and the instrument that you use to see or characterize these.	3–9	General science	Size and scale
Percolation	Students will be able to determine how the size of particles affects the retention of water and how to determine the particle size of different soils. They will observe the absorption of water in different soils.	6–12	Earth science Biology	Surface area to volume ratio
Organic batteries	Students will use fruits and metals to build electrical circuits.	7–12	Physics Chemistry	Applications: energy
Surface area to volume ratio in nature	Students will explore the chemical nature and the smoothness of leaves responsible for the lotus effect in materials.	7–14	Chemistry	Surface area to volume ratio Applications: lotus effect

Glass or crystal?	Students relate the physical properties with the molecular structure of different samples and learn how to distinguish between glass and crystals	7–14	Physics Chemistry	Structure of matter Quantum effects
Fuel cells	Connect and build upon students' knowledge of oxidation–reduction reactions in order to introduce the chemistry involved in fuel cell technology. Students will also explore the concept of surface-to-volume ratio.	7–14	Physics Chemistry	Applications: energy production
Volta cell	Students will explore reduction–oxidation reactions and will build an energy device.	7–14	Physics Chemistry	Applications: energy production
Solar cells	Students will be able to learn about the conversion of solar energy into useable electrical energy by simulating photosynthesis	7–14	Physical science Physics Chemistry	Applications: energy production
What color am I?	Students will explore how the structure of organic dyes affects absorption of dyes during photosynthesis	7–14	Physical science Physics Chemistry	Structure of matter
Synthesis and characterization of gold nanoparticles	Students will prepare gold nanoparticles and will use UV-Vis spectroscopy to characterize the particle size.	7–14	Physics Chemistry	Quantum effects

11.4 Implementation of Developed Lessons: Implementing Electrochemical Concepts in the Curriculum

Christian Menéndez, a graduate student whose research was in electrochemistry, developed a general chemistry experiment that introduced electrochemical principles in fuel cells into the curriculum, thus bringing research into the classroom. This experiment was designed to address common deficiencies students have in understanding redox reactions and its application and engage students in a real-life problem [20–27]. This module explores the phenomenon of surface area to volume ratio, which is central in energy production. In the proposed experiment, undergraduate students explored the role of an electrolyte in an electrochemical cell when building a Volta cell [28]. A Volta cell is an electrochemical cell used to produce small amount of electricity utilizing reduction-oxidation reactions.

In this experiment, participants explore the use of the Volta's cell in order to measure the voltage generated using different electrolytes and cells with different surface areas. Participants build an electrochemical cell and apply those concepts to understand the operation principles of a fuel cell, which is an important device that uses nanotechnology as an alternative source of clean energy production. The arrangement of the cell consists of using six pairs of copper and zinc strips separated by a filter paper with the same geometry as the metal strips (see Fig. 11.2a). The student measure the voltage generated by the cell when they use different electrolytes or solutions—$CuSO_4$, NaCl, vinegar, n-hexane and a commercial Gatorade® drink—and classify these as strong electrolyte, weak electrolyte, or nonelectrolyte. Students also measure the voltage of the cell when using only one pair of electrodes so they can probe the effect of the area of the electrodes on the outcome voltage of the cell.

This experiment was introduced in the general chemistry and nanotechnology courses. This engaging experiment contrasts with the Daniel cell typically studied in general chemistry (see Fig. 11.2b), as both are galvanic cells. The Daniel cell is the typical experiment completed in general chemistry courses as students explore the basic principles of galvanic cells as it is taught in textbooks. The

Daniel cell contains all the elements of galvanic cells (electrodes, salt bridge, electrolytes, metal wire and resistance), yet is a limiting model as it does not show how electrochemical devices such as batteries and fuel cells are built in real life. By constructing the Volta cell, students gain knowledge in construction of batteries and other electrochemical devices. Students' learning was assessed by using the pre- and post-test to explore the effectiveness of the experiment in improving the students' knowledge in electrochemical concepts. Results obtained revealed that students have a significant knowledge gain and in areas of higher learning domains, results showed students had the biggest improvement. This type of module is promising to continue the development of upper undergraduate level modules that reinforce science concepts and bring key nanotechnology concepts in the curriculum.

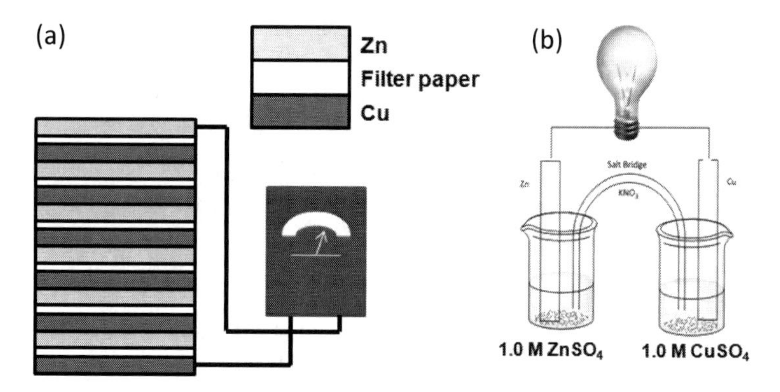

Figure 11.2 Copper–zinc galvanic cells. (a) Volta cell (adapted from www.funsci.com). (b) Daniel cell.

11.5 Demistifying NASA Technology: Water Quality

Melissa Dávila, a graduate student in curriculum development in science, developed a module in water quality. The purpose of this module was to demystify NASA technology in water quality and purification. This project is designed to spark students' interest in the STEM disciplines through multidisciplinary explorations of Space Exploration in Life Support Systems and Sustainability. The main goal of this module is to strengthen the message of space exploration by providing K-12 schools with a combination of science and engineering conceptual background and real world applications in Space Exploration and everyday life on Earth.

This module was designed for students to explore the qualities of water that makes it acceptable to drink and then design a purification system based on the International Space Station Environmental Control and Life Support System developed in NASA Marshall Space Flight Center. In this module, students use graphic calculators and sensors to determine the presence of common contaminants in water [10,28]. The parameters used are turbidity, acidity, salinity, and color (to model organic contaminants in water). After comparison with known samples, they must determine the contaminants present in unknown samples and determine which sample is acceptable for drinking. The second part of this module is still under development. It will be an engineering challenge in water purification that will include the use on nanotechnology to purify water, among other strategies.

Students and teachers impacted by this module were introduced to the use of technology in the classroom. They were exposed to the use of technology and reported that it promoted deep thinking, teamwork, and development of problem solving and analytical thinking skills.

11.6 Conclusion: What Are We Really Teaching: Concepts, Skills, or Both?

Although anecdotal data have only been collected, this initiative has been highly successful in effective in achieving the goals: (1) foster the pedagogical and communication skills of undergraduate and graduate fellows, by disseminating interdisciplinary inquiry based educational material developed by fellows inspired by NASA research and (2) infuse NASA mission in the K-16 curriculum by implementing educational materials developed by fellows.

Requiring graduate and undergraduate students to participate in the development and implementation of education modules has been instrumental in their professional development. All students interviewed reported that this process help them (1) understand their research better; (2) review the basic science concepts developed in the modules; (3) improve their communication skills; and (4) improve their pedagogical skills. In most cases students reported that this experience transform their pedagogical outlook and taught them the importance of the use of effective pedagogical tools in the classroom. Teachers and students impacted in the K-12 level have also benefited from this program. Teachers in the Master Teachers Program, where these materials are used, reported that they have transformed their teaching styles and have incorporated more inquiry-based hands-on activities in the STEM curriculum. They have increased their science concept knowledge and have been familiarized with current research events. Their students have had the opportunity of experiencing engaging activities that have helped them become more motivated in STEM careers.

References

1. Cooper, M. M. (1995). Cooperative learning: an approach for large enrollment courses, *J. Chem. Educ.*, **72**(2), 162–164.
2. Emenike, M. E., Danielson, N. D., Bretz, S. L. (2011). Meaningful learning in a first-year chemistry laboratory course: differences across classical discovery, and instrumental experiments, *J. Coll. Sci. Teach.*, **41**(2), 89.
3. Farell, J. J., Moog, R. S., Spencer, J. N. (1999). A guided inquiry general chemistry course, *J. Chem. Educ.*, **76**(4), 570–574.
4. Mazur, E. (1996). *Peer Instruction User Manual*, Addison-Wesley Press.
5. O'Hayre, R., Cha, S., Colella, W., Prinz, F. B. (2006). *Fuel Cell Fundamentals*, John Wiley & Sons New York, p. 3.
6. O'Sullivan, D. W., Copper, C. (2003). Evaluating active learning: a new initiative for general chemistry curriculum, *J. Coll. Sci. Teach.*, **32**(7), 448–452.
7. Rogers, F., Huddle, P. A. and White, M. D. (2000). What next? *J. Chem. Educ.*, **77**, 7.
8. Rocco, M. C. (2003). Converging science and technology at the nanoscale: opportunities for education and training, *Nat. Biotechnol.*, **21**(10), 1247–1249.
9. Stevens, S. Y., Sutherland, L. A., Krajcil, J. S., *The Big Ideas of Nanoscale Science and Engineering: A Guidebook for Secondary Teachers*, NSTA Press Book.

10. Su, L., Jia, W., Schempf, Ding, A., Le, Y. (2009). Free-standing palladium/polyamide 6 nanofibers for electrooxidation of alcohols in alkaline medium, *J. Phys. Chem. C.*, **113**(36), 16174–16180.
11. Bentley, A. K., Crone, W. C., Ellis, A. B., Payne, A. C., Lux, K. W. (2003). Incorporating concepts of nanotechnology into the materials science and engineering classroom and laboratory. *Proc. Am. Soc. Eng. Educ. Annu. Meeting*, **1462**(2024), 11.
12. Condren, S. M., Breitzer, J. G., Payne, A. C., Ellis, A. B. Windstrand, C. G., Kuech, T. F., et. al. (2002). Student-centered, nanotechnology-enriched introductory college chemistry courses for engineering students, *Int. J. Eng. Educ.*, **18**(5), 550–556.
13. Winkelman, K. (2009). Practical aspects of creating an interdisciplinary nanotechnology laboratory course for freshmen, *J. Nano Educ.*, **1**(1), 34–41.
14. Zheng, W., Shih, H.-R., Lozano, K., Pei, J.-S., Kiefer, K., Ma, X. (2009). A practical approach to integrating nanotechnology education and research into civil engineering undergraduate curriculum, *J. Nano Educ.*, **1**(1), 22–33.
15. Keeney-Kennicutt, W., Gunersel, A. B., and Simpson, N. (2008). Overcoming student resistance to a teaching innovation, *Int. J. Scholarsh. Teach. Learn.*, **2**(1), 1–26.
16. Sanger, M. J., and Greenbowe, T. J., (1996). Science-technology-society (STS) and ChemCom courses versus college chemistry courses: is there a mismatch? *J. Chem. Educ.*, **74**(6), 532.
17. National Committee on Science Education and National Research Council (1996). *National Science Education Standards*, National Academy Press.
18. Anderson, L. W., Krathwohl, D. R., Airasian, P. W., Cruikshank, K. A., Mayer, R. E., Pintrich, P. R., Raths, J., Wittrock, M. C. (2000). *A Taxonomy for Learning, Teaching, and Assessing: A Revision of Bloom's Taxonomy of Educational Objectives*, Abridged Edition.
19. Marzano, R. J., Kendall, J. S. (eds.) (2008). *Designing and Assessing Educational Objectives: Applying the New Taxonomy*, Corwing Press.
20. Ménendez, C., Díaz-Vazquez, L., Mayol, A.-R., Cabrera, C. R. (2012). An educational module to explore what is the role of an electrolyte in an electrochemical cell, *Proc. 2012 MRS Spring Meeting*. Vol. 1446, Cambridge Journals Online (DOI: http://dx.doi.org/10.1557/opl.2012.953).
21. Bard, A. J., Faulkner, L. R. (2001). *Electrochemical Methods: Fundamentals and Applications*, 2nd ed., John Willey & Sons: New York, USA.

22. Bianchini, C., Shen, P. K. (2009). Palladium-Based Electrocatalysts for Alcohol Oxidation in Half Cells and in Direct ALcolo Fuel Cells, *Chem. Rev.*, **109**(9), 4183-4206

23. Gilbert, J. K., De Jong, O., Justi, R., Treagust, D. F., Van Dreil, J. H. (eds.) (2003). *Chemical Education: Towards Research-based Practice*, Jeon, D., Treagust, D., Chapter 14 "The teaching and learning of electrochemistry," (Kluwer Academic Publisher, Netherlands) pp. 317–337.

24. Landolt, D. (1994). Electrochemical and Materials Science Aspects of Alloy Deposition, *Electrochimica Acta.*, **39**, (8–9), 1075–1090.

25. Liang, Z. X., Zhao, T. S., Xu, J. B., Zhu, L. D. (2009). Mechanism study of the ethano oxidation REaction on palladium in alkaline media, *Electrochimica Acta*, **54**(8), 2203–2208.

26. Liu, J., Ye, J., Xu, C., Jiang, S. P., Tong, Y. (2007). Kinetics of ethanol electrooxidation at Pd electrodeposited on Ti, *Electrochem. Communi.*, **9**(9), 2334–2339.

27. McFarland, A. D., Haynes, C. L., Mirkin, C. A., Van Duyne, R. P., and Godwin, H. A. (2004). Color my nanoworld, *J. Chem. Educ.*, **81**(4), 544A.

28. Cortés-Figueroa, J. E., Moore-Russo, D. A. (1999). Using CBL technology and a graphing calculator to teach the kinetics of consecutive first-order reactions, *J. Chem. Educ.*, **76**(5), 635–638.

Chapter 12

Future Directions in Nanotechnology R&D at NASA

Michael A. Meador

NASA John H. Glenn Research Center at Lewis Field,
21000 Brookpark Rd., Cleveland, OH 44135, USA

michael.a.meador@nasa.gov

12.1 Introduction

Nanotechnology has the potential to broadly impact future NASA missions. Lightweight, multifunctional materials derived from nanoscale reinforcements can significantly reduce aerospace vehicle weight not only by the virtue of their improved mechanical properties but also by eliminating or reducing parasitic weight through the consolidation of structural performance with other capabilities such as radiation protection. Use of nanostructured electrode materials can radically improve the specific power and energy of batteries and fuel cells leading to enhanced capabilities for future missions. Nanostructured materials can enable the development of sensors for the detection of chemical and biological species with high sensitivity and selectivity and low power demands for use in future planetary exploration missions. Use of

Advanced Nanomaterials for Aerospace Applications
Edited by Carlos R. Cabrera and Félix A. Miranda
Copyright © 2014 Pan Stanford Publishing Pte. Ltd.
ISBN 978-981-4463-18-8 (Hardcover), 978-981-4463-19-5 (eBook)
www.panstanford.com

nanostructured materials and nanofabrication techniques can lead to new electronics for data processing and communications that are more radiation and fault tolerant than conventional CMOS electronics and have lower power demands.

NASA has been actively engaged in the development and application of nanotechnology in NASA missions for more than a decade. Over this time period, NASA has funded more than $350 M in nanotechnology research at NASA centers, universities and industry.[1] The bulk of these investments have been in the development of nanostructured materials and nanoscale devices, including sensors and electronics. Because of its potential for pervasive impact, nanotechnology R&D has been supported by each of the NASA Mission Directorates: Aeronautics, Space Science, Human Exploration, and Space Technology.

Nanotechnology-derived products have flown on NASA missions. In 2007, a NanoChemSensor unit, developed at the Ames Research Center, flew on a Midstar-1, a US Naval Academy satellite, and demonstrated the capability to detect trace amounts of nitrogen dioxide. A compact trace gas sensor ("electronic nose") consisting of a nanoparticle impregnated polymer primary sensor and a carbon nanotube-based auxiliary sensor successfully flew on the International Space Station in 2008. Carbon nanotube–based composites were used in an engine cover and struts for the Juno probe for the mitigation of electrostatic charging.

In order to better plan its investments in nanotechnology R&D, NASA has recently drafted a 20+ year roadmap for the development and insertion of nanotechnology in future space missions. The Nanotechnology Space Technology Roadmap,[2] one of 14 technology roadmaps developed under the auspices of the NASA Chief Technologist, identified opportunities for nanotechnology in four main areas: engineered materials and structures; propulsion and propellants; energy generation, storage, and distribution; and electronics, sensors and devices. Fourteen key capabilities enabled by nanotechnology developments were identified along with five grand challenges, high payoff nanotechnology research thrusts for future NASA investment.

This chapter will discuss the Nanotechnology Roadmap and provide a glimpse into future opportunities for research and development activities in nanotechnology for aerospace applications. A particular emphasis of the chapter will be on a discussion of the grand challenges identified in the roadmap and on the developments necessary to meet these challenges.

12.2 Discussion

12.2.1 Grand Challenge: 50% Lighter Composites

Vehicle weight is a major concern for most NASA missions. Reductions in vehicle weight enable increased payload capacity that can be utilized to carry more instrumentation, supplies, and/or power systems. For aeronautics, vehicle weight reduction leads to reduced fuel consumption and emissions. The current trend within the aerospace community is to maximize the use of lightweight composites in both aircraft and spacecraft. Composite utilization in the Boeing 787 is 50% compared to 7% in the Boeing 777. NASA is actively pursuing technologies to enable wider use of composites in launch vehicle dry structures and cryogenic propellant tanks.

Replacement of conventional carbon fiber reinforced polymer composites with materials that are 50% lighter would lead to as much as a 30% overall reduction in the dry mass (vehicle without propellant) of aircraft and spacecraft. While reducing the density of composites by 50% seems to be an impossible goal, recent results in nanotechnology R&D suggest that it is achievable.

In conventional composites, it is desirable to have a fully densified matrix since voids can act as mechanical defects resulting in premature failure and can also act as sites for environmental degradation of the composite. However, by introducing nanoscale pores or voids into a matrix in a controlled manner, it may be possible to make matrixes that can sustain the same types of mechanical loads and with the same durability as conventional fully dense polymer matrixes. Researchers at the NASA Glenn Research Center have been working on structural polymer aerogels for use in multifunctional insulation.[3] These nanoporous polymers have densities on the order of 0.2 g/cm^3, roughly 1/5 that of a fully dense polymer. At this density level, replacement of a conventional polymer matrix with one of these aerogels would reduce the density of a composite by about 25%. While the mechanical properties of these materials are quite impressive compared to conventional polymer foams (tensile strengths as high as 8 MPa and compressive moduli up to 100 MPa) they are not sufficient for use as a composite matrix. Further improvements in mechanical properties could be achieved by adding nanoscale fillers (carbon nanotubes, nanoclays or graphene) to strengthen nanopore walls in these materials.

Replacement of conventional carbon fiber with carbon nanotube–derived fibers having higher tensile properties will also enable significant reductions in composite panel and, ultimately, component weight. Systems analysis indicates that utilizing a carbon nanotube fiber with a tensile strength of 10 GPa, twice that of conventional intermediate modulus fibers, would reduce the dry weight of a launch vehicle by up to 25%. Fibers comprising solely carbon nanotubes or nanotubes dispersed in a matrix, e.g., poly(acrylonitrile), have been produced from a variety of techniques. Jiang[4] was the first to demonstrate that pure carbon nanotube yarns could be drawn from vertically aligned carbon nanotube arrays (VANTA). While the initial mechanical properties of these yarns were poor, improvements in tensile properties have been made by utilizing VANTA with longer and better aligned nanotubes[5] and through modifications to the spinning process.[6] Zhang[5] reported the production of low density (0.2 g/cm^3) yarns with tensile strengths as high as 3.3 GPa spun from 1 mm-tall VANTA. Tran and coworkers[6] achieved more than a doubling of the tensile strength of CNT yarns (from 0.5 to 1.4 GPa) by optimizing the spinning process.

A second method for producing carbon nanotube yarns involves direct spinning from a carbon nanotube aerogel produced by high temperature catalytic decomposition of hydrocarbon feedstocks. Windle[7] has reported fibers prepared by this method have specific tensile strengths on the order of 1 GPa/SG. Both processing and post-processing conditions are important in obtaining good tensile strength. Tensile strength increases with winding rate–fibers wound at a faster rate tend to have a higher degree of nanotube alignment. In addition, densification of the fibers after winding by passing them through acetone vapor improves the tensile strength by increasing the contact area between nanotubes and van der Walls attractive forces that hold the yarns together. A similar process has been developed by Nanocomp Technologies, a New Hampshire–based company, to produce not only carbon nanotube yarns but also sheets and tapes. Nanocomp is producing these materials in reasonable quantities and has recently expanded their production capabilities to meet demand from the defense and aerospace communities.

Another approach that is being explored to increase fiber tensile strength is to connect individual nanotubes within the fiber via covalent chemical bonds (cross-linking). Theoretical studies by Cornwell and Welch[8] suggest that this approach could yield carbon nanotube fibers with tensile strengths as high as 60 GPa. Several approaches to cross-linking have been explored. Espinosa[9] reported that cross-linking of double wall carbon nanotube bundles by irradiation with 2 KeV electrons in the cavity of a transmission electron microscope increased their effective tensile strength from between 2 and 3 GPa to over 13 GPa. Windle and coworkers[10] have shown that infusing small amounts of 1,5-hexadiene, HDE, into a carbon nanotube fibers followed by UV-initiated polymerization of the HDE in situ produced a fiber with a specific tensile strength as high as 2.5 GPa/SG–nearly twice that of the as drawn fiber. While significant progress has been made in increasing the tensile properties of dry spun carbon nanotube fibers, more work needs to be done in both processing methods to produce highly aligned, ultralong nanotubes and post-processing methods, such as cross-linking, to further increase the strength and modulus of these materials.

Carbon nanotube-based fibers have also been produced by gel spinning—either from neat carbon nanotube solutions or from carbon nanotube/polymer blends. Fibers have been gel spun from lyotropic solutions of single-wall carbon nanotubes in strong acids, such as fuming sulfuric[11] and chlorosulfonic acids.[12] While good alignment is obtained in these fibers, the tensile properties reported to date have not been very good. Better success in getting good tensile properties has been achieved in fibers spun from carbon nanotube/polymer blends. Kumar has reported tensile strengths as high as 4 GPa for carbon fibers prepared by pyrolysis of gel-spun carbon nanotube/polyacrylonitrile fibers.[13] Further work is needed to produce fibers with better tensile properties and with consistent quality and mechanical properties.

12.2.2 Structures with Integrated Energy Generation and Storage

Developing multifunctional structures with integrated energy generation (photovoltaics, fuel cells) and storage (batteries, ultracapacitors) capability will reduce vehicle weight and provide power on-demand for use in future exploration missions and aircraft. This capability would require developments such as flexible photovoltaics with high efficiency and good radiation stability, better fuel cell and battery electrode materials and battery electrolytes. Nanotechnology can play a critical role in satisfying these needs.

For aerospace systems, photovoltaic efficiency and stability as well as systems weight are important concerns. While efficiencies above 40% have been reported for monolithic multijunction solar cells,[11] the efficiencies of flexible photovoltaics are less than half of that. Significant advances in nanotechnology R&D, primarily supported by the Department of Energy, have enabled large gains in the efficiencies of nontraditional photovoltaics within the past 5 years. Work on flexible solar cells falls into three broad categories: organic photovoltaics, dye-sensitized solar cells, and quantum dot solar cells. The use of carbon nanostructures both in the active layer and as electrode materials has been explored. A bulk heterojunction solar cell utilizing a mixture of semiconducting single-wall carbon nanotubes, reduced graphite, and PCBM, a fullerene derivative, in the active layer has a reported efficiency of 1.3%.[15] The authors claim that efficiencies as high as 9–13% are possible by utilizing semiconducting carbon nanotubes with smaller diameters to better facilitate hole transport by matching the energy levels of the nanotubes and PCBM. Efficiencies as high as 11.5% have been achieved for dye-sensitized solar cells through the use of nanoporous titanium dioxide electrodes.[16] Further enhancements have been reported with the addition of quantum dots to not only harvest more of the light from the solar spectrum but to also enhance efficiency via Forster resonance energy transfer.[17] While considerable progress has been made in developing high-efficiency flexible photovoltaics, more work is needed to improve their performance. Another area that requires more attention and research is the stability and performance of these systems in extreme environments, which could be problematic especially for dye sensitized and organic photovoltaics.

Nanotechnology has also had a significant impact on improving the performance of batteries, fuel cells, and ultracapacitors. Lithium ion batteries have received increased interest for many terrestrial applications, including electric and hybrid automobiles, and have been used in NASA missions including the Curiosity

Mars rover. Recent research has led to significant improvements in battery performance (power and energy density, discharge rates) and durability (cycling performance). For example, use of hybrid cathode materials comprised of $Li_2MnO_3 \cdot LiMO_2$ (LMNCO) and graphene in lithium batteries increased the discharge capacity of a lithium ion battery from 256 to 290 mA hg^{-1} over a battery using an LMNCO cathode.[18] Improvements in battery performance have also been achieved by improving the ionic conductivity of the electrolyte through the use of nanoionics.[19] Nanoionics has also been applied to the design of fuel cell membranes with enhanced ionic conductivity[20] and to prepare membranes for PEM fuel cells that can operate at high temperatures without the need for external humidification.[21]

Incorporation of batteries into structural components will provide some challenges in terms of their inherent durability and the effects that their incorporation will have on the host structure. Flexible, solid-state zinc-carbon batteries have been fabricated using carbon nanofiber mats in the current collector and as a cathode support, carbon nanotube cathodes, and titanium oxide nanoparticles in the electrolyte.[22] While these batteries were flexible, the performance of the battery decreased as the extent of bending (radius of curvature) increased. In addition, the effect of low temperatures and thermal cycling on battery performance was not reported. Batteries incorporated into rigid structures must have good dimensional stability under charge and discharge. One approach to provide this stability that is currently being explored is to utilize networks of carbon nanostructures (carbon nanotubes, graphenes) or nonporous carbon membranes as the electrode material in lithium ion batteries.[23] These networks also function as a structural buffer to mitigate against to counteract expansion and contraction of the batteries during charge and discharge.

Energy harvesting through use of thermo- and piezoelectric materials and devices has the potential to provide some power by harnessing excess thermal or mechanical energy. Embedding these devices into structures would provide some power that could be used to supply sensors, displays, or other low power demand devices.

In thermoelectric materials, electrical current is generated by temperature gradients in a material. For optimal power generation, the material must possess a low thermal conductivity, in order to maximize the temperature gradient within the material. In addition, the material should have a high electrical conductivity in order to produce high carrier diffusion currents. For bulk materials, these properties are interrelated–increases in electrical conductivity are usually accompanied by increased thermal conductivity. However, in nanoscale materials it is possible to optimize both properties independently. Further increases in thermoelectric performance are achievable through phonon scattering from the large number of interfaces in these nanocrystalline materials.[24] While the most efficient thermoelectric materials are the phonon-glass/electron-crystal (PEGC) materials, these are fragile and not suitable for most structural applications. There has been a significant amount of activity focused on carbon nanotube–based thermoelectric.[25] While carbon nanotubes have thermoelectric figure of merit, ZT, values that are much lower than PEGC thermoelectrics, they are more mechanically robust. Recently several labs have reported the fabrication of flexible, carbon nanotube thermoelectric devices.[26,27] While these studies demonstrated the potential for CNTs in structural thermoelectric devices, more work is needed to optimize ZT and power output.

Improvements in the performance of piezoelectric energy scavenging devices have been achieved through the use of nanostructured materials. Conventional bulk piezoelectric materials such as barium titanate suffer from high impedance characteristics, high permittivity, and poor mechanical properties that could limit their application in energy harvesting devices. Piezoelectric nanogenerators have been produced from electrospun nanofibers,[28] zinc oxide nanowires,[29] and doped graphene.[30] Recently a "flexible energy cell" has been reported comprised of thermo-and piezoelectric nanogenerators integrated with a flexible thin film solar cell.[31] The device generated enough energy to power four small LED lights. Further work is needed to optimize the performance and power output of these nano-generators, develop effective schemes for integrating them in structural components and evaluating their stability and performance in space environments.

12.2.3 Graphene: From Molecules to Devices

Future NASA missions will require electronic devices, including sensors, data processing and memory, that are fault tolerant, radiation hard, mechanically robust, require less power and are capable of operating in harsh environments. Replacement of conventional CMOS electronics with devices fabricated from graphene could provide such devices. Graphene has a number of properties that make it attractive for use in electronics. The carrier mobility in suspended exfoliated graphene has been measured as high as 200,000 $cm^2 V^{-1} s^{-1}$, the highest of any known material.[32] This mobility drops depending upon the substrate on which the graphene has been deposited. Graphene has a thermal conductivity on the order of 1000 $Wm^{-1} K^{-1}$, also the highest of any known material. Graphene also has excellent mechanical properties and durability, with a Young's modulus of 0.5 TPa[33] and a tensile strength of 130 MPa.[34]

Use of graphene in transistor applications is problematic since it has a near-zero bandgap energy and, hence, cannot be switched on and off. One approach that has been explored to address this is to reduce the dimensionality from 2D to 1D and utilize graphene nanoribbons rather than large-area graphene films. Graphene nanoribbons have been prepared using a variety of different approaches. Tour and coworkers demonstrated the synthesis of graphene ribbons from carbon nanotubes using a strong oxidizing agent[35] or under reductive conditions using potassium vapor.[36] Direct synthesis of graphene nanoribbons has been reported by several groups.[37,38] Another approach to opening the bandgap in graphene is through functionalization.[39]

Graphene is a very attractive material for use in RF transistors due to its durability and the combination of high carrier mobility and transconductance. For this reason, there has been a considerable amount of activity in the development of graphene field-effect transistors (GFETs).[40] Challenges in preparing such devices include the ability to grow large-area, defect- and wrinkle-free graphene on a variety of substrates, identification of doping schemes to control the bandgap of the graphene, and the development of interconnects that do not adversely affect the chemical structure of the graphene substrate. The current limitations of GFETs were recently demonstrated in the fabrication of GFETs from both wafer-scale CVD-grown graphene and graphene grown epitaxially on a SiC substrate.[41] These devices had intrinsic cutoff frequencies above 300 GHz and voltage and power gains up to 20 dB. The authors suggest that further increases in performance could be achieved with improvements in materials and design.

Graphene has an optical transparency between 97% and 98%[42] and has attracted significant interest as a transparent conductor for use in solar cells and LEDs. Tour and coworkers recently reported the fabrication of flexible transparent conducting films from graphene/metal grids deposited on PET. These films had sheet resistances as low as 4 Ω, optical transmittance between 70% and 91%, and did not experience any cracking or changes in resistance after 500 bending cycles.[43] Transparent conductive electrodes based on chemically reduced graphene oxide have been demonstrated on flexible organic-LEDs.[44] Gradecak[45] recently reported the fabrication of zinc oxide nanowire/graphene hybrid solar cells with performance (efficiency, J_{SC}, V_{OC}) equivalent to or better than that of comparable devices constructed with an ITO cathode.

12.2.4 Nanopropellants

Conventional propellants present special challenges in their handling and storage. Cyrogenic propellants require insulation and cryo-coolers to prevent their premature evaporation (boil-off), both of which add vehicle weight and complexity to ground-based operations. Noncryogenic liquid propellants, such as hypergolics, are toxic and require special storage and handling procedures that complicate launch operations and add cost and safety concerns to a mission. Use of nanoscale particles as either fuel or gelling agents for conventional propellants could lead to new propellants that are less toxic and easier to handle than conventional propellants.

There has been a considerable amount of activity funded by NASA and the Department of Defense in the development of propellants and energetic materials based on nanoscale metal particles in an oxidizing medium. A propellant based on a slurry of nanoscale aluminum particles in ice was used to propel a small rocket to a height of about 1800 feet.[46] While the high surface area and increased chemical reactivity of nanoscale metal particles makes them excellent candidates for use in nanopropellants, they also present safety risks since these particles can spontaneously combust. In addition, the formation of an oxide layer on the nanopropellant during storage is undesirable since it can add weight to the propellant and can also lead to incomplete combustion. As a result, a major thrust of nanopropellant R&D is the development of surface passivation approaches that improve the handling characteristics of the nanoparticles without adversely affecting their combustion behavior. For example, aluminum nanoparticles generated in solution by decomposition of aluminum hydride complexes with titanium isopropoxide were surface passivated with perfluoroalkyl carboxylic acids to provide nanoparticles with good stability in air.[47] The combustion behavior of these materials was not reported. Lewis and coworkers report that oleic acid passivated aluminum nanoparticles are more reactive than uncoated nanoparticles, suggesting that some types of passivation chemistries could improve combustion characteristics.[48] Another approach that has been investigated to control nanoparticle reactivity is to utilize core-shell structures comprised of a highly reactive oxidizer core coated with a less reactive shell.[49] Nanopropellant morphology can also play a role in combustion performance. Electrostatic self-assembly of a metal nanoparticle and metal oxide nanoparticle oxidizer mixture has been shown to lead to a propellant with better energy release than could be obtained by mechanically mixing the nanoparticles.[50]

More work is needed on the development of passivation chemistries and on understanding of how surface modification affects both stability and combustion behavior. Methods are needed to develop tailored nanopropellant morphologies to control combustion behavior. In addition, the development of "smart" nanopropellants whose reactivity and burning behavior can be controlled by applied external fields (electrical, magnetic, electromagnetic) will enable control of propellant ignition.

12.2.5 Hierarchical Integration

Insertion of nanotechnology into future NASA missions requires the ability to integrate nanotechnology components into macroscale systems in a manner that preserves the attributes and benefits derived from working at the nanoscale. Among others, the development of advanced manufacturing technologies is critical for this integration. Utilization of graphene in transistors requires methods to fabricate interconnects that do not adversely affect the electronic properties of the graphene. Production of ultrahigh strength, light-weight nanocomposites from carbon nanotube reinforcements requires the development of functionalization chemistries to maximize wetting and load transfer between the reinforcements and the matrix materials and fabrication techniques tailored to processing of thin gauge composite structures. Integration of energy generation and storage systems into structural components will require the development of fabrication and packaging schemes that maximize the energy output of these structures, preserve their mechanical properties and durability, and minimize structural weight. The recent emphasis on the development of advanced manufacturing technologies, in particular those related to additive manufacturing, will have a significant impact on successful integration of nanotechnologies into macroscale systems.

While integration of nanoscale materials and devices into macroscale systems may require new approaches and manufacturing techniques, it is important to focus on integration schemes that are compatible with existing manufacturing and assembly methods. Aerospace companies and their suppliers have significant capital investments in manufacturing facilities and in the development and qualification of manufacturing approaches to reliably produce components, systems, and vehicles. Technologies that can fit into existing manufacturing lines with little or no modification will be more readily embraced by the industry than those that require substantial retooling and requalification.

Hierarchical integration will also require the development of efficient and accurate multiscale modeling tools that will aid in the design of nanotechnology integrated structures and their manufacturing and testing. There is a growing interest in the development of these tools under both the Materials Genome Initiative and the Nanotechnology Knowledge Infrastructure Signature Initiative recently initiated by the National Nanotechnology Initiative.

12.3 Conclusions

Nanotechnology has the potential to have a pervasive impact on NASA missions and on the aerospace industry as a whole. While nanotechnology has found its way into some aerospace applications, these examples are somewhat limited. Much work is needed to improve the efficiency, performance, and durability of nanotechnology-derived materials and devices to make them suitable for future applications in aerospace. In some cases, nanotechnology that has been developed for terrestrial applications could be directly applied to support NASA mission needs. However, these devices and materials will need to be tested in relevant mission environments to determine if they have adequate durability in the harsh environments that would be encountered in many NASA missions.

Through the Nanotechnology Space Technology Roadmap, NASA has developed a long-term strategy for the development of nanostructured materials and devices and their insertion into future missions. NASA is actively pursuing technologies identified in the roadmap and is working with academia, industry and other Federal agencies to develop and qualify these technologies.

References

1. Source: National Nanotechnology Initiative, www.nano.gov.
2. Meador, M. A., Files, B., Li, J., Powell, D., Siochi, E. J. *Nanotechnology Roadmap (Technology Area 10)*, National Aeronautics and Space Administration, April 2012.
3. Guo, H., Meador, M. A. B., McCorkle, L., Quade, D. J., Guo, J., Hamilton, B., Cakmak, M., Sprowl, G. (2011). *ACS Appl. Mater. Interfaces*, **3**, 546–552.
4. Jiang, K., Li, Q., Fan, S. (2002). *Nature*, **419**, 801.
5. Zhang, X., Li, Q, Holesinger, T. G., Arendt, P. N., Huang, J., Kirven, P. D., Clapp, T. G., DePaula, R. F., Liao, X., Zhao, Y., Zheng, L., Peterson, P. E., Zhu, Y. (2007). *Adv. Mater.*, **19**, 4198–4201.
6. Tran, C. D., Humphries, W., Smith, S. M., Huynh, C., Lucas, S. (2009). *Carbon*, **47**, 2662–2670.
7. Koziol, K., Vilatela, J., Moisala, A., Motta, M., Cuniff, P., Sennett, M., Windle, A. (2007). *Science*, **318**, 1892–1895.
8. Cornwell, C. E., Welch, C. R. (2011). *J. Chem. Phys.*, **134**, 204708.
9. Filleter, T., Bernal, R., Li, S., Espinosa, H. D. (2011). *Adv. Mater.*, **23**, 2855–2860.
10. Boncel, S., Sundaram, R. M, Windle, A. H., Koziol, K. K. K. *ACS Nano*, (2011). *ACS Nano*, **5**, 9339–9344.

11. Zhou, W., Heiney, P. A., Fan, H., Smalley, R. E., Fischer, J. E. (2004). *J. Am. Chem. Soc.*, **127**, 1640–1641.
12. Davis, V. A., Ericson, L. M., Parra-Vasquez, A. N. G., Fan, H., Wang, Y., Prieto, V., Longoria, J. A., Ramesh, S., Saini, R. K., Kittrell, C., Billups, W. E., Adams, W. W., Hauge, R. H., Smalley, R. E., Pasquali, M. (2004). *Macromolecules*, **37**, 154–160.
13. Liu, Y., Chae, H. G., Kumar, S. (2011). *Carbon*, **49**, 4487–4496.
14. *Source:* National Renewable Energy Laboratory *Best Research Cell Efficiencies.*
15. Brenardi, M., Lohrman, J., Kumar, P. V., Kirkeminde, A., Ferralis, N., Grossman, J. C., Ren, S. (2012). *ACS Nano*, **6**, 8896–8903.
16. Nazeerrudin, M. K., DeAngelis, F., Fantacci, S., Selloni, A., Viscardi, G., Lisca, P., Ito, S., Takeru, B., Gratzel, M. G. (2005). *J. Am. Chem. Soc.*, **127**, 16835–16847.
17. Buhbut, S., Itzhakov, S., Oron, D., Zaban, A. (2011). *J. Phys. Chem. Lett.*, **2**, 1917–1924.
18. Jian, K.-C., Wu, X.-L., Yin, Y.-X., Lee, J.-S., Kim, J., Guo, Y.-G. (2012). *ACS Appl. Mater. Interfaces*, **4**, 4858–4863.
19. Wagemaker, M., Wulder, F. M. (2013). *Acc. Chem. Res.*, **46**, 1206–1215.
20. Matsui, J., Miyata, H., Hanaoka, Y., Miyashita, T. (2011). *ACS Appl. Mater. Interfaces*, **3**, 1394–1397.
21. Tezuka, T., Tadanaga, K., Hayashi, A., Tatsumisago, M. (2006). *J. Am. Chem. Soc.*, **128**, 16470–16471.
22. Hiralal, P., Imaizumi, S, Unalan, H. E., Matsumoto, H., Minigawa, M., Rouvala, M., Tanioka, A., Amaratunga, G. A. J. (2010). *ACS Nano*, **4**, 2730–2734.
23. Xin, S., Guo, Y.-G., Wan, L.-J. (2012). *Acc. Chem. Res.*, **45**, 1759–1769.
24. Dresselhaus, M. S., Chen, G., Tang, M. Y., Yang, R., Lee, H., Wang, D., Ren, Z., Fleurial, J.-P., Gogna, P. (2007). *Adv. Mater.*, **19**, 1043–1053.
25. Prasher, R. S., Hu, X. J., Chalopin, Y., Mingo, N., Lofgreen, K., Volz, F., Cleri, F., Keblinski, P. (2009). *Phys. Rev. Lett.*, **102**, 105901.
26. Hewitt, C. A., Kaiser, A. B., Roth, S., Craps, M., Czerw, R., Carroll, D. L. (2012). *Nano. Lett.*, **12**, 1307–1310.
27. Yu, C., Choi, K., Yin, L., Grunlan, J. C. (2011). *ACS Nano*, **5**, 7885–7892.
28. Wu, W., Bai, S., Yuan, M., Qin, Y, Wang, Z. L., Jing, T. (2012). *ACS Nano*, **6**, 6231–6235.
29. Hsu, C.-L., Chen, K.-C. (2012). *J. Phys. Chem. C*, **116**, 9351–9355.

30. Shin, H.-J., Choi, W. M, Choi, D., Han, G. H., Yoon, S.-M., Park, H.-K., Kim, S.-W., Jin, Y. W., Lee, S. Y., Kim, J. M., Choi, J.-Y., Lee, Y. H. (2010). *J. Am. Chem. Soc.*, **132**, 15603–15609.

31. Yang, Y., Zhang, H., Zhu, G., Lee, S., Lin, Z.-H., Wang, Z. L. (2013). *ACS Nano*, **7**, 785–790.

32. Du, X., Skacho, I., Barker, A., Andrei, E. Y. (2008). *Nat. Nanotechol.*, **3**, 491.

33. Frank, I. W., Tanenbaum, D. M, van der Zande, A. M, McEuen, P. L. (2007). *J. Vac. Sci. Technol. B*, **25**, 2558–2561.

34. Lee, C., Wei, X., Kysar, J. W., Hone, (2008). *J. Sci.*, **321**, 385–388.

35. Kosynkin, D. V., Higgenbotham, A. L., Sinitski, A., Lomeda, J. R., Diiev, A., Price, B. K., Tour, J. M. (2009). *Nature*, **458**, 8723–8876.

36. Kosynkin, D. V., Lu, W., Sinitski, A., Pera, G., Sun, Z., Tour, J. M. (2011). *ACS Nano*, **11**, 968–974.

37. Pisula, W., Feng, X., Mullen, K. (2011). *Chem. Mater.*, **23**, 554–567.

38. Cai, J., Ruffieux, P., Jaafar, R., Bieri, M., Braun, T., Blankenburg, S., Muoth, M., Seitsonen, A. P., Saleh, M., Feng, X., et al. (2010). *Nature*, **466**, 470–473.

39. Niyogi, S., Bekyarova, E., Hong, J., Khizroev, S., Berger, C., de Heer, W., Haddon, R. C. (2012). *J. Phys. Chem. Lett.*, **2**, 2487–2498.

40. Avouris, P. (2010). *Nano. Lett.*, **10**, 4285–4294.

41. Wu, Y., Jenkins, K. A., Valdes-Garcia, A., Farmer, D. B., Zhu, Y., Bol, A. A., Dimitrakopoulos, C., Zhu, W., ZXia, F., Avouris, P., Lin, Y.-M. (2012). *Nano. Lett.*, **12**, 3062–3067.

42. Nair, R. R., Blake, P., Grigorenko, A. N., Novoselov, K. S., Booth, T. J., Stauber, T., Peres, N. M. R., Geim, A. K. (2008). *Science*, **320**, 1308.

43. Xhu, Y., Sun, Z, Yan, Z., Jin, Z., Tour, J. M. (2011). *ACS Nano.*, **5**, 6472–6479.

44. Matyba, P., Yamaguchi, H., Eda, G., Chhowalla, M., Edman, L., Robinson, N. D. (2010). *ACS Nano*, **4**, 637–642.

45. Park, H., Chang, S., Jean, J., Cheng, J. J., Aurojo, P. T., Wang, M., Bawendi, M. G., Dresselhaus, M. S., Bulovic, V., Kong, J., Gradecak, S. (2013). *Nano Lett.*, **13**, 233–239.

46. Risha, G., Connell, Jr., T. L., Yetter, R. A., Yang, V., Wood, T. D., Pfeil, M. A., Pourpoint, T. L., Son, S. F. *Proc. AIAA/ASME/SAE/ASEE Joint Prop. Conf. Exhibit*, AIAA 2009–4877 (2009).

47. Jouet, J. A., Warren, A. D., Rosenberg, D. M, Bellitto, V. J., Park, K., Zachariah, M. R. (2005). *Chem. Mater.*, **17**, 2987–2996.

48. Lewis, W. K., Harruff, B. A., Gord, J. R., Rosenberger, A. T., Sexton, T. M., Guliants, E. A., Bunker, C. E. (2011). *J. Phys. Chem. C*, **115**, 70–77.
49. Prakash, A., McCormick, A. V., Zachariah, M. R. (2005). *Nano. Lett.*, **5**, 1357–1360.
50. Malchi, J. Y., Folewy, T. J., Yetter, R. A. (2009). *ACS Appl. Mater. Interfaces*, **1**, 2420–2423.

Index